TRAVEL

...have but recently re-
...ed from a southern ram-
...Washington for several
...s, then Florida, Sarasota,
...Pcksburg & Tallahassee, .
...n New Orleans. Then a ramble
...up the riverside to West Feliciana Parish.
...Then Oxford, Mississippi. Then Jackson,
...Miss. Then Tuscaloosa, Ala. Then Carbon
...ally rotten, Hill, Ala. Then Birmingham. Then Washington
...I don't mind it. once more, then N.Y.

INCIDENTALS

I no longer have to worry about
binding Elie Faure or other unbound
art books as I no longer have any
of them, except the Maillol you gave me.
  I have now published two books
and 53 shorter papers and all of them are
lousy. I owe apologies to posterity.
Souvenirs de Voyage:

LANGUAGES

I am rather lan-
guidly studying Span-
ish, making some
progress. It's easy to
pronounce and to spell,
but full of tricks never-
theless. I am about
far enough to try to
read a novel in it, which
I shall soon attempt

...ed chicken and grits with the
...te Geologist of Fla. and his
...ousey but tyrannical wife.

...night in an enormous four-
...ter in a room 15 feet high
...an old plantation mansion
...the wilds of Louisiana, and
...charming and extremely
...uble New Orleans French hostess,
...more fried chicken and grits,
...d scrambling about stream
...nks with an incomprehensible negro

...day spent weirdly
...d unsupported by the
...place of tobacco in
...the depths of a coal
...mine, looking for the
...footprints of animals
...ead since 250,000,000 B.C.
...freezing cold, a man dropping
...my door.

German dialect recitations by the chief engineer of a
steamboat in New Orleans Harbour.

Driving furiously at night along
the invisible Mississippi, with
dense corporate clouds of fog
oozing over the levee and flowing
singly off over the dark marshes

A very wild party
all night in the room
next to mine in
an otherwise very
good hotel.

A long drive
with a high-school boy
for chauffeur who explained
the personal attractions and
morals of every young lady
in town and nearly landed
us in the ditch several times.

A heavy snowstorm
in Alabama &
dead just outside

Lunch in the open court
of an old house in the
Vieux Carré, & tea on
the balcony of another on
the cathedral square.

A stay in a town where I was supplied
with a house and a servant and literally
could not spend any money, all the
store keepers informing me that they
were ordered not to let me pay for
anything.

A very suspicious Professor of
Geology in a tiny college who
thought me a city slicker come
to steal his lousy specimens.

A very jovial oil geologist who
put (literally) a barrel of
whiskey in his car and then drove
me unsteadily but widely through the
surrounding country, at his company's
expense.

and many others.

And so to — a meeting of   ...gy and
Mineralogy of the New York Acade...
Honi soit qui

SIMPLE CURIOSITY

W.D.M. G.GS.   MASTODON   PLIAUCHENIA   PLIOHIPPUS   HYPPARION   HETHRCOFORDIA
D AMNATA

Quelques uns de mes amis
de la Floride

What some pterodactyls
looked like really.

Cynognathus

Skull of Triceratops

Dimetrodon ((an another of mine)

Camarasaurus   (still another dinosaur)

SIMPLE CURIOSITY

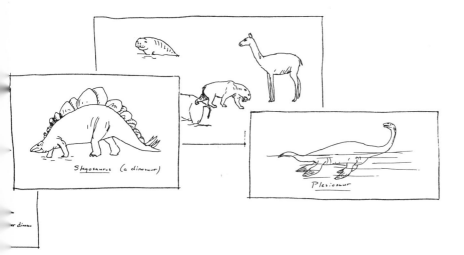

Letters from George Gaylord Simpson
to His Family, 1921–1970

♦

EDITED BY LÉO F. LAPORTE

UNIVERSITY OF CALIFORNIA PRESS · BERKELEY · LOS ANGELES · LONDON

University of California Press
Berkeley and Los Angeles, California

University of California Press, Ltd.
London, England

Library of Congress Cataloging-in-Publication Data
Simpson, George Gaylord, 1902–1984
  Simple curiosity.

  Includes bibliographical references.
  1. Simpson, George Gaylord, 1902–1984—
Correspondence.   2. Paleontologists—Correspondence.
3. Vertebrates, Fossil.   I. Laporte, Léo F.
II. Title.
QE707.S55A4   1987   560′.9   [B]   86-25046
ISBN 0-520-05792-9 (alk. paper)

Printed in the United States of America

1   2   3   4   5   6   7   8   9

LETTERS are valuable and entertaining in proportion to the wit and ability, and above all to the imprudence, of those who write them. For the merit of a really good letter is always colloquial; it is full of news or gossip, it is personal, unstudied and indiscreet. It shows the writer without his guards or defences, uncovers all his thoughts and feelings; and that is why honest letters are more informative, more amusing, more pathetic, more vital than any considered autobiography. Of all documents these are the most essentially human.

C. E. Vulliamy, 1945

CONTENTS

PREFACE

THIS COLLECTION of family letters was written over a
period of fifty years by George Gaylord Simpson (1902–
1984), the "greatest paleontologist since Cuvier."[1] Not only did Simpson
describe, classify, and interpret myriad fossils of ancient life, especially
mammals, but he was also one of the founders of the modern evolutionary
synthesis. His classic *Tempo and Mode in Evolution*, published in 1944,
demonstrated that the genetic principles of living populations could, contrary to much paleontological opinion of the time, explain the major patterns and rates of evolutionary history as inferred from the fragmentary record of fossil remains stretching backward hundreds of millions of years.[2]

The quantity, diversity, and especially the quality of Simpson's published writings—almost eight hundred bibliographic entries—secure him a
place in the history of biology and geology. Future paleontologists and evolutionists will be interested in this man of genius not only as a scientist but
also as a human being. Simpson himself no doubt had some sense of this
interest when he began making autobiographical notes as early as 1933;
these eventually culminated in the published work a decade ago.[3] Yet any
autobiography, particularly one written late in life, gives only a retrospec-
tive view of the life in question. Unless accompanied by contemporary
documents recording attitudes, opinions, points of view, and states of
mind, such a view of the past is inevitably colored by later experience and
subsequent emotional and mental development. In fact, most memoirs are
more like memories of memories than statements of the way it "really"
was.

We are therefore fortunate that Simpson was an inveterate letter writer,
especially to his older sister Martha (1898–1984) and his parents. Martha
recognized very early the special gifts of her brother and as a measure of her
affection and admiration managed to save some 156 of his family letters.
These she eventually entrusted to the archives of the American Philosophical Society in Philadelphia. Another 54 World War II letters to his par-

1. This judgment was made by Simpson's doctoral thesis adviser at Yale, Prof.
Richard S. Lull, according to Norman D. Newell, professor emeritus of Columbia
University and a colleague of Simpson's for more than a decade at the American
Museum of Natural History (NDN to LFL, May 1979).

2. Laporte, L., "Simpson's *Tempo and Mode in Evolution* Revisited," *Am. Phil. Soc.
Proc.* 5:127 (1983), pp. 365–417.

3. Simpson, G. G., *Concession to the Improbable* (New Haven: Yale University,
1978).

ents and sisters were discovered among Simpson's personal papers after his death; these are also included here.

I learned of the existence of the letters several years ago from conversations with Martha Simpson Eastlake in Tucson, Arizona, when I was doing background research on Simpson's professional life and scientific work. As soon as practicable I went to Philadelphia and spent several days reading the letters. I was immediately convinced that they would interest a much wider audience, so I had them photocopied and sent to me in California.

Since then I have transcribed the letters, all but a few of them handwritten. Although Simpson's handwriting is more legible than most people's, when in a whimsical mood he would write in mirror writing, or in tightly coiled spirals around and around the page, or in French, Spanish, or doggerel German; some sentences even are in his own made-up code. Because of the informality of the letters, Simpson was not always careful or consistent about punctuation. His French and Spanish, too, are often colloquial or ungrammatical. Except in instances of obvious errors that would mislead readers, I have let these idiosyncrasies remain. Such idiosyncrasies, however, slowed transcription considerably.

I have arranged the letters in chronological order. In most cases Simpson dated them, but for those undated I have relied on internal evidence for their chronology. Dates and places in brackets are my inferences; a question mark indicates any doubt about such inferences.

The letters are divided into somewhat arbitrary, yet logical sections. For each I have written a short introduction to provide the biographic background necessary to understand the letters. I have included brief footnotes to identify or clarify persons, places, and events that might be unknown to the general reader.

I thank the American Philosophical Society for permission to publish the letters housed in their archives. I thank too Dr. Anne Roe Simpson, who has encouraged the project from the beginning in many ways yet has been scrupulous in giving me complete editorial freedom.

Finally, I would like to dedicate this volume to two older sisters, Martha Simpson Eastlake and Joanne Laporte Sheridan, whose affectionate support could always be counted upon by their younger brothers.

Simple Curiosity

George Gaylord Simpson
Blantyre,
North Carolina

[ca. 1915]

YE EYEBROWS

There was a princess, very fair
   Whose mouth was made to kiss;
Above her eyes there were two brows
   Which eyebrows went like this:

    In ye ordinary manner.   ⌒ ⌒

Ye princess' father was a king;
   He liked to have his say.
He told ye princess whom to wed.
   Her eyebrows went this way:

    Ye haughty disdain.   ⌃ ⌃

That night a prince, by king abhorred,
   Did stand beneath ye wall.
Ye princess coyly gazed at him
   And let one eyebrow fall:

    Ye wink.   ⌒ —

Next day ye king did seek ye maid.
   I fear ye king did fuss
For he did find ye window wide,
   And eke her eyebrows thus:

    Ye princess had taken them with her.

"Tell you what—I'll toss you for it. Heads we'll name it Simpsonia tuberosa, tails it'll be Schwartzia tuberosa."

INTRODUCTION

P REEMINENT AMONG paleontologists and evolutionary
biologists for the two decades following World War II,
George Gaylord Simpson achieved such fame in his field—a discipline as
arcane as paleontology—that his renown spilled over into public consciousness. In the 1950s and 1960s Simpson appeared on the cover of the
Saturday Review of Literature, was the subject of a full-page cartoon in the
New Yorker, was featured in a radio broadcast by Lowell Thomas, and was
periodically mentioned in the national newspapers, especially his home-
town papers, the *New York Times*, the *Herald-Tribune*, the *Sun*, and the
World-Telegram. He even appeared on television, guiding a commentator
through the fossil displays at the American Museum of Natural History in
New York City.

Despite this public exposure and despite his many writings, both professional and popular, Simpson was not an easy man to know. Most people,
even those colleagues with whom he worked closely, viewed him as
reserved, guarded about his private life, and capable of sharp critical comment. He did not make friends easily, and some of those whom he mentioned as good friends in his autobiography were surprised to be so considered. Because Simpson made no special effort to cultivate friendship
among his many acquaintances, he put most people off. Clearly more brilliant and renowned than most around him, he accentuated the distance by
his apparent lack of warmth. By the time he died, some colleagues of his
own generation had long given up knowing him; others, through hearsay,
assumed he was cranky, difficult, even embittered.

If Simpson had been an ordinary scientist, the accuracy of these judg-
ments would probably not make any difference. But in Simpson's case we
do want to know more about the private person lying within the public
man. What was he *really* like? What motivated him?

These private letters provide remarkable insights into the Simpson per-
sona. The letters show a teenager finding his life's direction, an enthusiastic youth from the West going East to be educated. We see Simpson dazzled
by the history and culture of Europe. We learn how difficult his first marriage was and how he eventually married the woman he truly loved. By
current societal standards his marital problems were hardly remarkable, but
the way he handled them reflects American morals and culture of an earlier era. We also see Simpson's growing recognition of his own talents as he
found approval and support on every side.

The letters also reveal Simpson's wit, his keen sense of observation, especially of nature, his honesty and personal modesty, his devotion to his immediate family, and in later years his periodic illnesses.

The public face of Simpson is a quintessential American success story in science; his private face is very different, more richly complex and emotional. We are reminded of Tolstoy's opening statement in *Anna Karenina* that although all happy families resemble one another, each unhappy family is unhappy in its own way.

By the late 1930s and early 1940s the relatively young Simpson was already a distinguished paleontologist at the American Museum of Natural History in New York City. His accomplishments had been many: a Yale doctorate in geology and paleontology, a post as visiting research scholar at the British Museum, the leader of two year-long fossil-collecting expeditions to Patagonia, the author of two books and more than a hundred scientific articles and monographs, and a newly elected fellow of two of the most distinguished honorary and scholarly societies in America, the American Philosophical Society and the National Academy of Sciences. As discoveries and new ideas and theories in genetics were being published, Simpson sensed their importance and kept himself informed. Before entering military service in late 1942, Simpson completed a major revolutionary text published two years later as *Tempo and Mode in Evolution*.

Simpson's book applied the concepts and conclusions of recent discoveries in genetics to the large body of fossil evidence of life's long history, and he demonstrated that the "microevolution" of the geneticist could indeed be extrapolated to explain the "macroevolution" of the paleontologist. According to Simpson, the mechanisms of generating and accumulating inherited variation, as described by laboratory geneticists and field naturalists, provided a parsimonious explanation of the adaptations, specializations, and evolutionary trends of the paleontologists, as measured in their fossils. In this respect, then, *Tempo and Mode* became one of a half-dozen books of this era which formed the basis for what came to be called the modern evolutionary synthesis. The word *synthesis* betrays the origins of the new body of theory in a variety of fields—genetics, ecology, anatomy, field biology, paleontology, embryology, botany as well as zoology, and biogeography—which were integrated into a unified whole.

Single-handedly, Simpson brought the discipline of paleontology into the mainstream of biological research; he validated the use of fossil evidence in resolving evolutionary questions. Before Simpson, what fossils had to say to biologists as articulated by paleontologists was at best confusing, at worst contradictory. Once and for all, he debunked previous evolutionary explanations of fossil phenomena which depended on inherent or internally directed forces, like "momentum and inertia," which argued that once organisms began to evolve they continued to do so because of the momentum of past evolution; or "racial senescence," a theory that organisms in a given line may exhaust their evolutionary reserves and become extinct; or "orthogenesis," that organisms evolve toward some future goal and thus intermediate stages exist only as steps toward that goal rather than as perfectly viable ends in themselves; or "aristogenesis," that organisms are driven forward by their striving for perfection. Simpson demonstrated that such explanations were not consistent with modern genetic theory. He also delivered the coup de grace to lingering notions of the inheritance of acquired characteristics through use and disuse. Thereafter, if paleontologists were to carry conviction, they had to ground their interpretations of macroevolution in terms of microevolution.

Another equally important contribution of *Tempo and Mode* was Simpson's identification of significantly varying rates of evolution—very fast, average, and very slow—and his explanation of how such differing rates yielded characteristic patterns of evolution within the fossil record (hence the title *Tempo and Mode in Evolution*). Simpson attributed a special importance to the environment, in all its physical, chemical, and biologic manifestations, in influencing the patterns of rates of evolution.

Toward the end of Simpson's life his reputation dimmed somewhat, in part because his contributions had been so thoroughly assimilated into current theory and practice that the identity of the originator was forgotten. In addition, the succeeding generation of evolutionists, especially among paleontologists, began to question some of Simpson's conclusions. Some recent paleontologists have even challenged the idea that macroevolutionary events portrayed by fossils are merely long-term extrapolations of the short-term microevolutionary processes seen by the experimentalists and naturalists. It remains to be seen if the challenge will succeed. Whatever history's final evaluation of Simpson's contribution may be, during his lifetime he was judged *the* major paleontologist and one of the founders of modern evolutionary theory.

George Gaylord Simpson was born in Chicago on 16 June 1902. He was the third and last child of Helen J. (Kinney) and Joseph A. Simpson; two sisters, Margaret (b. 1895) and Martha (1898–1984), preceded him in the world. At the time of George's birth his father was an attorney handling railroad claims, but he soon became involved in land speculation and mining in the West, which resulted in the family's resettlement in Denver while Simpson was still an infant. His mother had been born in Iowa, but because of the premature death of her mother she had been raised in Hawaii by her grandparents, who were lay missionaries there. The family's Scots ancestry and missionary background led to a strict Presbyterian upbringing for the young Simpsons, but George turned his back on the church by his early teens.

As a boy Simpson was curious about everything. He talked his parents into subsidizing his purchase of the now classic eleventh edition of the *Encyclopaedia Britannica*, which he then read straight through. It became the foundation of what was to become a huge personal research library, and he was still using it at the end of his life. He also kept a notebook in which he recorded random facts, including such dubiously useful information as the densities of various materials.

Simpson had only a few close friends in childhood, chiefly a neighborhood chum, Bob Roe, and Bob's sister, Anne, whom Simpson would marry years later. In old age Simpson reminisced about his childhood and noted that being more intelligent, shorter, and redheaded had guaranteed him antagonism from his peers. He was also afflicted with an eye condition that made it difficult for him to follow the flight of a ball—a serious handicap preventing his participation in virtually all team sports. His father, his sister Martha, and his friends Anne and Bob Roe all enjoyed the outdoors, so Simpson spent much of his time exploring the Rocky Mountain landscape, which undoubtedly fed his interest in natural history.

Simpson attended grammar school and high school in Denver. Despite losing a year or so because of eye ailments and appendicitis, he managed to skip grades and graduate from high school close to his sixteenth birthday. In the fall of 1918 he entered the University of Colorado at Boulder. He was undecided about his major, although he thought he wanted to be a creative writer. In his second year he enrolled in a course in geology and was quickly converted, in part because of the enthusiasm of his instructor and the mutual respect each developed for the other. It was in this geology class that Simpson learned the importance of trusting his own observations, even those contrary to the received truth. On one particular field trip, when he found a bone of a terrestrial dinosaur in a deposit that was clearly marine in origin, he was told that he could not have found the bone in the location he described. Simpson quietly stuck to his guns, reasoning that the bone must have been washed in from a nearby river. And, of course, he was right.

In his senior year Simpson transferred to Yale, because he was told by a professor that if he wanted to be a geologist and paleontologist, Yale was the best place to study. Other factors may have played a role in his decision: he had lost his girl to his best friend and his position on the literary magazine he had helped found was in dispute. Yale's requirements for graduation were more stringent than those at Colorado, so Simpson had to make up several general education courses, among them a foreign language. As a result, Simpson spent the summer of 1923 in France, boning up on French so he could pass a test enabling him to graduate retroactively with the class of 1923.

In February of his senior year Simpson married Lydia Pedroja, whom he had met at the University of Colorado and who was now attending Barnard College in New York City. Yale at that time did not allow its undergraduates to marry, so Simpson and Lydia were married secretly. His parents apparently did not learn of the marriage right away either. This betrayal of his parents seems to have plagued Simpson all his life and may explain his deferential attitude toward them well into middle age.

The marriage thus got off to a troubled start. Simpson continued at Yale as a graduate student in paleontology, but Lydia's mental instability soon became apparent and compounded the normally difficult circumstances of graduate students burdened with young children and meager incomes. Despite their problems, Simpson and Lydia produced four daughters in their first six years of marriage: Helen, Patricia Gaylord, Joan, and Elizabeth.

At Yale Simpson discovered a large collection in storage of primitive mammals from Mesozoic-age rocks of the American West. Despite the initial lack of enthusiasm on the part of his doctoral advisor, the distinguished paleontologist Richard Swann Lull, Simpson decided to work on these fossils for his dissertation. Lull's hesitation revealed some uncertainty about Simpson's qualifications for the task. Moreover, Simpson had earlier annoyed Lull by his elaborate preparations for Lull's lectures on vertebrate paleontology. In class Simpson would visibly tick off in his own notes various points as Lull covered them. Lull found this unnerving and asked Simpson to stop. Simpson continued the practice, but unobtrusively.

Simpson received his doctorate in 1926 and moved on to the British Museum of Natural History in London to continue his study of primitive mammals and examine British and European specimens. Lydia decided that London was unpleasant and took their two daughters to spend the year in southern France, which further strained family finances. Yet the separation gave Simpson the freedom of a bachelor in an interesting foreign capital. He thoroughly scouted London, the English countryside, and made occasional side trips to the continent. He was able to meet the leading scientists in his field as well as indulge his growing interests in art, sculpture, architecture, foreign languages and customs, and he imbibed a cosmopolitanism hitherto denied the provincial youth from a Western cowtown.

On his return from England in the fall of 1927 Simpson joined the scientific staff of the American Museum of Natural History in New York City as assistant curator of fossil vertebrates. He accepted the museum's offer after rejecting an offer made by Yale University, which had vacillated because of malicious gossip about Simpson's treatment of his wife. Simpson replaced William Diller Matthew, who left the New York museum to assume the directorship of the museum of paleontology and the chairmanship of the reestablished department of paleontology at the University of California, Berkeley. Matthew was impressed with Simpson's studies of Mesozoic mammals and with the endurance and diligence the young scientist had displayed as his field assistant in the baking summer heat of Texas when he was a first-year graduate student at Yale.

Having established himself as a paleomammalogist, Simpson wanted to continue his fossil mammal studies by examining the South American strata in Patagonia, which had yielded unusual and important fauna that had evolved in isolation from the rest of the world during the long time that South Africa was an island-continent. Simpson befriended a museum patron, who put up the money—so critically short during the Great Depression years—for two expeditions to Patagonia, one in 1930–31 and another in 1933–34. In later life Simpson remarked that he spent many hours drinking with the patron while persuading him that the expeditions were necessary; Simpson quipped, "I only regret that I have but one liver to lose for my museum." Simpson's first book, *Attending Marvels*, is a travel journal of the first expedition. It was quite successful (it is still in print) and brought Simpson notice from the world outside of paleontology, including a radio interview in New York City and front-page coverage in the *New York Times Book Review*.

In the late 1930s and early 1940s Simpson's work turned more theoretical as his focus shifted from fossil mammals to problems of evolution in general. Just before America's entry into World War II, Simpson published *Quantitative Zoology* and completed two book-length manuscripts, which were published as *Tempo and Mode in Evolution* (1944) and *Principles of Classification and a Classification of Mammals* (1945).

When Simpson left for South America in 1930, he had broken with Lydia, although they would not be legally separated until two years later and were only finally divorced in 1938. Lydia had a long history of mental problems which had begun even before she married Simpson; some were serious enough to require hospitalization. The daughters had been constantly moved about, at times with Lydia, at other times with their grandparents, and occasionally with their father. Domestic life was by no means tranquil and one wonders if part of Simpson's motivation for his two South American ventures wasn't the prospect of escape from an increasingly difficult situation at home.

During his last year of graduate work at Yale Simpson had chanced to meet Anne Roe, his childhood friend from Denver, who was working for her doctorate in psychology at Columbia. Great friends as children and teenagers, they had lost touch over the intervening years. She had gone to Denver University while he had moved East to Yale. Moreover, Anne had had a brief adolescent interest in fundamentalist evangelicism, which had rather discouraged Simpson. Thus, while remaining friends, they had drifted apart. When Simpson saw Anne again in New York, their friendship was revived. Both were now mature adults with fully developed, interesting personalities. Their renewed relationship quickly developed into a love that could not be acknowledged. Within a few months Simpson left for his postdoctoral research in London; he and Anne managed to correspond surreptitiously through his ever-reliable and supportive sister, Martha. Anne came to realize the impossibility of their situation and impetuously married while Simpson was in London. Rather than making things simpler, the marriage only created further complications, for Anne and Simpson began seeing each other again when he returned to New York in the fall of 1927. By 1930, when Simpson left for a year of fossil hunting in Patagonia, he was no longer living with Lydia, and Anne was questioning her own marriage.

When Simpson returned in 1931 he immediately filed for a legal separation from Lydia, which he obtained in early 1932. For the next six years Anne, by now separated from her husband, and Simpson lived together whenever their respective jobs permitted them to do so. For a while Martha even joined the ménage, and all three happily lived together in a New York City apartment. Simpson eventually moved to Connecticut (and commuted to the museum in New York) so he could file for divorce from Lydia on the grounds of mental cruelty, which was not permissible in New York State. His eldest daughter, Helen, was living with him, attending a nearby private school as a day student; Patricia Gaylord (called "Gay" in the family) was living with her maternal grandmother in Kansas, and the two youngest daughters, Joan and Elizabeth, were in the temporary custody of their mother, pending the outcome of the divorce and final custody arrangement.

In April 1938, after a rather long court battle, Simpson was awarded a divorce from Lydia and custody of the two older children. (Not long afterward, he received custody of the other two girls as well.) A month later Simpson and Anne were married. Later that year they went to Venezuela on a fossil-collecting expedition; Anne put aside her own work in clinical psychology and collected various mammals for the museum. The following summer the happily married couple set up house in New York City and brought together the far-flung family for the first time in years, except for Gay, who remained in Kansas where she had lived practically all her life. The next three years were tranquil: Anne was caring for the family and working part-time as editor and researcher, the girls were in junior and senior high school, and Simpson was working on the manuscripts of *Tempo and Mode* and *Classification of Mammals*, among many other projects.

In 1942 the director of the museum was contemplating reorganization of the various departments, which included combining the paleontologists with the zoologists, that is, putting scientists working with living groups with those studying fossils. Simpson resisted this change and became depressed about his prospects. Before he was forced to make a decision, however, he enlisted in the armed services and in December 1942 joined military intelligence as a captain in the U.S. Army. He must have flustered his superiors by completing a six-week course in intelligence methods in a single week. By spring of the following year he found himself in North Africa as part of the Allied Forces Command led by General Eisenhower. He was transferred to Sicily and Italy during the invasions in the summer of 1943 and returned later to North Africa, where he remained until the fall of 1944, when he was shipped home with a severe case of hepatitis. He held the rank of major and two Bronze Stars.

When Simpson returned to the museum, the earlier reorganization plans had been scrapped and a different plan proposed, which included a department of geology and paleontology of which Simpson was to be named chairman. Simpson also accepted appointment as professor of vertebrate paleontology at Columbia University. In 1949 Simpson published a popular account of modern evolutionary theory with reference to the fossil evidence, *The Meaning of Evolution*, which was subsequently translated into ten languages and sold some half a million copies. This book further familiarized the educated public with Simpson the scientist. A few years later Simpson completely reworked *Tempo and Mode* into *Major Features of Evolution*, considered by some a more ponderous, detailed version of what had been a succinct and brilliant monograph.

In 1953 Simpson published a small volume, *Evolution and Geography*, which climaxed a series of writings published over more than a decade, all of which addressed the principles for explaining the distribution of fossil land animals, especially mammals. Reacting to the ideas of Alfred Wegener, the German scientist who argued that drifting continents had carried land animals far and wide from their original distribution, Simpson made a cogent case for the natural dispersal mechanisms of organisms over long geologic time intervals to accomplish the same results on continents that did not move. So great was Simpson's authority and so persuasive his reasoning that the theory of continental drift slipped still further in its credibility with North American geologists. Wrong for the right reasons, Simpson only converted to the new theory of plate tectonics when the geophysical evidence from the oceans provided compelling proof that the seafloor was spreading from mid-ocean ridges that carried the continents—and their resident flora and fauna—farther and farther apart. Simpson, like most other scientists who live their allotted three score and ten, was forced to acknowledge that what had been learned earlier in his professional career was now either wrong, obsolete, or irrelevant. He did not relish the opportunity for recantation the way a number of his colleagues did, born-again enthusiasts for the new geology.

The centrality that Simpson enjoyed in the earlier revolution, that of the modern evolutionary synthesis, was visible beyond his contribution of learned treatises. Just before the war he helped found the Society of Vertebrate Paleontology and became its first elected president. After the war he helped to organize the Society for the Study of Evolution, which brought together a whole range of life scientists, including paleontologists, working within the new paradigm. He was its first president and, later, when the society wanted to begin publishing its own journal, *Evolution*, Simpson managed to raise initial funding from private sources.

In the summer of 1956 Simpson undertook yet another South American expedition, this time to the headwaters of the Amazon in Brazil to collect fossils to fill in gaps between Argentina to the south and Colombia to the north. The area was so covered with vegetation that the best prospects for exploring outcrops were along the river banks exposed during seasonal low-water. Simpson headed a small party of North and South American scientists, assisted by several Brazilian workmen. Near the end of their river trip, camp was being cleared along the river's edge and during its preparation a tree fell right on Simpson, causing multiple injuries, the most serious of which was a bad break of his lower right leg. After a painful and dangerous trip back down the river and a series of transfers from boats to planes, he arrived a week later in New York City. He spent the next two years in and out of hospitals, undergoing twelve separate operations, one of which was necessary because of a surgical sponge left behind from the preceding operation. Simpson resisted all advice to have the leg amputated and finally recovered the use of the leg, although it always pained him and he was somewhat lame.

During this painful time Simpson's second daughter died. Gay had been born with a congenital heart defect and had been shielded from most of her parents' marital problems by living with her maternal grandmother in Kansas. Recently married and working as a librarian, Gay died of a brain abscess at age 31.

In 1958 Simpson gave up the chairmanship of the department of geology and paleontology at the American Museum, under pressure from the museum's director. Shortly thereafter Simpson resigned from the museum altogether and was appointed as one of the Alexander Agassiz professors at the Museum of Comparative Zoology, Harvard University. There seems to have been some resentment of Simpson's periodic absences from the museum during his two-year convalescence and the delegation of his administrative chores to others. The director, intending to lessen Simpson's museum responsibilities, suggested that he accept a sinecure, but Simpson, feeling that he had been carrying on satisfactorily, was so insulted by the idea that he made up his mind to seek a position elsewhere. Despite the turmoil, Simpson managed to produce a major biology textbook, *Life: An Introduction to Biology*, in collaboration with two other biologists, see into print a number of articles started before his accident, and write a number of book reviews.

The centennial of Darwin's *Origin of Species* was celebrated in 1959, a year that not only signaled a fresh start for Simpson at Harvard but also brought him once again into the spotlight as a leading evolutionist. Conferences, symposia, and special events marked the centennial, and Simpson was often present either as a contributor—as in Chicago where he gave a keynote public lecture at the annual meeting of the American Association for the Advancement of Science entitled "The World into Which Darwin Led Us"—or as an honoree—as in London where he received the Darwin-Wallace Commemorative Medal from the Linnean Society in whose meeting rooms a century earlier Charles Darwin and Alfred Russel Wallace first announced their theory of natural selection as the mechanism for evolution.

Darwin always held special interest for Simpson, not only for the obvious reasons but also because Darwin represented the quintessential liberator of the human spirit, seeking to find answers to the puzzle of human existence without recourse to supernatural explanation. Simpson's own penchant for a naturalistic, positivistic view of the world was well reflected in his reading of Darwin. Thus Simpson regularly took the opportunity throughout his career to speak and write about Darwin for nonscientific or nonspecialist audiences. He was thus in his glory during the Darwin revival of the late 1950s.

Simpson and his wife Anne settled in at Harvard. He was busy writing once again, but because of his accident his fieldwork days were over. Anne was eventually granted a full professorship in the graduate school of education. They were thus the first married couple to receive professorial appointments at Harvard. They spent winters mostly in Cambridge, summers at their simple home in the New Mexico mountains, with a number of trips abroad, including England, France, and Spain as well as Africa and Australia. These trips were as much opportunities to renew scientific friendships and receive further honors and awards as they were to examine fossil collections in museums and universities. The Simpsons had conversations with the duke of Edinburgh, the granddaughter of Darwin, the Huxleys, and the Leakeys among others. Nor was Simpson ignored at home: in February 1966 he received the National Medal of Science from Lyndon B. Johnson, then president of the United States.

Yet not all was unmitigated satisfaction. Simpson and Anne both suffered heart attacks and were hospitalized. In fact, some of their traveling was in part intended as recuperation. And although, strictly speaking, Simpson's Agassiz professorship at Harvard did not explicitly require residence at the museum, his frequent absences once again made for some difficulty with the administration.

In 1967 Simpson and Anne decided to move to Tucson, where they had earlier bought a house in preparation for retirement. He was given an appointment at the University of Arizona, which involved some teaching at first and later evolved to weekly luncheon meetings with interested students and faculty. By 1970 he had severed all formal ties with Harvard. Simpson continued to publish books—on South American mammals, penguins, Darwin, fossils and the history of life, collections of essays, and his autobiography—as well as the usual monographs and articles on Cenozoic mammals, obituaries of deceased colleagues, and book reviews. He worked in a small building he built next to his house, surrounded by his research files and extensive personal library, walls and surfaces scattered with honorary degrees, photographs from the past, and replicas of the many medals he had been awarded.

Despite the change in locale and greatly increased domestic tranquillity, Simpson's life did not alter much. In fact, his Arizona years reveal an intensification of the activities he really loved: close, daily contact with Anne; small, informal get-togethers with close friends; occasional visits from children and grandchildren; trips of varying length in the U.S. and abroad; association with the academy without too much formal responsibility; and long stretches of time to devote to his research and writing with minimal interruption or distraction. Of course, there were periods when his or Anne's health was poor, which then required a slow rebuilding of energy and reserves to reestablish the routine.

In the last decade of his life, as his friends and colleagues died, Simpson himself seemed to become more a memory from the past than a prime mover in the present. Most thoughtful students were aware of his contributions but took them for granted. They looked to the writings of younger paleontologists and evolutionary biologists for new ideas. In a way, Simpson had outlived his fame. He had become a living monument of past discoveries.

Simpson's father lived to age eighty, his mother to ninety. His sister, Martha, to whom he was so attached and to whom he had written the bulk of the letters contained here, died in July 1984, at age eighty-six, of complications resulting from prolonged emphysema. Martha had moved to Tucson some years before and lived nearby. Brother and sister saw each other regularly, and Martha often accompanied Simpson and Anne on their trips. When Martha died, Simpson himself was suffering the aftereffects of a serious attack of pneumonia that he had contracted on a South Pacific cruise. His condition worsened over the summer and he succumbed on Saturday, 6 October 1984, at age eighty-two. His remains were cremated and then dispersed in the Arizona desert.

Colorado
1921-1922

GEORGE GAYLORD SIMPSON entered the University of Colorado as a freshman in the fall of 1918, a few months after his sixteenth birthday. He spent three years at the university without choosing a major field, although he pursued his interest in writing, including poetry. One outcome of this interest was the founding of a college literary magazine, *Dodo*, together with two other classmates, one of whom, Mort Lippman, later wooed away Simpson's girlfriend, Helen Stewart, who also worked on *Dodo*. Another young woman, Kate Smith, was sweet on Simpson, but he wasn't at all interested in her.

Simpson's first geology course converted him to the discipline. His earlier interest in nature and the outdoors no doubt influenced his conversion. Simpson's geology instructor, Arthur Tieje, was an appropriate role model. He not only had a doctorate in geology but a prior one in English. Tieje had taught scientific composition at the University of Minnesota for several years before deciding to pursue his doctorate there in geology. Simpson always acknowledged in his later years the importance of Tieje's influence on him, especially his high standards and his refusal to condescend to his students. Simpson found this challenging, and it may also be that Simpson liked the idea that Tieje was a literate scientist.

Toward the end of his freshman year Simpson was hospitalized with Spanish influenza. The following year he stayed out of school because the family finances were, as he put it, "at the lowest ebb ever." His father had speculated in a gold-mining venture in Colorado which failed in the postwar years. Simpson took a job first in Chicago, later in New Orleans and Port Arthur, Texas. Although his parents were fairly strict disciplinarians, they obviously gave their children considerable freedom, enough to let their seventeen-year-old-son traipse about the country for a year on his own. His sisters Martha and Margaret had also left home in their teens. Martha went off to the Art Institute of Chicago to study painting, Margaret to travel and soon to marry.

Simpson returned to the University of Colorado in 1920 to complete his sophomore and junior years. During this time Tieje left the university for the Los Angeles Museum of Science and two years later joined the geology faculty at the University of Southern California, where he became professor and department chairman. Tieje advised Simpson to transfer to Yale if he planned to specialize in paleontology, for Yale was then one of the strongest schools, both in geology and paleontology.

As a youngster in Denver as well as a college student, Simpson held a wide variety of jobs, part-time during the school year and full-time over the summers. These varied from handyman, park guide at a summer resort, bellhop, mule skinner, housepainter, trail-builder in a national forest, runner for a bond broker, and advertising copywriter for a piano company. This no doubt gave him plenty of opportunity to see the outside world and to judge where he might fit in.

By the time Simpson left for college, he had given up the strict fundamentalist Presbyterianism of his family. For a while his parents had taken him to triple Sunday services and midweek evening prayer meetings. By his early teens, Simpson had given up being a Christian, although he had not formally declared himself an atheist. At college he began the gradual development of what might best be called positivistic agnosticism: a belief that the world could be known and explained by ordinary empirical observation without recourse to supernatural forces. Ultimate causation, he considered unknowable.

Simpson met his future wife, Anne Roe, when both were quite young. They lived on the same street in Denver, and one of Simpson's best chums was her brother, Bob. Thereafter, Simpson and Anne, as well as several other neighborhood youngsters, spent much time together, especially singing songs around the piano as Anne played. Anne and Simpson kept up a correspondence while he was away at college, although it cooled somewhat when Anne flirted briefly with Christian evangelicism, and their communication stopped when he married at Yale.

The letters in this section were all written during Simpson's last two years at the University of Colorado: five to his sister Martha, who was six years older than he, and one to his father. It was apparently when he was first away at college that Simpson began a lifelong routine of weekly letter-writing, usually on Sundays, to his sister and parents.

Calc Class—Monday
[Univ. of Colorado, Jan. 31, 1921]

Dearest Martha Lee—

I hear very indirectly (from mother in other words) that my only famous relative (outside of Cromwell & Louis XVI & they're dead) is considering accepting $500 a week from Redpath. Today's the 31st—you'll know tomorrow. Let me know immediately as I am quite as thrilled as a hard-boiled Associate Editor dares allow himself to become. The suspense resembles the sensation of rapid sinking in an elevator or that awful moment when you come to class unprepared & the prof. starts putting quiz questions on the board. (The latter simile's better but too provincial for my broad public.)

I was very sorry to hear this morning that Dad is sick & has had an operation. Of course, mother wrote you & you know as much about it as I do.

Of course, also, either mother or myself has told you that we finally had our magazine proposition [for *Dodo*] authorized & that I am duly appointed Associate Editor with a bright airy office & regular office hours as well as three assistants & a secretary. I burst with importance & have given up wearing a cap (altho I had before because (1) I like to (2) my cap is worn out (3) it pays to advertize.) I appointed a beauteous ΔXZ damsel whom I fuss some 6 or 7 times weekly, as my secretary & so my office hours never seem too long. Very good idea.

I am, while we're on the subject, in the midst of the struggle of puppy-love & every night I dream of the following squib which never will appear in the Silver & Gold (our [college] newspaper). "Delta Chi Zeta announces the engagement of Helen Stewart to George G. Simpson. Mr. Simpson is a prominent independent, very active in literary circles." Isn't this pathetic?

I do get my name in every issue of the S.&G. because I write our magazine news stories & am so unblushing as never to fail to mention the associate editor one or more times.

I'm afraid I'd better listen to the prof. for a few mins. & finish this tonight.

NOW—in my office with the seething mob locked out. Secretary not present. I just finished a heller of a psych quiz. It's terrible the way lessons do bust in on a fellow's time here. Some days I have very little time to devote to anything else.

You remember Kate Smith of course. She's here now but she & I get along like chickens & hawks in friendly amity (to be redundant). I am finally leaving my boarding house, as Kate has taken up there. She's a nice girl—but oh my!

Recently heard The Devil's Disciple read. You've read it, of course, & I agree with you that the characters are not people but are symbols of people & that it probably means something or other.

Also heard Sherwood Eddy[1] who (figuratively) patted us all on our nicely combed heads, beamed out at our shining faces, & told us what to think. Whereupon we went & didn't. Some of the freshmen who still have to put sand in their shoes to feel at home enjoyed him.

Also heard the New York Chamber Music Society or Ensemble or something. Very good for Boulder. First *music* I've heard since I left Chicago.

And I crown my intellectual achievements with the digestion of Sara Teasdale's "Love Songs" & the current "Poetry" & "Poet Lore", and am beginning on the "Dial" you so kindly sent. So far—fine. Thanks. The

---

1. Sherwood Eddy was an official of the YMCA, a Christian internationalist, and author of the book *Everybody's World*. He was visiting Boulder and giving a series of talks about the postwar situation worldwide.

Sarg program is good too—how were his pups? Kindly don't attempt comparison with yours, start on a lower level.

Of course I read a lot of college magazines too.—I have here now a "Pelican", a "Crimson Bull," two "Punch Bowls" & a "Cornell Era", but wouldn't class these with the intellectual refreshment. Also Judge & Life, not, as customarily taken, as a pabulum, but as a stimulant. Not that *we'd* accept that sort of drivel, but you never can tell what's likely to give me an idea which I can turn into something good, something worthy of the "Dodo" (for such is the name of that gem of humor & literature which, had we not hurriedly named it, was in danger of becoming universally known as Mort-George-and-Dick's Dirty Sheet, altho, since we've not issued any yet, how on earth did they know?).

This week-end I go to the player's club production of Pinero's "*Thunderbolt*" & to another play-reading Wednesday. Are not my pastimes elevating? Tonight I dine with Dr. Williams.[2] He has some proposition to make to me & I am all agog (forming a frenzied mob all by myself as 'twere) to see what now. He can't surprise me tho. I refuse to be surprised. I've thot of everything from an attempt to rejuvenating my soiled & frayed but still struggling soul, to making me editor of *Atlantic Monthly*. What do you bet the answer is? I hope he takes me to the Boulderado. That's our good hotel—and it really is good—& I won't even object to the soul-saving if we go there. Mercenary—that's what this life does to a fellow.

Well, I must write a couple of more letters then if no one drops in I'll go home & have it out with the landlady. (You see, I'm moving tomorrow & have not told her as yet.)

Yours till the ink gums,
G

(I indulged, nearly to repletion, my love of parentheses & —'s in this letter; a member of the family won't mind.)

[Univ. of Colorado, 1921]

Chère Martha Lee—

To one of my natural higher sensibilities & possibilities the student life is deadening is it not? That is the cue to my thots this evening and on that topic I would discourse somewhat as follows—

An example might first be given, such as when I was experimenting on etching glass with fluorite[3] & instead of inscribing an intricate scientific figure etched the following little ditty.

> This dainty little flower
> So pleasing to the sight
> Was etched upon the glass
> By means of *fluorite!*

2. Dr. Williams was the minister at the Capitol Heights Presbyterian church in Denver who was apparently keeping an eye on his wandering sheep.

3. Fluorite is a fluorine oxide mineral standardly used in beginning geology courses to determine the relative hardness of various other minerals.

And was forced to do the whole experiment over by a professor who thinks frivolity in the presence of science the utmost sacrilege.

I might then continue to tell you how my soul abhors the petty things & longs to roam, in fancy, with yours, mon amie, among the realms of the eternal—

—as I said, I would say all that but I haven't time.

When did I write you last? Seems to me I wrote you a couplatimes since Xmas, but, strangely, I just found one of the letters floating around here & can't remember mailing the other. Anyway there's no news. I'm back at school but not yet studying noticeably. Kate Smith is up here too & leading me a merry life. She rooms & eats at the house where I eat & she tries to be my little guardian angel, aged advisor, & fiancée all rolled into one sweet package but doesn't make a noticeable success in any of those rôles. Let the good work continue.

I don't know all my marks for last quarter yet but those I know are Chemistry Lectures—97. Chemistry Lab. 92. Physiography 97. Psychology 90. All of which are very satisfactory.

Also my room-mate got put out of the house by the landlady (He was a ΣN & she hates 'em) so I have a new one—John Stahl—a wild boy from El Paso whose faults are human and whose—etc. He's a darn nice fellow and the bottom part of his trunk contains 2 decks cards, 1 p'r dice, 1 bottle (quart) gin, 1 bottle Tiquila [sic] (distilled Pulqué), 4 bottles whisky, 1 six-shooter—a most interesting chap and quite naïve, for all the accoutrements. But I expect he'll be bounced out soon too unless he uses discretion where he gets stewed.

Well, must go walking to clear my head & gimme a chance to figure out a few of my social difficulties & will mail this on the way.

Your uncouth brother
George Gaylord[4]

There's something else I want to say but I can't think of it. Consider it said

G

[Estes Park, Colorado]
Bear Lake Camp
July 10, 1921

Dearest Marty:

Mother just forwarded me a letter to her on the cerulean Chicago, Duluth, and Georgian Bay Transit Company informing me of the disgrace that has fallen on our family. My sister a waitress. Horrors! (Do you get good tips?) However I am not jealous. I have succeeded in descending even lower. I am a mule skinner. My mulesmanship is superb. It would surprise you—It does me. What I lack in vocabulary I make up in enthusiasm. In my odd moments I convey tourists from hither to yon for the Lewis Byerly

4. Simpson used his full middle name of Gaylord from this time to differentiate himself from all the other George Simpsons in the world. Gaylord was the surname of Truman Gaylord, a lawyer friend of his father's in Chicago.

Tours. This is an ideal place to work. "Glaciers, gorges, magnificent alpine scenery, wildflowers, easy (!) one day hikes with competent (!) guides, happy hours in the heart of the glorious Rockies"—to quote the prospectus. I live all over the landscape, but if you address the reply you are about to pen to George G. Simpson, Estes Park, Colorado, c/o Frank W. Byerly, he'll see to it that I get it before the summer is over.

I have just returned from a trip to Grand Lake & return by a different route—several days of walking staying at $6 a day lodges—expense of Frank W. I am now engaged in piloting a New York young lady on a week tour of the park and am spending a couple of days at Bear Lake—my home camp. The young lady is about three-fourths dead, so the trip may end suddenly. However, as Omar used to say Kismet, Allah, I enjoy life whatever happens.

I rather expect to go over to Grand Lake Lodge to stay for quite a while soon. It is great over there. Sumptuous hotel & sylph-like waitresses for the cheap help (C'est moi) to kid with. The people everywhere are fine & I find friends wherever I go.

You will be intrigued to learn that I am now equipped with a pale pink mustache which, in the Roe-Simpson exhaustive classification, falls into the class of Junior Walrus. It is extremely artistic and attracts more attention from tourists than does the less wild scenery, such as the Peaks, Gorges, & Glaciers.

There are about a dozen individuals in camp, each a perfect jewel in this perfect setting. We have native Estes-Parkians whose conversations, during those supremely esthetic moments that come to all of us, turn to the "Tryolyan" yodelers or the "Peerless" saxophone trio which they heard in Niwot in '74. We have the captain of the football team of a miniature college in Crete, Nebraska. We have a retired cowboy more stunningly homely & more *intensely* humorous than Will Rogers. We have a rolling stone who has just been relieved from an inforced cessation of rolling involving extended incarceration. We have a vocabularious mule skinner (all supplies are packed in on burros). And we have others. My education progresses.

There is a miniature riot going on outside & I must go out.

Yours,
The blunt manly son of the
hard-riding west.

Eagle, Colo.
July 18, 1922

Dear Marthé—

I couldn't stand the quietness & small town ways of Denver so I have flitted up here to the center of Style & Beauty. The Chamber of Commerce population of Eagle is 358, but that includes the deceased and the expected.

After coming up I spent five or six days on a ranch with Harry Fuller. The ranch consists of 30 acres of potatoes & one medium sized house. The logs fail to meet by a foot & are unchinked. The doors haven't been hung yet and are nailed up so entry & exit are by way of the windows which are innocent of glass. We cooked for ourselves & lived on corn fritters and rice pudding. I was sincerely sorry when we had to hitch up the potato planter & ride to town across the sage brush flats. I should hate to tackle anything rough on one of those planters. We both fell off regularly every hundred yards coming across the flats.

Eagle itself is quite Spoon Riverish.[5] Each inhabitant has a deep and interesting (but not unknown) history. And stylish! As a rising young designer you should never fail to keep in touch with the tendencies of Eagle modistes! Here, for instance is the very latest model for afternoon outdoor wear: In summer of course the designers turn their attention to sports models which are here exemplified in their most artistic and modish development.

Here is pictured the latest thing in a stylish bathing suit, which is attracting much attention along the river now at those exclusive watering places of the smart set. While here is the most popular gown for evening wear after eight o'clock. I am going to prepare a fashion letter from an Eagle correspondent for a few of the more refined magazines.

Last Sunday Harry & I issued ourselves free passes to Glenwood Springs.[6] For some reason the conductor did not collect the passes—altho, to give him credit, he tried hard to find us in order to do so. And coming back, a number of uncouth yokels, known professionally as railroad bulls, raised silly obstacles & objections to our boarding our private car, which we had hitched on immediately behind the engine. However, by a simple expedient we managed to avoid contact with these vulgar persons and had a very pleasant ride. The Canyon of the Colorado *in* Colorado is smaller than the Grand Canyon but of the same type & very beautiful. It was moonlit on the other side of the gulch, & our side would be lit up fitfully when the firebox was opened. We were sincerely glad that our car had no roof on.

5. Eagle is a small Rocky Mountain village in central Colorado, some thirty miles west of Vail. The short poems of Edgar Lee Masters in *Spoon River Anthology* (1915) described life, love, and death in small-town America.

6. Glenwood Springs is a Rocky Mountain resort town in central Colorado with hot springs, some twenty-five miles west of Eagle.

Mother says that the book arrived OK in Denver but I probably shan't have it forwarded, for safety's sake. Thanks—I look forward to it.

We don't know yet what we shall do here, but probably shall work in a mine about 20 miles from here.

I'll have to have mother forward this as I don't know your address.

George

I am going to Yale, & plan to get out there about the 20th of September or a few days after. You won't be gone by then will you?

GS

Box 146
Eagle, Colorado

[Eagle, Colorado]
July 28 [1922]

Dear Dad—

I haven't had a renewal of your address for a long time & can only hope that 4422 will reach you eventually if not now.

Harry & I have been painting a house & just finished day before yesterday. It is a wooden house, of course, with General Grant style wood scroll work stuck on everywhere so that it looks like the abortive child of a cheap lacemaker and a lame carpenter. However it was quite a joy to feel the creative thrill and development of the delicate technique in bringing out most effectively each hideous little angle and ugly little curve of the arabesques and curlecues in beaver brown against the slate gray of the body. It was formerly forest green on canary yellow, so we did improve it somewhat.

I guess the house in Denver is coming all right & mother also, but her letters are not at all explicit. I begin to feel the reason for remonstrance when I am less explicit than I might be. But I have written a number of times—much better than once a week.

Harry's father is County Assessor & has only two joys left in life—politics and being unreasonable & mean with everyone. "Well, sir," he began as soon as I was seated the first time I entered this home, "I am told that I have an incurable disease." And thence from inception & symptoms thru many diagnoses & misdiagnoses along with patent medecines [sic] and other failures at a cure to the conclusion that he is an old man "and its pretty tough sitting here and dying by degrees thisaway." And then his wife will come and he'll spend the next five minutes growling and swearing at her. She is a pleasant but unattractive old lady who has been thru all the pioneer trials & hardships out here in the West and is difficult to comprehend but easy to like & get along with.

This is a farming town but everyone in it thinks that he would make a wonderful miner. Passing thru the country with a prospectors [sic] pick in my belt I seldom pass a farmhouse or stop for a drink but some grizzled old fellow does not drag out a piece of ore or a sack of ore from his mine or claim up the creek. The ore is always poor & usually consists of barely visible specks of pyrites.[7] Going up the Muddy[8] about 15 miles from here yesterday we stopped at a little godforsaken hovel in the exact middle of as flat & drap [drab?] & hot & hopeless a sagebrush prairie as can exist anywhere. The hut had one door with no hinges & an unconceivably [sic] old woman with two very old men, her sons, stumbled out. The younger son, who talked to us, was clad in a pair of what had been socks, & a pair of what had been pants. The latter had slipped below his flabby rotundity and were scarcely retained at a level of some decency by a string over one shoulder. He told us that he was a poet, but even on urging would recite none of his verse so he was surely not a typical poet, being either a liar or crazy. Then he spotted our picks and brot out his ore. Fool's gold, of course, & almost none of that even. But he'll live another 50 years in the surety of wealth by the time he's, say, 125. Mr. Fuller says he appeared from nowhere way back in the 80's with $90 in his pocket. This he spent in the nickle [sic] in the [sic] slot gambling machines, going from one place to another without pause to eat, drink, or sleep until it was all spent. Then he went out and settled down in an abandoned cow barn & later his mother & brother drifted in. He raises enough potatoes to live on. His name, by a queer circumstance, sounds exactly like *beaucoup*!

There are lots of characters like that.

In town are a few real prospectors who nightly & dayly [sic], when they can corner us, tell us of lost mines and strange rocks and rich veins. There is a young revival in process at a camp about 20 miles from here, Fulford. We may get work there later.

Tomorrow we are going back in the depths of the forest primeval to build trail for the Forest Service in Holy Cross Forest, Eagle Division. We will be alone & camp out, moving with the work. We are to receive $3.50 per day until between us we have earned $150 which has been appropriated. After that we hope to do a little mining. I think we will enjoy it very much, this trail work I mean. No pick & shovel work, just saw & axe and a healthful 8 hour day outdoors in nice but not grand surroundings. Also we will have quite a chance for a little geologizing by the wayside.

Love,
George

Box 146,
Eagle, Colo.

7. Pyrite, or "fool's gold," is a relatively worthless mineral of iron sulfide.
8. Muddy Creek is on the western edge of the continental divide and drains into the headwaters of the Colorado River.

Out in the Sticks
Sometime in August
[1922]

Dear Marthé—

Mother forwarded me a letter from you so at last I know where you are and what you are doing and why. Apparently you are well informed as to me.

One of my little playmates & I are out in the deep dark forest primeval sawing logs and swinging a pickaxe & grubbing hoe & axe & shovel & a few other sordid instruments which it is too painful to mention here. We are repairing trail. No one ever goes over these trails but the people who work on them. The finished part is handy to get to work on! I hadn't seen a soul except Harry [Fuller] for two weeks until the Ranger came up tonight to sample some of my rare bisquits & put us on a new trail.

We have the sweetest little horse. He has a Roman nose & a melancholy eye. We call him Manuel because he follows after his own thinking and his own desires. We spend more time chasing him than working on trail. He thinks it's a game and enters into it in a gentle but uncontrollable spirit of joyousness.

We live in a tent, which is so full of paraphenalia [sic], from a pack saddle to a portable stove, that we have to go outside every now & then, turn around & carefully insert ourselves again in order to keep our lameness equal on both sides.

This is country such as a colorist might dream of. The atmosphere is heavy & hazy, so the lavenders are here, and of course the sky is unoriginally blue. But the canyons are deep gashes, two or three thousand feet deep, cut thru varying beds of intensely red sandstone. The walls are precipitous, castellated & buttressed. In the bottom grow the pines & spruces, silvery to deep green, & aspens & willows supplying a lighter green which contrasts vividly with the red where they climb up little gulches in it.

I am sending off my final registration papers to Yale by the ranger along with this, so I guess I am really going. Are you still planning to dash over to Paris? and you're coming right back or at least soon aren't you? Let's both get jobs on a passenger boat & go over next summer after school is out & I'm finally authorized to step forward and look the world over.

Living in the woods is conducive to mental simplicity. But that's the only thing unpleasant about it.

Love,
George

Harry & I have read a book "Mayfair to Moscow" by Clare Sheridan—English Sculptress with a drag.[9] She was coming to America. What about her?

Box 146, Eagle, Colo.

9. Clare Sheridan (1885–1970), Irish sculptor, traveler, and writer. She visited the Soviet Union in 1920 and did busts of various Soviet leaders; she later wrote of her experiences in Russia. The phrase "with a drag" means with influence or pull.

YALE
1923-1926

SIMPSON STARTED his senior year at Yale in the fall of
1922. His performance at Colorado had earned him a
scholarship from Yale, which, supplemented by part-time tutoring, allowed
him to meet all his expenses. Soon after reaching the East, Simpson
encountered Lydia Pedroja, whom he knew from Boulder and who was
attending Barnard College at Columbia University in New York City.
Despite Yale's prohibition of undergraduate marriages, Simpson and Lydia
secretly married in February 1923.

Yale's graduation requirements forced Simpson to make up courses he
had missed at the University of Colorado, including American history,
economics, and two years of a foreign language. To fulfill the language
requirement, his only obstacle to graduation in 1923, Simpson traveled to
Paris that summer and became sufficiently competent in French to pass the
second-year examination on his return and thus graduate retroactively
with the class of 1923. Although Simpson states in later autobiographical
notes that Lydia accompanied him to Paris, the letters from France do not
mention her. Apparently, Simpson kept the news of his marriage from his
parents as well as from Yale.

While in France Simpson (and presumably Lydia) rented rooms from
Jean Van Dongen, a Dutch-born sculptor who was the brother of the
rather well-known Fauvist painter, Kees Van Dongen. From Simpson's letters it is obvious that he plunged headlong into the French language and
culture, developing a special interest in medieval art and architecture.

At this time Martha was living with Simpson's parents, who had relo-
cated in Washington, D.C., where his father was employed as an attorney
with the Federal Trade Commission. Shortly thereafter Martha herself
went off to France, as did so many young Americans in the 1920s, the generation of Hemingway and Fitzgerald.

After their return from France in fall 1923, the Simpsons settled in New
Haven and he began his graduate studies in geology and paleontology.
Their first child, Helen, was born that winter. The following summer
Simpson spent in the field in the Texas Panhandle, collecting fossils with
William Diller Matthew, a distinguished vertebrate paleontologist from the   27

American Museum of Natural History in New York City. It was on this expedition that Simpson made his first important fossil discovery, a complete skull and partial skeleton of a Pleistocene ancestor of modern horses, *Equus simplicidens*.

After Matthew returned to New York, Simpson spent the rest of the summer collecting additional fossils in New Mexico for another vertebrate paleontologist at the American Museum. Simpson apparently misunderstood—if he knew at all—the conditions under which he was working; for he published a brief abstract describing the results and earned the annoyance of the senior man, who had paid for the expedition and who obviously felt proprietary about the results. Simpson seems to have thought that the person doing the work should be the person to publish it.

Simpson returned to Yale and soon decided to study a collection of primitive early mammals in storage at Yale. He was at first discouraged from doing so by his major professor, Richard Swann Lull, but Simpson quietly persisted and when Lull relented he completed his dissertation within two years.

Although still a graduate student in his early twenties, Simpson did not hesitate to behave like a professional in his field. He went to scholarly meetings, read papers dealing with his research, and associated with the leading figures. He also began publishing a stream of technical papers that would continue for the next sixty years, accumulating a bibliography that would eventually total almost eight hundred separate items: more than a dozen books, scores of monographs, and many articles, book reviews, and letters to the editor.

On the domestic side, however, all was not well. Bills piled up, another child was on the way, misunderstandings with the in-laws occurred, and the first signs of estrangement became evident. And at just this time Simpson and Anne Roe renewed the friendship that had lapsed for several years—he was twenty-three and she was twenty-one. Anne had completed her undergraduate work and a master's degree in psychology at Denver University and was now working on her doctorate at Columbia, just uptown from the American Museum, which Simpson often visited from New Haven. Thus they took the first steps toward a love affair that persisted until they could marry a dozen years later.

Toward the end of Simpson's graduate work, Yale offered him the opportunity to stay on as a research fellow. But he had applied for an overseas fellowship to allow him to follow up his research on North American early mammals by studying the collections from Europe and Britain in the British Museum of Natural History in London. The Yale offer was thus put on hold until his return.

Major publications resulting from this period include a series of a dozen papers in the *American Journal of Science* on various aspects and features of American Mesozoic mammals that Simpson was studying for his doctoral dissertation. Because he was quickly becoming a world expert on Mesozoic mammals, he coauthored an article on several tiny, primitive mammal skulls that had been discovered on the American Museum's central Asiatic expeditions of 1924 and 1925. At the tender age of twenty-three Simpson also wrote a popular article on his Mesozoic mammals for the magazine section of the *New York Times*, which was to foreshadow a lifelong interest in bringing the results of paleontological discovery to the interested public. Simpson also published a short article on the fauna of one of the quarries at Como Bluff in Wyoming which had yielded a series of spectacular dinosaurs. He also discussed the primitive mammals discovered there, and in particular he attempted to reconstruct the ancient ecosystems, aquatic and terrestrial, represented by all the fossils. This was an early paleoecological study of the sort not to become routine until several decades later.

[New Haven?]
18th [January 1923]

Dear Marthe*—

Here it is ten days since I've returned from Washington, and it seems to me that I've done nothing. Examinations start the 26th of January. I have one on Saturday the 27th, one Monday, one Tuesday, one Wednesday, and one Friday the 2nd of February. Classes don't begin [again] before the 6th, and I don't know what I'm going to do on the days off—sleep probably.

Mother said that you gave her a linoleum print to give me and that she lost it. That would be "Beauty" wouldn't it? I certainly hope that you will send me another print on that subject and also some others that you have made since my departure. "Cleopatra and Her Slave" is on my table where it is the cynosure of all eyes at Yale.

Here are some proverbs that are in the Petit Larousse [French dictionary]—

That is the question (zat is ze kouess-tcheun): Cela est la question
The right man in the right place (ze ra-it man in ze ra-it ple-se):
L'homme qu'il faut dans la place qu'il faut, and so on.

I have received a book from the National Museum which contains ". . . descriptions of all stones and minerals used as gems with illustrations, and catalogue of gems, gems of the Bible, and the mystical properties of each separate gem." The stone of your zodiacal control is Emerald. The angel of emerald is Muriel, the planet Mercury ☿, and among about two pages of other things the Emerald is "Emblematic of Happiness . . . antidote for poisons . . . bestows contentment of mind . . . nourishes the soul . . . its wearer never becomes poor . . . strikes serpents blind . . ." The Holy Grail was carved from an emerald. My it's a beautiful book. I will carry it to Wash. when I go there.

Here are some little verses that I just did. Neither good poetry, nor good grammar, but something of both.

Instant de Joie
Dans un jardin de fleurs, lointain,
Tout parfumé

*Original in French; translation by editor.

La nuit venait, et y errait
　　Mon âme troublée.

Les oiseaux chantaient heureusement
　　Chansons de rêve.
Elles me touchaient, mais me semblaient
　　Pas assez brèves.

Revenait à moi une chose perdue
　　En fort chantant;
Derrai-j'oser la supporter
　　Qu'un seul instant?

　(You well know, don't you, that in French the letters "ai" are pro-nounced like the "e" in the English word "pep" except in the words "ai" and "mais.")
　Otherwise the first verse rhymes *too* much—the double rhymes in the third line becoming the same as the main rhyme if "ai" is pronounced like "é."
　So there. Enough and even too much for a single time.
Redheadedly yours,
George Gaylord

———————————————————————

[Previous page(s) of this
letter missing]
[Paris]
[Summer 1923]

[In French, translated by editor]
　One can truthfully say that it is impossible to travel and to write about that travel at the same time. Some eight days have passed since I wrote the above. The things I have seen! If I have seen one thing, I could easily write about it, but having seen a thousand and one things, I can scarcely describe a single one. Of the great views of Paris—it's perhaps Notre Dame (especially the interior) which has moved me the most. How indescrib-able! The somber aisles—the high vaults, the lighted windows. What an atmosphere of age, of glory, of piety, of mystery—But take my advice and see the other churches before going to Notre Dame. After that, the others lack power. Add everything else that everyone sees—the streets: the grand boulevards, Champs Elysées, Rue de la Paix (where the pretty dresses and jewels are! I also saw, somewhat against my will, the summer fashion show) and so on. The parks and the squares: Place Vendôme, de la Concorde, the Luxembourg gardens, the Jardin des Plantes, the Bois de Boulogne, a thou-sand small squares and enclosed parks. The buildings, the museums, the galleries—all endless! The Louvre is a small city all by itself. I spent sev-eral days there. And yesterday I spent the afternoon there, from 11 A.M. to 6 P.M., in the Salon. I so wanted you to escape there [with me]! There were Matisses which I liked perhaps the most, but there were also at least a

thousand other paintings, half of which I liked. The sculpture seemed rather bad to me—but, of course, I'm hardly familiar with the subject. I bought two photographs that were on sale for my room. I also bought several prints at a little bookstall of which there are a thousand along the length of the Seine. They are of the city's churches. These bookstalls by the way are one of the great joys of Paris. I'm going to send you a catalogue of part of the Salon, and also a little book I bought on the Quai Malaquais (where lived that charming Sylvestre Bonnard about whom Anatole France has written).[1]

And the palaces—the Grand Palais (where the Salon was held), the Palais de la Ville, the Palais Royale, Palais du Luxembourg, and a dozen more.

I also stole a few hours for myself and spent them looking at the Museum of Natural History—the halls of paleontology, mineralogy, geology, and the zoological gardens. I made the acquaintance of Professor M. Boule,[2] a renowned paleontologist, and I am going to chat with, the first part of this week, Mr. P. Lemoine, one of the great geologists. In the hall of paleontology I met a young man who was sketching some fossils. He was a student from the French School of Mines. We chatted together for some time, while tasting a pint of beer.

But it isn't only those things that make it worthwhile to stay in Paris. There are so many other things, too small to describe, for example the "Bonsoir, M'sieur, Merci, M'sieur," the shop girls, the delicious cheeses (I'm going to send some to Dad), the strawberry tarts, the old streets—the life of Paris.

We are going to leave Paris this Saturday, probably. But if you continue (is that the word? I've hardly heard any news from you) to write to General Delivery, Paris, your letters will be forwarded to me.

> And now back to the fray—
> All [my best] to you,
> Georges

[Marly-le-Roi]
[Summer 1923]

My dear sister*—
I don't have a dictionary or a grammar of any sort here. Because I do have to write to you, you'll have to forgive my mistakes.

Here is yet another change in my living situation. Having been given several days to spend as I wish, I wanted to spend them near Paris, but in a

1. Anatole France, pseudonym of Jacques Anatole Thibault (1844–1924), who was awarded the Nobel Prize for Literature in 1921. His novel *Le Crime de Sylvestre Bonnard* (1881) tells of Bonnard's abduction of the daughter of a former lover from a boarding school where she had been maltreated and how the elderly scholar sold his precious library to provide a dowry for her.

2. Marcellin Boule (1861–1942), French geologist and paleontologist, director of the Institute of Human Paleontology at the National Museum of Natural History, Paris. He had done extensive studies on fossil humans.

*In French; translated by editor.

rather small town. So I took a map and I put my finger randomly on Marly-le-Roi.[3] Fine, I went there. I looked for a long time for a convenient room—there weren't any. All rented, or else completely unsuitable. I began to walk slowly to the train station. Then I thought—but this is ridiculous, I was a fool for coming here, [but] would be a still greater fool for leaving, and I began ringing the door bell of the first pretty garden that I passed. A woman came to the door. —Madame, can you help me? I've come to Marly looking for a room. There aren't any. However, I do not want to leave. It's far too pretty here. So, would you be so kind . . . She laughed, and said, but I have a room for our house guests. I've never rented it, it's hardly furnished, but if you would like it.

And that's how I've come to occupy this very room in which you see me right now. It's a pretty room, isn't it? Look at the walls. Japanese woodcuts. A bronze plaque. Look through the window—the little wooden balcony, the garden in bloom—one can even just make out the magnificent portrait from the 12th century which is in a corner, a piece for which my hosts have been offered 50,000 francs. And inside once more, look over the fireplace. That pretty vase! It's Persian, very old. Look how iridescent it is! And here, the tiny little statue of a seated woman—Egyptian, a real archeological artifact. The reproduction over there is Chaldean and comes from the Louvre. But look at the wardrobe especially. It isn't at all ugly, with simple and pure lines, but the real treasure is inside. More than one hundred pieces, all original artifacts. Notice the Persian vases, the iridescent ceramic, the Tanagras, the Egyptian bronzes, the old Etruscan and Roman glass whose iridescence is as magnificent as ever—and more and more.

And it's like that throughout the house. We are on the second floor. On the third is a sculpture studio. On the ground floor there is, besides the kitchen and dining room, a beautiful salon—soft sofa, drapes on the walls, sculptures from the Middle Ages and from Egypt, books (worth a whole lot, these are. I'm in the process of reading them).

As a matter of fact, it's the house of Mr. Van Dongen. He's the brother of the Van Dongen of whom you've heard, isn't that so? He paints well-known portraits, Van Dongen, a Dutchman.[4] Mr. Jean V. D. is an artist too, a sculptor, of course. He is Dutch, thoroughly affable. His wife is French, charming, vivacious, talkative, a real French woman. Wasn't I lucky? It is real close to Paris, yet keeping the ways and walls of a small town. The house is beautiful; with a French ménage, something one doesn't often see while traveling, isn't that so? Here we speak French, of course. I speak it too. It is necessary. Consequently, each day my accent gets better, and each day I can speak more quickly, more easily.

And the town? It's a dream. Old, from the Middle Ages at least and not at all changed. The streets which always go up and then down, all along

3. Marly-le-Roi is five miles north of Versailles and just on the southwestern outskirts of Paris. Louis XIV had a famous château built there by Mansart which was destroyed during the Revolution.

4. Kees Van Dongen (1877–1968), Dutch-born French painter; one of the founders of the school of Fauvism, which flourished between 1905 and 1908. Other Fauvists were Matisse, Rouault, Vlaminck, Braque, and Dufy.

their length, are old, old high walls, crowded with roses. When there is any news, there is a drummer who comes into the streets drumming and saying what there is to say! Incredibly old women who jabber away in the doorways. Fat peasant women with bare feet walking in the narrow streets. How picturesque it is!

Monsieur has a bicycle that I use. I have been in all the little towns and hamlets in the surrounding area. Tomorrow I will go to St. Germain-en-Laye, and Tuesday I'll go to Chartres (pretty far away, that) to see the cathedral, the number one monument to original art.

But this is already enough.

I have to leave in a little while, of course—the Carters are in Paris with friends. They will leave in several days—I'm not certain when just now. Until then I'm going to stay here.

I've heard nothing from you—neither father, nor mother, nor sister although I have received lots of other letters. What's going on?

Georges

[Marly-le-Roi]
[Summer 1923]

Dear sister*—

I just finished *Religious Art of the 12th Century in France* by Emile Mâle—a beautiful book of which you've heard, haven't you? And more than that, I've just seen what "represents 12th century art in its full development." That is to say the Royal entrance of the cathedral at Chartres. Of the three entrances of this, the prettiest cathedral of all, this one is the smallest, the most ancient, and by far the most beautiful. How I love it! I went, two days ago, with M. Van Dongen, the sculptor at whose place I spend almost all my free time. We went by bicycle—eighty kilometers, a good long way for me, who isn't used to that. We stayed there two days—I saw such beautiful things indeed. But what was most entertaining (in the artistic sense always) was the moon which rose, completely full, and then the glory of the cathedral in full moonlight, all golden, the night, the centuries, the old houses, deep shadows, ghosts speaking from an epoch where one constructed this dream in stone, which seems to spring right up before us, from the fertile soil, among the houses, here these eight centuries, so much has been lived. A solemn thing! The stones of the flying buttresses, of the buttresses, seem to melt, to blend, only the golden lines outlined on the mass remained—a mass that is the symbol of an idea completely ineffable. The clocks strike midnight. Oh my god, if one could tell of it!

But also that Monsieur Coué[5] is not the least of philosophers with his "that passes!" [For] the next day, which was yesterday, we had kilometers to cover. And we came back to Marly. Today, a small trip to Paris, not too amusing, that, when one is tired and only wants to rest. Tomorrow to Ver-

*In French; translated by editor.

5. Émile Coué (1857–1926), French psychotherapist who used autosuggestion as a method of treatment.

sailles—I've already been there, entertaining, always entertaining, but there are so many things that are!—and then one more time to Chartres, but with the whole world. It's well worth the trouble to stay there one's whole life, but just the same I will forget less quickly my first impressions. And one leaves (absolutely at that) for the provinces Friday (today is Sunday). First to Clermont-Ferrand.

At last! Some mail—or rather one letter! As a tip (and everything here is tips, especially when it's a matter of a restaurant check)—as a tip for your very detailed impression of Paris, I'm giving you a little, very complete map—everything you'll need to find your way around Paris.

And so you've cut your hair, eh? You said ["chevaux" for "cheveux"] "horses"—be careful, the French will laugh—but all the same it's funny, that. As for me, however,—I have to confess that I have a beard in the French manner! I'll shave it when I return, of course, but here where hot water is like fine gold, it is indeed convenient.

And my love to mother, too. I'm going to buy her a photograph of a saint, or, rather, a King of Judah for her birthday—don't say anything to her about it. It comes from Chartres, from the twelfth portal. How thoroughly lovely it is, and even religious. Just like here but (of course) much, much prettier. The lines are superb, one couldn't do better!

Ah well! I hope everything goes okay. And Dad, and little John.[6] I will write again once more in a few days, because I don't know if you're receiving my letters.

Best to you,
Georges

[Marly-le-Roi]
Le 1ᵉʳ Septembre

My dear sister*—

Here it is several weeks that I haven't written. That's because I've been indeed occupied.

I left for the provinces. I went to Bourges, to Clermont-Ferrand (a dirty city, that one), to Lyons (even dirtier, but at least interesting), and finally to Grenoble. From Grenoble I saw the French Alps, including Mont Blanc. Then unfortunately I was sick, so I had to return to Paris, where I stayed for several days, and then after returning to my pension at Marly, I went a little everywhere that wasn't too far away from Paris. Thus, I saw Amiens[7] (the only trip in the [war-]devastated countryside—in Amiens nothing more remains of the devastation, except the fact that the windows are still not replaced). And, in returning from Amiens, to Beauvais, one of the oldest and prettiest cities, with a cathedral, or rather a half-cathedral because the nave is entirely missing and also a church of the 11th and 12th

6. This may refer to his oldest sister's son, John, who perhaps was staying with his grandparents in Washington.
    *In French; translated by editor.
    7. Amiens, a French city in the valley of the Somme, eighty miles north of Paris.

PETITE CARTE PRATIQUE DE PARIS ET SES ENVIRONS

PAYS SAUVAGE ET INCONNU

ESPRIT DE PARIS

PAYSAGE PARISIEN

ENCEINTE

XIIe

XIIIe

XIVe

XVe

XVIe

SACRE COEUR

MONT MARTRE

INCONNU

INCONNU

QUARTIER

PANTHEON

JARDIN DU LUX

PALAIS

ETOILE

TROCADERO

LA TOUR

ENCEINTE

AGENT DE POLICE

LA SALE SEINE

ST CLOUD

COURSES

VOIE PNR

CHAMPS

LA CAMPAGNE

PAYSAN

PAYSANNE

CHAUMIERE

GRANDS TOURS T. COOK

LA TOURISME

PAUVRE

centuries with magnificent sculpture—a decorative entrance and a rose window like a "wheel of fortune." Then I also went to Vézelay—a little isolated town, crowning a hill with a superb view, and everywhere encircled by fortified walls from the Middle Ages—here is found the most complete and also the most famous monument of Romanesque (*not* Roman) art, that of Vézelay Abbey, or the Madeleine Basilica. Upon returning I went to see Sens, to the old churches and to the cathedral equally renowned, especially for its windows and its unsurpassably pretty tapestries. I went all alone to Senlis where there's an almost forgotten cathedral but one very important for the history of religious art.

And always Paris, the Louvre still several times more, the Cluny Museum, with its collections of the Middle Ages, the Trocadéro where one can study in plaster reproduction the whole evolution of the plastic arts, both foreign and French. Also to the Luxembourg where among the paintings (all more or less modern) ones reads the names Degas, Whistler (his famous mother), Puvis de Chavannes, Manet, Monet, Pissaro, Gauguin, Sisley, Carrière, Renoir, even Van Gogh and a single Gauguin [sic]. To see Matisse as well as the real moderns it is necessary to go to the art dealers.

My how the boulevards are amusing just now! All of Paris has gone away, to the mountains, to the watering places, to the sea, and a whole world descends on Paris, where [if] there aren't Americans (and they are nearly everywhere), there are the English. And one can hear all the civilized languages and many which are not. Here, the Moroccans, all in white except for a kind of red hat, and speaking their guttural tongue. And there, the little Annamans frightened in front of the big Western world. And, of course, everywhere German, Spanish, Italian, Russian. It's only French that hardly isn't spoken in Paris. As for myself, I do not much like the American tourists. When one among them speaks to me, I pull my beard and say, "But, Monsieur, I spik not Eenglish" and they go away nine times out of ten.

Happily, Don Stauffer[8] is in Paris right now. I haven't yet seen him, but I received a card and I am going tomorrow to look for him at his hotel. He has been to Spain, Italy, Switzerland, everywhere. (Me, I don't like to dash about. I came back to see this time what I hadn't been able to see before.)

Because, as a matter of fact, my sojourn is almost over. At the end of several more days I'm going to go back home to New Haven. You'll have to address your letters (rather infrequent, at that) to 82 WHALLEY AVENUE, NEW HAVEN, CONN.

Also I still have some very interesting books—For example, several volumes of the monumental "Histoire de l'Art" de M. André Michel, the volumes of the 12th, 13th, and 14th centuries (those which interest me the most, from the point of view of the plastic arts). A "Human Anatomy" among the most precious. "Architecture of the Orient" which is to say architecture that is Parthian, Sassanian, Syrian Christian, Aegean, and

8. Donald Stauffer was a University of Colorado classmate of Simpson's, studying at Oxford University. He was later a professor of English Literature at Princeton University.

Armenian, Arabic, Coptic, Byzantine, Russian, Chinese, and so on. Several scientific books, and so on and on. I have a whole library, and I'm annoyed that I cannot have any more books.

What are you going to do this winter? I certainly hope I'll be able to see you soon.

> Mille petits baisers au cou
> (which is only too French)
> Votre frère,
> Le méchant Georges

May 26, 1924
Clarendon, Tex.

Dear Marthé—

I'm awfully sorry about the trunk. Mrs. B must have held out the telegram on me. We left the 12th—I had no word at all from you. It probably would not have done much good anyway as I did not come thru N.Y. but went up to Springfield & took the N.Y. Central [railroad] thru Albany & Canada to Detroit & Chicago—that way is quicker and cheaper from New Haven.

I'm tremendously pleased that you are really getting off to Europe. How do you get in on the Yale Excursion? Oughtn't I to get a commission or something? As Reverend Wilson, who stays at the hotel, is wont to say "Be good, my girl, and make some one a good wife"—this to a waitress with the eyes of a sphinx, the face of a slightly dissipated match girl, & and the mind of a subnormal pussy-cat!

However, pardon the digression. That is life—Digressions I mean. I've been digressing all over the landscape. I digressed around the midriff of a mountain yesterday, with everyone hugging the upper side of the Ford like a sailboat in a high wind, until I warped the tires right off the car—so disgusting the chief[9] that he couldn't help replace them. Now & then we find a fossil—every third day or so, if small fragments count. I am fortunate that I am paid whether or no, unlike the early collectors who were paid for what they got. Poor Dr. Matthew gets madder & madder "First Tertiary formation in which I couldn't find mammals." However, I am getting sunburned, & also acquiring an insouciance hitherto unattained in the presence of male cows & he-man cow-punchers. This country is built on two levels—a broad flat plain covered with mesquite & rattlesnakes, & deep gorges of canyons, quite unexpected till you almost run over the edge. The

9. William Diller Matthew (1871–1930), Canadian-born vertebrate paleontologist at the American Museum of Natural History for most of his professional life until he went to the University of California, Berkeley, where he became chairman of the reestablished department of paleontology and director of the museum of paleontology in 1927. He was especially known for his work on Early Tertiary mammals of the American West and his book *Climate and Evolution* (1915). Although not one of Simpson's formal teachers, he was one of his most influential mentors, as Simpson himself acknowledged at every opportunity. Simpson replaced Matthew at the American Museum.

W.D.M. G.GS. MASTODON PLIAUCHENIA

PLIOHIPPUS HYPPARION HATRACOFORDIA DAMNATA

Like all good illustrations, this one has nothing to do with the text.

canyons are vivid red, with a pink or yellowish rim, like frosting, on top. The people are hearty but suspicious. I am getting to understand Texan, but I fear I'll never speak it.

How I envy you! Every time I have to get out & dig the car out of the sand or mud, or walk ten miles for gasoline, or change tires in the boiling sun—while I'm walking miles over chalk-white beds in the noon sun feeling like a herring being kippered, munching my lunch of 3 graham crackers, pulling a drowned rat out of the cow trough before I drink the nasty luke warm "water"—I think of you in la belle France, careening about in a taxi driven by a pirate, strolling down the Champs Elysées or along the Seine, sipping a [bock (?)] or liqueur on the side walk! You surely are going to enjoy it.

I'm afraid I can't possibly get over for three or four years at least. Matthew scouts my idea of studying in Paris & says Munich[10] is the only place for me—the trouble is, he's right, and I'd ever so much rather stay in Paris!

I must dash along. We're going out onto the ranch to hunt the more elusive *Protohippus*.[11]

George

Gen. Del. Clarendon, Texas.

10. The University of Munich had a long and deserved reputation for excellence in paleontological studies.

11. *Protohippus* is a three-toed grazing fossil horse.

Crosbyton, Texas,
1924

Dear Mart—

Your welcome letter forces me to another attempt to reach you before you sail, especially as my last note was so brief and blue.

Topic I. Transferrence [sic] of thot waves by inanimate objects. One of those things which one may *say* one believes, for the sake of conversation or diplomacy, but one which only morons *really* believe.

Topic II. Negromancy[12] [sic]—another, but more interesting. The names of the demons and familiars is worth the price of admission, alone. I am very fond of Rosicrucians & regret that times have so deteriorated that only *clumsy* magic is really believed in nowadays.

Pertinent Detour:—Plump, preserved "Texian" [sic] Lady: "Oh, *how* do you find those petrified bones?" "We just walk along & look for them, lady." Lady (Much disgruntled & sure that nothing so full of common sense can be really scientific.) "Oh! I thot perhaps you were like geologists & went along with a little stick & dug where it dipped."

Further—Much discourse, pyramids, Pocahontas, original protoplasm, & our three-toed horses are branded "all prehistoric you know" by the lady, who also says "Now science says ———" follows discourse from Hearst papers.

Formerly the faker said "The great God Yubble-Yammer hath inspired me to say that black is white with green spots." Now he says "science says the lost Atlantis was an Atlantic land bridge on which man entered America." Impossible but true, by Yubble-Yammer; not so, but believed, thru faith in Hearstian Science.

Science says thumbs up.

Science the sacred cow, digesting (in a sterilized test-tube) the fresh herbiage of knowledge, chewing (in an automatic Jena glass masticator) the cud of contemplation, & occasionally solemnly saying something to W. R. Hearst, to be relayed to the people, who's [sic] humble servant he is.

Peter Whiffle does not defeat Paleontology. No paleontologist can be ignorant of anthropology but has in early youth looked "steatopygous" up in the dictionary, that well-stocked source of precocious sophistication, & come away sadder & wiser. I've known it for years, & would have sprung it long ago had I suspected you ignorant of it.

Whoever interests himself in Magdalenian art[13]—as who does not?— soon encounters not merely the word, but unforgetable [sic] examples of its signification in the female figurines common in Europe at that time. In fact many think them religious, & that the main character of the cult was this graven exemplification of Steatopygosity.

There Art, Anthropology, Philosophy, & Religion mix. There is a

12. Necromancy is the pretended art of revealing future events by communicating with the dead; more generally, magic, enchantment, conjuration.

13. Magdalenian culture of the Late Paleolithic associated with early *Homo sapiens* in Europe; its crowning achievement was the cave paintings in France and Spain.

book, La Réligion de la Préhistorie. I forget the author—a priest, rather strangely. [Not] strangely in France as it would be here. In France the church sponsors every scientific advance at the present time, thru its individual workers, thus swelling its prestige while in America the church is wrecking itself on that rock, instead of building its house on it.

In a gallery in the Museum d'His. Nat. in the Jard. des Plantes, no longer used for exhibition I saw the stuffed remains of the "Hottentot Venus" once resident in a sideshow in Paris.

She, dear readers, was steatopygous.

When you can't afford good wine get Vichy & rouge ordinaire & mix them 50–50; it turns purple.

Ordinarily drink white wine, or else a good "extra pay" wine.

A "quart" is *not* ¼ of a gallon, but it's a plenty.

You can feel fine sitting down yet have your knees give way when you get up.

Climb the tower at Notre Dame.

Go to Chartres before seeing any other cathedral then Paris.

Don't ask for butter, unless you eat radishes.[14]

Go to St. Cloud along the river on the boat & bus.

Take a St. Germain-en-Laye train from the Gare St. Lazare, get off at Marly-le-Roi, walk straight up the right hand side of the street to number 15, ask for Madame Van Dongen (the French call it Vanne Donguenne), tell her you are Monsieur Georges' sister, pretend to believe all M. Jean Van Dongen tells you.

Eat many Reine Claude plums & wild strawberries in season.

Look at the Pantheon from the *outside*.

Do as you please—as you will—& tell me about it—as you won't.

TOPIC—SANTA FE POEM—It *doesn't* mean anything, but it sounds nice. Sound—very nice sound. A mean criticism, meant well. A poem is anything to a reader, another to the author. I'm not talking of the poem you wrote, but the one I read.

One criticizes oneself as well. Especially when one sees deep significance.

Sante Fe—I'll be there soon.[15] May outfit there. Who[m] do you know there? I'd rather like to find someone who wanted to wander and paint, while I hunt bones—lovely country I'm going to, & I hate going alone. I could make an artist enjoy it. Recommend me someone, or someone who knows someone.

If I don't find someone I'd have to buy a dog, which would be more expensive!

I'll be out 2 to 3 months; sheepherders go crazy. I'm in no mental condition for introspection.

There's been a lot of talk in France about "seeing with the skin"—no, it is *not* possible. More "science says ———" in my humble opinion.

And good for St. Louis. It means more of France in the end, of course.

14. Refers to the French tradition of serving butter with a garnish of radishes.

15. After his work with Matthew in Texas, Simpson was going on to New Mexico to collect fossils in the Santa Fe formation of Miocene age.

SMITH HOUSE
New and Modern
Owned and Operated by
Mr. and Mrs. J. Frank Smith

X is my window! as in Postcards from Uncle Silas at the
Hotel Pennsylvania.

Crosbyton, Texas,................................192.....

They call it Sainta Fee here, by the way. Their tongue is vilely unbeau-
tiful, wildly unlike English.

Especially to one who passes his days with Hipparion gracile, Pliohippus
simplicidens, Stegomastodon mirificus, Pliauchenia spatula, Borophagus,
Platygonus,[16] etc. White slivers of bone on the shimmering white outcrops—an animal caught in the deep white mud a million years ago.
Uncover it with an awl & fine brush, shellac it, lay rice paper on it, bandage it with cloth & flour paste, put on a wooden splint, bandage that, roll
it over carefully, repeat, carry it to the car, take it to town, make a box,
ship it to N. Y., unpack it, unwrap it, clean it carefully, restore any chipped
or rotten places with plaster, carefully rivet in place on a steel rod, put it in
a glass case.

"Is that bone?" "Hunh! Looks like a little old 'possum bone, le's go look
at the stuffed elephants."

You won't have read this far, so I don't have to introduce a "natural ending" like my teacher told me to.

George

New Haven, Connecticut
1 January 1926

Dear Mother and Dad:

Thank you so much for all the Christmas things. We appreciate and
enjoy them tremendously. The rug fits in beautifully—we are short on floor
coverings and that is certainly a beauty. The same applies literally to the
socks and stockings. The baby[17] takes her things to bed with her every
night—especially the striking blonde! She refuses, however to be mislead
[sic] by her parents' assertion that the little animals are lions—she knows
puppy dogs when she sees them!

16. Late Cenozoic beasts, including fossil horses, proboscideans, and ancestral
camels, dogs, and pigs.
17. Helen, born in December, 1923, now just two years old.

The little one has almost died with excitement too great for one tiny being to hold. She has three new dolls, half a dozen books, four strings of beads (she has a distinct predilection for them), a tea set, a kiddie car, enough chocl'te canndy [sic] to last her the rest of her life at the present rate of permitted consumption, and I don't know how many other toys and playthings as well as piles of new clothes which she definitely appreciates but, in the manner of kids, does not rate as gifts! We had a nice big Christmas tree which she helped to decorate and which we have not yet taken down.

My own Christmas was followed by much excitement and now that it is all over I feel rather ill. The meeting of the Geological and Paleontological Societies was held here[18] and I have been attending as many sessions as possible. I have met a great many people whom I wanted to know—Scott, Merriam, Loomis, Sinclair, Osborn, Gregory, Wood, to mention only vertebrate paleontologists. I also had a chance to visit with Matthew, Gidley, and Gilmore among my old paleontological acquaintances.[19] Monday night I went to the smoker and there had a long talk with Osborn. He expressed the greatest interest in my work and made one or two valuable suggestions. He is so engrossed in administration that he goes around very little in scientific circles and I had never happened to meet him before although I have done so much work in and for the American Museum. Tuesday night Lydia and I went to the banquet. She attended the Anthro-

18. The new wing of the Peabody Museum of Natural History was being dedicated at Yale; the annual meetings of the Geological Society of America and the Paleontological Society were held in New Haven to coincide with the dedication. Osborn (see fn. 19 below) gave the address at the dedication in which he developed once again his view of the internal forces within organisms which led to their striving for perfection, so-called "aristogenesis," a theory subsequently debunked by Simpson in his *Tempo and Mode*.

19. William B. Scott (1858–1947), who founded the Princeton geology department; particularly known for his study of horse evolution.

John C. Merriam (1869–1945), expert on living and fossil reptiles and mammals from the Carnegie Museum in Pittsburgh.

Frederic B. Loomis (1873–1937), paleomammalogist from Amherst College.

William J. Sinclair (1877–1935), a protégé of Scott who was a vertebrate paleontologist at Princeton; he left a large estate of oil-derived wealth to Princeton's geology department.

Henry Fairfield Osborn (1857–1935), the dean of American paleontology at the time, who was a non-Darwinian evolutionist. Longtime president of the American Museum of Natural History and expert on fossil proboscideans. Active against antievolutionists, he wrote *The Earth Speaks to Bryan* in 1925 as a riposte to William Jennings Bryan's antievolutionary stand as manifested that same year in the Scopes trial.

William King Gregory (1876–1970), vertebrate paleontologist at the American Museum and Columbia University, especially known for his work on human origins and vertebrate evolution.

Horace Wood, II (1901–1975), vertebrate paleontologist at the University of Newark (now Rutgers-Newark), who was a student of fossil rhinoceroses.

William Diller Matthew, Simpson's mentor at the American Museum whom he succeeded when Matthew went to Berkeley the following year.

James Gidley (1866–1931), curator of fossil mammals at the National Museum of Natural History in Washington, D.C.

Charles W. Gilmore (1874–1945), vertebrate paleontologist at the National Museum who studied fossil vertebrate tracks and fossil reptiles.

pological Society sessions and had planned to go to their dinner but at the last minute I persuaded her to come with me and we had a jolly time.

Wednesday morning I gave my two papers[20] before the P.S. [Paleontological Society]. Dr. Matthew who is very much inclined to kid anyone he knows well is still teasing me about my launching into the field of fishes for one of them. It provoked a decided response from Raymond of Harvard from whom I disagreed and who, to my great surprise, was present, but Patten of Dartmouth defended me, to my relief.[21] Lydia and I had lunch with the latter and found him very pleasant indeed. When I came to my second and more serious paper, that on Pre-Cretaceous Evolution of Mammalian Lower Molars, Professor Lull, who was in the chair, surprised me by introducing me with a long and embarrassingly complimentary presentation to the society on my first appearance before it and after I had spoken Dr. Matthew got up and pronounced an even more unexpected eulogy from which I am recovering but slowly. Dr. and Mrs. Matthew later came up to my office and spent some time. He went over considerable of my material with great care and as always, his suggestions were most stimulating and helpful. He also relieved me by agreeing on some of the essential points on which I differ from my predecessors in this field.

Tuesday afternoon the museum was dedicated and is now open to the public at last. It seems very strange to have such crowds in the halls where I have been accustomed to roam alone.

Wednesday evening Lydia and I went to a dinner and party for returning Yale geologists and had a very nice time. The faculty club here has a very jolly arrangement for keeping babies when their parents want to be free and this enables us to go about almost as if we did not have one. Don't think we abandon her there however! She spends not more than one or two afternoons a week there ordinarily and likes it very much as they have a jolly nursery and several of the little boys and girls she knows also go.

I have several other letters to write and must close. With all love,

George

I am sorry that I cannot give you any more copies of the Times article.[22] I have only one myself and would like to keep that on file here. Would you mind letting me know just what of my articles you have? I have lost track and want to be sure that you have at least two copies of each, if you wish

20. These were formal oral presentations. In the first paper Simpson presented a new interpretation of a very primitive fossil fish. In the second he discussed some results from his dissertation on Mesozoic mammals.

21. Percy E. Raymond (1879–1952), Harvard paleontologist and curator at the Museum of Comparative Zoology; subsequently president of the Geological Society of America.
William Patten (1861–1932), zoology professor at Dartmouth College who had done research on Paleozoic fishes.

22. "Mammals were humble when dinosaurs roamed" (New York Times, 18 October 1925, part 10, p. 11). The first of Simpson's popular-science articles underscored the fact that mammals coexisted with the dinosaurian reptiles for millions of years as small, rodentlike animals. Later, mammals diversified from these earlier forms once the dinosaurs became extinct. "Theirs was the promise of the meek; they were to inherit the earth," wrote Simpson.

them. I have none coming out in January, but there will be another the first of February, and at least half a dozen more this year that I am sure of. If I stay here at Yale I have decided to start work on a book in the Fall, as that is the quickest way to advancement (!), but don't know yet what I would write about.

Oh, yes. I neglected to mention that I also at last met Riggs, the vert. pale. there at Chicago at the Field Museum. He is a big slow-moving and -thinking Scandinavian. If you are ever at the Walker Museum at the Univ. there introduce yourself to Romer[23] and remember me to him. He wasn't at the meeting.

I will send reprints to Truman Gaylord.[24] Did you give me his address? I seem to have mislaid the last letter.

G

New Haven, Connecticut
Jan. 5, 1926

Dearest Mother & Dad—

I am afraid something else in the nature of a tempest in a teapot has been stirred up, but perhaps I can allay it.

In the first place regarding our condition we have a hundred a month to live on, but living here is expensive & we have many unusual expenses such as tuition, books, etc. which are unavoidable as well as dues in several organizations, reprints, etc. Some of these may seem extravagant to you, but I assure you that they are as essential as food & shelter. We have to watch expenses very closely, but on the other hand we are far from being in want. We have a pleasant appartment [sic] & plenty of nourishing food and sufficient warm clothing. Of course we would like to have & really need some things we cannot afford at present, but we are certainly not in want or in any degree lacking the essentials of civilized existence. Lydia's mother apparently has gotten a mistaken impression as Lydia has never told her that we are in real need. She has felt rather ill at times & has not written very frequently, from which Mrs. Pedroja has gathered the mistaken impression that we are in want.

The $2500 alluded to by Mrs. Pedroja was, as you inferred, Lydia's allowance since coming east. This was sent as an allowance, not as a loan, although it was charged up against Lydia with previous allowances to the total of $5000. We have consistently refused to sign a note for this amount or in any way to acknowledge it as a debt. We have no objection to its

23. Elmer S. Riggs (1869–1963), vertebrate paleontologist at the Field Museum of Natural History in Chicago; he was known for his work on fossil marsupial mammals. Alfred S. Romer (1894–1973), distinguished vertebrate paleontologist at the University of Chicago and later at Harvard University. A student of fossil amphibians and mammal-like reptiles, he educated several generations of zoologists and paleontologists through his various textbooks. Years later as director of the Museum of Comparative Zoology he arranged an appointment for Simpson there as Agassiz Professor when he learned of Simpson's resignation from the American Museum.

24. The lawyer associate of Simpson's father whose surname became Simpson's middle name.

being taken out of any share of Lydia['s] in their estate however. Mrs. Pedroja recently made a gift of $3000 to each of the three other children & sent us a legal receipt for that amount applied on the debt (as they call it) of $5000, so that considerably more than clears any amount which Lydia & I have both profited from. Most of the remaining $2000 was spent on Doctor bills by Lydia when she was ill in Colorado (she has had pneumonia three times), largely as a consequence of having to work too hard to keep herself in school.

The Pedrojas are not at all as hard up as they themselves honestly believe. They have the habit of over-extreme frugality but never have really been forced to work as hard as they believed necessary, nor have they been in want.[25] Just at present their income is considerable as may be judged by the recent gifts of $9000 to the other children. That was paid out of their *current income*. In passing, with the exception of Lydia's sister Louise none of the children had any really legitimate need for that money. One of them, Dan, didn't even want it. The other, Ferd, is unmarried, lives at home, & is doing very well in the cattle business. The nearest approach to hard times which the Pedrojas have had was when Mr. P. was swindled out of about $20,000, after which they had to borrow on their extensive land holdings. This sum was later compensated by a nearly equal amount made in a lump on an oil lease in 1922. The $800 which Mrs. P. mentions is, you notice, *taxes* on all their land plus interest. Their capital is all in land and the interest was not the pennies going to stave off foreclosure but in any sane view simply a matter of sober business dealing enabling them to utilize their wealth when needed in a more fluid form than land.

There is no reason why Lydia could not have been given an education without the struggle & loss of health which she has been through. The martyred & hard-up attitude of her parents may well be sincere, but it is sadly mistaken. In view of their attitude, the course we have followed has been the only one possible if Lydia was ever to marry & enjoy any happiness at all.

We do not need or want an allowance from either family, although we wish to thank you for considering it. Any gifts for the babies which the giver makes of his own free will & without stipulation & which the giver can *afford* we will, of course, accept. We have over $200 in insurance premiums every year and we are considering asking Mrs. Pedroja to help us out in that respect as we wish to provide adequately for the babies in any eventuality.

I hope you won't misunderstand our attitude towards Mrs. Pedroja. She does a great deal of sewing for Lydia & helps in that way very materially.

25. Lydia's parents, Joseph and Mary Pedroja, lived in Buffalo, Kansas, and were financially comfortable as a result of the discovery of oil on their land. They had earlier been hard pressed, however, because Joseph Pedroja had gambled away large sums of money. Ironically, Joseph had originally married Mary, many years his junior, to pay off a mortgage debt owed him by her parents. Simpson's youngest daughter, Elizabeth, later wrote a novel, *Stranger from Home* (1979), which is a partially fictionalized version of the Pedroja family story. The old woman in the book, in particular, is a well-drawn portrait of Mary Pedroja.

We understand the way in which hard marital conditions & years of (usually unnecessary) labor & pinching can mold a character.

My achieving my degree this spring will mark the edge of the woods for us and conditions hereafter will be more livable. We have decided to have another baby as soon [as] we are established & before Hélène is too old— also before Lydia begins graduate work. He (I hope the sex is correct) is on his way & will arrive next July.[26] I very probably will go out in the field regardless, as the doctor assures us that there is no need to anticipate trouble. The head of the department of obstetrics in the Yale Medical School is taking care of Lydia personally as a friendly favor altho he no longer handles cases of this sort as a general thing. Mrs. Benedict, a very charming woman who teaches anthropology at Columbia and whose husband is in biological chemistry at Cornell & a member of the American Academy, is coming to stay with Lydia in July, so, if I do go to the field, I shall be relieved of some worry at least.[27]

I hope this hurried discourse has settled some doubts & worries—

with love,
George

If there is time to enumerate we might save some expense by selecting the things we want from the stuff at Blantyre.[28] Most or all of it, however, sounds very usable & we should be delighted to have it.

G.

Excuse this scrawl, but I've not felt well the last few days and am resting in bed today with a cough & slight sore throat.

G.

[New Haven, 1926]
Jan. 25th

Dearest Marthe:

I'm sorry such a devilish long time elapses between communications. I always swear that it won't next time and yet it always does. I wish that we were nearer and could communicate more freely and oftener as I think we should understand one another very well and I have always been tremendously proud of your work and appreciative of your mind. If we couldn't understand life any better, and of course no one who really understands anything thinks that he can understand even a little of life, why at least we would know how to get a great deal out of it.

26. The Simpsons' second daughter, Patricia Gaylord, was born in July 1926.

27. Ruth Fulton Benedict (1887–1948), who subsequently wrote the classic anthropological study *Patterns of Culture*, was a student and later colleague of Franz Boas at Columbia University. It turned out that Simpson remained that summer in New Haven, where Patricia Gaylord was born.

28. Blantyre, North Carolina, where Simpson's paternal aunt and uncle had a farm. Simpson spent two summers there before World War I; his sister Martha went there in 1914 to recuperate from a tuberculosis infection. Simpson and Anne also stopped there on their honeymoon car-trip in 1938.

External circumstances here are rather solemn. I feel like a figurehead on the front of a tremendous and very pompous vessel under full way—save, alas! that no figurehead was ever called upon to do so much of the labour about the boat despite being quite unable to guide it nor to stay its majestic course. I pound away at my typewriter and squint industriously through my microscope. I publish articles and give papers and meet people and try not to laugh, but the moon is always over my shoulder. The spectacle at which I attend is vastly moving. There is an almost painfully epic sweep to the vastness of geologic ages which pass between my fingers in tattered fragments. The commonplace room is always filled with the mute cries of ages impossible to contemplate in which life has blindly toiled upward, or at least to further complication and further ability to realize that it cannot realize anything at all. It is all very strange and thrilling in a way which is, I am afraid, incommunicable.

All of which can only serve as a deeper contrast to the utter futility of it all. The reconstruction of the past, even so great a past as that which lies before me here, can add only a melancholy significance to the fact which we know but dare not realize that the present must become as truly past and perhaps even more irrevocably. As for science, one who is not engaged in it can hardly realize to what extent petty motives dominate even here. The highest possible scientific motive is simple curiosity and from there they run on down to ones as sordid as you like. And all our scientific interpretations and theories are simply meaningless. There are facts of course, in any workable use of the term facts, but with us as with artists and other impractical people here facts are considered as only so much mud and straw unless they can be piled up into a hypothesis, gaily stuccoed and concealed with theory. And like other futile edifices of man these are inhabited for a brief space giving glory to the proprietor of the most unusual or striking and then left to melt back to dust and be forgotten, or worse yet, to become curiosities for generations with other "latests".

Don't think I am bitter or unhappy about my work. I like it very much and get pleasure out of it. I am also achieving considerable success.

I see or hear very little that is movingly beautiful. My life seems to be turning in on itself almost viciously. I feel a desire for things which is too poignant to be called vague and yet which is utterly lacking in definite direction. Most acute of all is the desire for companionship. I literally haven't a single friend in the world. Of my many acquaintance[s] some think I'm merely damned queer and others don't even realize that much about me! In spite of the fact that I write and speak with great ease and sufficient fluency, some item of my personality, as you know, makes intimacy on my part even with those who could eventually comprehend me and enjoy me, very slow to come by. People simply talk. This runs pretty much like idiocy or morbid introspection, which is worse, but it will be clear to you I hope.

I've just come back from a trip to Washington, Philadelphia and New York. I enjoyed it perfectly, never slept, and got very nearly drunk the night before I had to return. I accomplished a lot scientifically too—that being the purpose of the trip, but that you can take for granted. In Washington, after finishing my work, which depended on daylight and hence stopped about four-thirty, I would go to the hotel, change my clothes in a leisurely fashion, have an early dinner at Harvey's or Wearley's, go to a show, then wander the streets until some hour in the early morning. The pavements were usually wet and made reflections. I only spent a single eve-ning in Philadelphia and passed that enjoying the novelty of soaking in a warm bath, reading la Vie Parisienne and listening to the radio all at the same time!

I saw Anna Roe in New York. She has left the parental roof—although the latter is something of a misnomer. Her mother has become one of these horrible persons who spends her time touring the country telling other women how to bring up children and letting hers do as they damn well please. Ed is a wild little devil just leaving high school I think, drunk half the time and in trouble with the girls. Mr. Roe who had only two interests in life, his wife and his business, has practically lost both and has become a pathetic figure keeping books for someone else at about a third as much as his wife earns, or rather, gets. Bob, who always was made of sub-stantial rotarian timber, has a fair job as a chemist for the government in Chicago and has a wife as sensible as himself.[29]

To return to Anna, she interests me tremendously. Her judgment and taste, once so far behind, are at last catching up with her remarkable intel-lect. She has abandoned her bigotted [sic] pseudo-religion and her ridicu-lous priggishness and is really human and sensitive,—all of this greatly to my surprise for I had not seen or heard from her for several years.[30] Physi-cally she has ripened and is really striking in appearance. She is supporting herself in New York and at the same time taking a Ph.D. which she will get next year. She doesn't plan ever to return to Denver except to visit— an excellent determination I am sure.

The baby is well and is a darling. Lydia does not feel very well. We quar-rel rather continually and she regularly threatens to leave me—sadly enough, I'm afraid I wouldn't care much if she did. (This I need hardly insist, is for no ears other than your own, without *any* exceptions). In spite of which we are going to have another child next July . . . Odd. I can't begin to express how the whole thing hurts and disgusts me.

I have only a little time for reading. I am now reading Sons and Lovers by D. H. L[awrence]. and I like it a lot. I spend most of my scanty spare time learning Egyptian—to the immense annoyance of my more literal-minded colleagues. It fascinates me: physically because of the beauty and decorative value of the writing and the frequent power of the things writ-ten in it, and intellectually because of the insight into human life in this language which was used with insignificant change for at least 5500 years—longer than any other whatsoever. I am getting along in it quite well and can read any but quite obscure passages now if given time enough.

29. Anne was studying for her doctorate in psychology at Columbia. Her mother was national traveling-secretary for the Parents-Teachers Association; her father had lost his transfer company business several years before and was now employed as a bookkeeper. Brother Bob was an analytic chemist and later administrator with the Food and Drug Administration in Washington, D.C.

30. Simpson is referring to Anne's brief adolescent interest in evangelical Chris-tianity back in Denver.

The Egyptians did some very fine things in sculpture and painting also and I think a knowledge, however slight, of the language helps in the appreciation of them also.

Thanks a lot for the Japaneses [sic] print which you sent for Christmas. I have it in my office and like it very much indeed.

Mother occasionally sends me word of you, but not very often. Write me if you ever have the time and the mood.

As ever,
George

New Haven, Conn.
April 5, 1926

Dearest Mother:

As usual I have been remiss, but as usual I have the best of reasons, if they are needed. Just twenty-five days to finish my thesis, and it's only about half typed! If that were not all I had to do it would not be so bad, but alas! it is not, indeed not. For one thing I have two artists working for me now and anyone who thinks that it doesn't take about six hours a day to keep two artists busy and to correct their work has never tried it. Then Friday Prof. Gregory and Dr. Raven came up from the American Museum in New York and spent two days with me going over these six new primitive mammal skulls which were found in Mongolia—if you read Asia you know about them, or perhaps you have seen comments in the papers.[31] Dr. Gregory and I are writing a joint paper on them, and I shall have to go to New York soon, which cuts out another two or three days of my precious twenty-five. They are well worth it, marvelously so, in fact, as they are generally considered the greatest paleontological discovery of the present century, so far, and my having a share in them is almost unbelievable good fortune as my reputation will be established at once. But I wish it had come about a month later!

Also I am putting in a restored group of Devonian fishes and have to do that in my spare time. It can't be put off either, as there is an artist here from California (that makes three!) and he must be kept busy. Also I have just sent off three articles to the [American] Journal of Science, one to Science (weekly magazine) and have another to go off to the Scientific Monthly if I ever get enough time to put it in an envelope. The ones for the Journal keep me rather busy as the editor is just down the street a couple of blocks and he calls me up every time he wants to ask about a comma. They're rather interesting, however, and I shall be glad to see them in print—the first won't be out till June, however, I suspect. One is on the supposed oldest American mammals, which I have just restudied (at Williamstown and Philadelphia) and which I attempt to show to be rep-

---

31. On two of the central Asiatic expeditions to central Mongolia in the 1920s a great variety of fossils was discovered, including several very tiny primitive mammal skulls of Cretaceous age. Simpson collaborated with W. K. Gregory on their description and interpretation in an article published that year by the American Museum.

tiles and not mammals at all. Another is a study of the lives, habits, environment, interrelationships, etc., of the group of animals, remains of which have been found in one of the quarries at Como Bluff in southwestern Wyoming on the old U. P. [Union Pacific Railroad]; and one is about the extinct American members of a group of lizard-like creatures now confined to New Zealand.[32]

Don't embarrass me! All faculty mebers [sic] are not professors, and I'm one who isn't. My title will continue to be merely Research Fellow, my rank will be that of Instructor. In the course of a few years, if I make good, my title will become Research Associate and my rank Associate Professor in the fullness of time, and finally, when I begin to turn gray, a Professor. There are only three full Professors in the entire Geology Department, and the youngest of them is nearly fifty, while the youngest Assistant Professor (Professor Longwell) is well up in his thirties.[33] So you see one does not scale Parnassus at a bound. One beauty of it, however, is that unlike the other youngsters only a few years removed from their Ph.D's I won't have to go through the drudgery of actually assisting and instructing—all I have to do is go on with my own work. I want to teach sometime, when I can run my own courses as it suits me, but not while I'm a neophyte and would be under someone's thumb.

I don't see how you got the idea my degree was already conferred. That happy day isn't [until] well into June—at commencement. I don't remember the exact date, but it's in the catalogue which I sent to Dad. I shall certainly be much hurt if you don't both manage to be here, as you will be in New York so near that date. I'm very happy that both you and Dad are going to be able to go over [to France], as I know how much you will both enjoy it. You talk as if I knew all the news, but I have not received a personal letter from anyone but you or Dad since Christmas.

I am overwhelmed by the prospect of a watch. I haven't even a tin one at present. From whom? I hate to accept valuable things much as I appreciate the feeling with which they are given. I don't feel that I very amply repay them and the obligation is too great already. Don't think this is unappreciative—and if it makes no impression, I much prefer one with a chain,—but I really can't help feeling a little uncomfortable.

One spells it Lydie, just like that, and accents the last syllable in pronouncing it but not in writing it, but her parents and everyone else call her Hélène, fits of temper to the contrary notwithstanding.

With all love,

George

We've all had colds but me, but we are slowly recovering again. What a beastly spring!

32. The Como bluff quarries in the Jurassic Morrison Formation of southern Wyoming have provided a wealth of fossils, especially those of large dinosaurs. Quarry No. Nine's fossils were reported by Simpson in the *American Journal of Science*, published at Yale in 1926.

33. Simpson is referring to Charles Schuchert, then 68, a distinguished invertebrate paleontologist; Lull (59), his adviser; William Ebenezer Ford (48), a mineralogist; and Chester Longwell (39), a structural geologist.

I am told, *sub rosa*, that the U. of California, at Berkeley, is considering giving me an offer to be their vertebrate paleontologist, but it will have to be an awfully good one to tempt me.[34] I haven't received official notification yet so don't know any details. I didn't apply for the place.

G.

New Haven, Conn.

June 15, 1926

Dearest Marthe:

First of all, a thousand most sincere thanks for the delightful gifts which you were so thoughtful as to send me. As you well know, nothing could have pleased me more. I have been happily immuring myself in my attic chamber (how appropriate!) and reveling in Verlaine and Beaudelaire [sic]. I confess that I become a little restive when the former becomes religious and even wish that he, too, had had no need for "cette hypothèse-là", for religion is such a controversial thing and brings up so many unpleasant associations, especially to one engaged in scientific work in this country, that it does not seem to be a fit subject for emotional verse. It seems to be the one thing which sincere people should not discuss—but no doubt that is a very non-gallic and narrow point of view. I agree with the Englishman who said, apropos of the Russians, I believe that it is an excellent thing to think about the soul, but not to talk about it. Usually, however, Verlaine is a delight, and Beaudelaire [sic] is so always, even with his bloody décorchés [skinned beasts?] grinning their way through his pages.

I think M. Séché is a most extraordinary and not a very admirable critic when, in essence, he apologizes for Beaudelaire [sic] by saying that after all he was only showing off!

And the work explaining the proper begetting of every old French verse form from epigram to epic is altogether delightful!

Thank you too for your long letter of some time ago. I followed your peregrinations in the south of France with the greatest pleasure, marred only by the desire to be there also. I am happy that you have been able to accomplish so much and hope to hear how everything has since turned out.

And next the most loving and soaring birthday wishes for a measure of happiness and for the achievement of beauty. I hope that in your case, as in so few others, they may be more nearly synonymous than antagonistic. You seek beauty as I seek truth, and I fear, or rather hope, that you have the better of it. For beauty is plainly everywhere, while truth is—heaven knows where—and one has to seek it with the tongue in one cheek to keep from going mad. But most of all I send wishes for more than the usual human share of happiness.

---

34. Simpson is probably referring to the vacancy at the Berkeley campus, eventually filled by Matthew, which in turn opened a place for Simpson at the American Museum. It is not known if, in fact, Simpson was ever formally offered the position at Berkeley.

I have an unusually large budget of external news. Next week I terminate some eighteen years of going to school and become one of the learned doctors whom the populace so distrusts. This distinction, which has overshadowed my more juvenile efforts with awe for their goal, has shrunk sadly in dimmensions [sic] now that one sees it closely, but still is pleasant to contemplate.

I am rapidly acquiring a bibliography, which is the scientific equivalent of the little bag in which the untutored indian keeps his bits of shells, odd hairs, rabbits' feet, and other garbage of obvious magical virtue. The bibliography is the fetish to which the scientific world bows and, what is more important, it has the power of opening doors, hardly less prosaic than that through which our old friend Ali peered at his peril. Indeed I can but laugh at his simple sesame—had he but known the vastly more far reaching effect of the equally irrelevant terms such as "Brachydiastematatherium," "rhipidistian crossopterygian"[35] and other admirable elements of the slightly significant hocus-pocus of which I shall soon be a doctor! I have published six papers and a number of book reviews, etc., and have about a dozen more in print or nearly ready. I have been honored by the most extraordinary attentions of our friend Figgins at the Denver Museum who devoted a whole number (beautifully illustrated and lavishly supported by taxpayers' money) of the bulletin of that institution of human curiosity to pointing out the congenital defects in my character.[36] I am proud to announce the birth, at his hands, of a new adjective, "Simpsonesque" which means an abundance of unpleasant things—could one ask more of fame? A paper from a Fräulein Hertz just now received closes with, "Mit dieser Aufgabe is[t] gegenwärtig G. G. Simpson beschäftig[t]" ["With this task G. G. Simpson is now occupied"]. Not having my German dictionary by me I do not yet know whether to be pleased or otherwise.

I was going to go to Alberta this summer, but at the last minute the money gave out, as money will, and it was called off. They now have found a little and want me to go out for a month and pay my own expenses and salary out of $300—a proposition which I was forced to decline with thanks not altogether sincere, since about $275 is the least my expenses could come to. So I shall stay right here until Fall. I have an appointment here in the Museum which is permanent and I was about resigned to staying right here next winter also. I am also offered a place on the geological faculty, and I might have combined that with the museum job and done quite well. After three hearty refusals from three several foundations, however, I at last found someone soft hearted enough to give me the money for

35. Although a mouthful, these are real animal names. *Brachydiastematatherium* ("short tooth-space beast") is an early Tertiary distant cousin of rhinoceroses, horses, and tapirs. A rhipidistian crossopterygian is a freshwater, lobe-finned fish of Devonian age on the main line of evolution to the first amphibians.

36. Jesse D. Friggins (1867–1944) was director of the Colorado Museum of Natural History in Denver. He is perhaps best known for his work on the excavations of prehistoric sites where Folsom spear points of Paleoindians were found with extinct bison bones in the mid-1920s. His museum work included the mounting of fossil vertebrate skeletons for display and he disputed with R. S. Lull at Yale over priority for a new technique of using plastic in such displays. In rebuttal Lull quoted Simpson's testimony on the issue, which Friggins in turn lambasted.

a year in Europe, so that I will shortly become a National Research Council Fellow in the Biological Sciences and shall sail to join all my other compatriots who like America best when not in it about October 1st. Although hitherto a dyed-in-the-wool geologist, I have dabbled in biology and I now become a biologist with all the ease of the graceful python shedding last season's skin. I only shed the title, however, for like the python my ordinary and ornery nature remains the same or even worse—I shall carry on with exactly what I am doing now, under the guise of a geologist. Happy is he who is all things to all men, for I had exhausted the geological possibilities. No doubt some bona fide young biologist is making wax images of me and sticking needles in them for spoiling his chances!

Although moderately deluded, do not get the impression that I am flippant toward my work. I slave at it—and with good results, for the chairman of the Ph.D. committee read my thesis in person and informed the other members that it placed me at once among the foremost members of my profession—which, little as you may be inclined to believe it, actually makes me feel rather humble.

So much for all that. Again, the very best for your birthday. I hope to see you again very soon.

Your affectionate brother,
George

[New Haven, ca. 1923?]

Dear Marthe—

The enclosed primitive manuscript is more in your line than ours, & I forward it to you. It is an exact copy of drawings & an accurate translation of rhymes found incised upon plates of phlogopite-phyllite[37] found fossilized together with three phalanges & a humerus, radius, & supra-occipital bones of a dinosaur of an unidentifiable genus, allied to *Styracosaurus*. The degree of ossification of the humerus indicated a young specimen. Evidently all the plates were not found, & one of those found was broken & one half lost, but altogether they constitute a memorial of no little scientific interest. In fact so impressed was one of my colleagues on the faculty of Extinction that he acclaimed: "Simpson, had I known that you possessed or were likely to possess such a piece of work, I am sure that you would not be at Yale now"—

G.

37. Phlogopite is a platy mineral in the mica family; phyllite is a scaly metamorphic rock rich in mica. The phalanges, the humerus, and the radius are forelimb bones; supraoccipital bones belong to the skull. *Styracosaurus* is a Cretaceous horned dinosaur.

LONDON
1926-1927

HAVING FINISHED his doctoral study of American Mesozoic mammals, Simpson took up residence in London in 1926 for a year of postdoctoral research on Old World Mesozoic mammals housed at the British Museum of Natural History. He had two fellowships to support him: one from the U.S. National Research Council and another from the International Education Board, which together paid him $2,500. Lydia refused to stay in London, preferring to settle in the south of France in the village of Grasse, the center of the French perfume industry, about ten miles north of Cannes. She had the two girls with her—Helen, aged two-and-a-half, and Patricia, barely four months old. This was the first of several, increasingly longer separations of the young couple. Simpson clearly was ambivalent about the separation; on the one hand, it permitted him the freedom to work and to move about without domestic restriction, but, on the other, it meant his finances were halved and he lived alone.

Martha also ended up in southern France for part of the year, living and painting in Les Arcs, some thirty miles west of Cannes. She later returned to the United States, spent some time in Santa Fe, New Mexico, then went to Hawaii where there were a number of relatives from her mother's side of the family. Martha continued to paint and give an occasional show. She also began making puppets and presenting puppet shows locally.

At Christmas time Simpson divided his holiday between his wife and his sister in southern France. The Simpsons' third daughter, Joan, was conceived during this visit.

After the beginning of the New Year, Simpson began getting mixed signals from Yale about his position there on his return to the United States. He had had an understanding that he would come back as a research fellow. But in the meantime Lydia had complained to several Yale faculty wives, who raised the question of Simpson's character, so when Lull was traveling through England he stopped to discuss the situation with Simpson. Lull was soon satisfied that Lydia's chief complaint—the lack of financial support—was unjustified. After Lull reported back to Yale, Simpson was offered the appointment. During this awkward period, however, W. D. Matthew had also visited Simpson and encouraged him to fill the position at the American Museum that Matthew was vacating to move to the University of California. When a formal offer came from the New York museum, Simpson, disappointed in Yale, accepted. Matthew's departure from the American Museum was no doubt softened by his ability to recruit the best young vertebrate paleontologist to replace him.

Simpson's year abroad was a kind of *Wanderjahr*. Besides consolidating his hold on the field of primitive fossil mammals, he had the opportunity to travel to the major research institutions to study their fossil collections firsthand and meet the leading scholars in the field. Equally important, Simpson immersed himself in the culture of Britain and Europe and, eager-beaver that he was, took full advantage of all there was to learn. He mastered French and German, read widely in French, especially poetry and nineteenth-century romantic novels, developed his interests in medieval sculpture and architecture, and experienced the nightlife of the Western world's great cities.

When Simpson returned, although still a young man in his mid-twenties, he was a mature scientist, initiated into the highest circles of his field, and a genuine cosmopolitan, knowledgeable and comfortable in cultured circles.

Simpson's closest friend in London was Arthur Tindell Hopwood (1897–1969). Hopwood was born of artist parents and was himself interested in painting, especially on porcelain. He served in World War I as a Royal Naval Air Service navigator. After receiving advanced degrees in zoology and geology at Manchester University, Hopwood joined the British Museum in 1924. Although his previous research had been on marine invertebrates, he studied fossil mammals at the British Museum, especially elephants and early humans. He was one of the first—along with Louis Leakey and Hans Reck—to work at Olduvai Gorge in East Africa. In 1933 Hopwood coined the term *Proconsul* for the Miocene fossil ape discovered in Africa (he derived the name from a famous London zoo ape named *Consul*). *Proconsul* is related to the group of fossil apes from which our early human ancestors descended.

Simpson and Hopwood discussed fossil mammals and roamed London during Simpson's stay, but they drifted apart in later years, particularly when, as Simpson remarked, Arthur T. Hopwood became A. Tindell Hopwood. Years later, when Simpson tried to arrange a meeting with Hopwood during a London visit, Hopwood declined, much to Simpson's disappointment. Hopwood's obituary in the *London Times* noted that although he was "rather solemn and occasionally pompous he was actually merry at heart."

Simpson's chief publication resulting from his work in the British Museum was a monograph describing and interpreting Old World Mesozoic mammals. Although completed after his Yale dissertation, the monograph was published before his monograph on American Mesozoic mammals, which was drawn from the dissertation. Today both monographs are classics and are still in print. Although superseded in some ways by more recent work, more than sixty years later they still provide a valuable source of detailed information.

London
Oct. 24, 1926

Dearest Marthe:

Here I am established at The Hall, 29 Trebovir Road, Earl's Court, London S.W. 5—which is oh! how much less aristocratic than it sounds, although not half bad. It's some quart hour's walk from the museum, in a fairly respectable neighborhood and with food neat but not gaudy.

Chauffage centrale [central heating] is even more distant as a rumor here than in la belle France and even in my nice new double weight pure wool two piece ones I am cold all the time. Each moring [sic] at 8:30 the slavey depositsa [sic] pitcher of luke warm water at the door and pounds vigorously—I take it in and superheat it on the gas ring (each room has one, on a shilling meter), go down at 9 to my liver and bacon, or fish, or kidneys or what not that constitutes an Englishman's way of beginning the day right, with liberal tea and marmalade, and then tool off to the Museum at 10. I am supposed to stop work there at four, but by special dispensation I stay till five, so that what with time out for lunch I get in 5½ or 6 hrs a day.

Everyone is delightful. The two men in charge of most of the fossil vertebrates are both under thirty—I have a place in their office. I am finding out many new things about my little beasties. At 1 Hopwood and myself (he's their fossil mammalogist) lunch together and then go for a half hour airing, and in the evening he walks home with me and then takes his train at the station near here. All of which is very pleasant indeed, but I still find myself in the evenings painfully close to the nervous depression which hounded me all spring and summer.[1]

The thing which perhaps amuses one most is that the soap and towels, not to mention more intimate things, at the museum all theoretically are the personal property of the king and are flamboyantly marked with his crest and monogram! It gives my democratic soul much pleasure to wipe my face on the king's towel!

Last Sunday I went to the art museum (Brit. Mus. at Bloomsbury) but came away feeling that I hadn't seen a thing worthwhile. One Greek vase or marble may be thoroughly inspiring, but a whole room or building jammed full of them is simply depressing. One can't see the forest for the trees. I always have maintained that the really good things should be scattered among ths [sic] small museums, but of course if they were[,] neither I nor anyone else would bother with them probably.

I so regretted leaving Paris—I haven't felt so gay and carefree for years as

---

1. Simpson perhaps is alluding to the emotions stirred by his relationship with Anne whom he had begun seeing the previous winter.

I did there and it was almost the first time we have been together for a decent visit since we have been grown up. I hope that everyone's departure has not made you too blue. Don't fail to give me your southern address. Do tell me, too, about the salon. I am sorry to miss it and hope I can manage to see your exhibition in the spring.

> With lots of the best—
> George

[London]
Nov. 20, 1926

Dearest Marthe—

I'm dining with D. M. S. Watson[2] in an hour, but I'll try to dash off a line. I was most depressed & grieved to hear of your troubles, dog-bite, tour, burn, shinlessness and all. I do hope that you are better now & beginning to get along. It's so hard to be sane in this world; how do you manage, or do you?

Gloom spreads as a pall. It has rained every day & most nights, & I cannot possibly get away at Christmas—lack of £. S. D. [pounds, shillings, pence]—even Paris would require at least seven pounds, which I haven't. If there's anything worse than being here at all it will be Christmas here.

People are most kind however & this week has been fairly bright. Monday eve I dined at the Zoo (i.e. Dinner Club of the Zoological society) with Hopwood & C. Tate Regan—the latter a famous fish (student). Wednesday evening I dined with Dr. Butler at his club. Sir Arthur Woodward, Sir Francis Ogilby [sic], & Sir Thomas Holland sat with us & we wined well & all most jolly.[3] Towards the shank of the evening Dr. B. promised me, as an old pal, to publish all my European results as a descriptive catalog of the Museum—fortunately he still remembers doing so & everything is lovely. Much *kudos*, as one says. Thursday a young Australian name of Overell & I went to the theatre & a number of whiskey & sodas, while Friday (yesterday) Hopwood & I spent the evening eating cocidos & other conglomerations & drinking Spanish sherry & port at the Casa Mar-

2. David Meredith Seares Watson (1886–1973), vertebrate paleontologist at University College, London. He demonstrated that rhipidistian fish are true ancestors of terrestrial vertebrates. He also showed that the Permian vertebrate fossil *Seymouria*, discovered in Seymour, Texas, was an almost perfect intermediate between amphibians and reptiles.

3. C. Tate Regan (1878–1943), director of the British Museum of Natural History; expert on fish evolution and biogeography.

Arthur S. Woodward (1864–1944), vertebrate paleontologist and museologist; student of Mesozoic fossil fishes. Coauthored with C. D. Sherborn *Catalogue of British Vertebrata* (1890); scientific describer of the Piltdown skull that was later shown to have been faked by unknown perpetrators.

Francis G. Ogilvie (1858–1930), Scots physicist, engineer, and geologist; chairman of H. M. Geological Survey; known for his studies of geomorphology and of the Cretaceous strata of Surrey.

Thomas H. Holland (1868–1947), geologist with many papers on the geology of India; rector of Imperial College, London.

tinez—you see like all of us Englishmen my thots of an evening's entertaining are rather alcoholic. (as whose wouldn't be in this trice-bitterly damned eternal clammy nasty climate). Tonight to Watson's.—All this to prevent your feeling sorry for your poor little brother all alone in that snug little isle.

In Hopwood I have, I believe, struck one of the few kindred spirits of a lifetime—for he is nearly my age (4 years older) & is both silly & a vert. paleontologist. I know too many silly people & too many vertebrate paleontologists but no other person who satisfactorily combines both. We get on most well.

I must dash. Do get better & get some rest. I suppose your coming here for the holidays is as hopeless as my going there, but I do wish you could. I could manage a place for you & nourishment of a sort very nicely, & Lord! (says Pepys) how glad I should be to see you. More anon—G.

[London]
Nov. 30, 1926

Dearest Marthe—

Of course I'll come down. As they say in the hair-oil (or organ-grinder, or patent shoelace or what have you) advertisements, "send positively no money, & you will be surprised at what you receive." In other words, keep your filthy lucre, or in pure London English, retain your sanguinary two quid. It's bad taste to flaunt your wealth before a poor boy trying to get along. We'll let some creditors fret; what, I ask you, else are creditors for?

Of course you know that if I come south I'll have to spend the major part of my time at Grasse, for I am still struggling along. But I should be able to manage, say a week, not more, and perhaps a day or two on my way back. Christmas itself I must spend at Grasse. Don't think this means only crumbs for you, for it doesn't. You know I have to reconcile different loyalties and make a dozen different ends meet. It's largely because others don't recognize that fact that I have what few troubles I do have.

Where were you going to move on to if I couldn't come? Perhaps it is not worth staying there for the little while I could stay. If you think otherwise (as I hope you will) suppose I come about the 9th & stay till the 15th, *possibly* returning for New Year's unless you want to dash off before then? If by any chance letters don't connect again, expect me about then and leave a note at Poste Restante, Les Arcs, telling how to reach you.

Cheerio!
Georges

Ballades are meant for France's sunny skies
Where winds and trees and smiling hills unite
To spread, before the weary, heart's delight;
It takes less wit a sonnet to devise
In Britain where the sun doth never rise
And black and grey divide the day and night.

Ballade in dainty measure can invite,
And clumsily the sonnet now replies:

> Mon Dieu! and else my God! were it not there
> But to the frigid pole you bid me speed,
> I yet would come at once, my only care
> Your slightest uttered wish (and mine!) to heed.
>   I shall arrive *aux Arcs* today a week—
>   And now no longer writes who soon will speak!

Thursday, December Second
[1926]

[London]
Feb. 20, 1927

Dear Dad—

Thank you for your letter of the 6th and its enclosure, which arrived safely.

Mother told me she & Marthe were going to Omaha & Marthe later told me she was going to Santa Fé, but I didn't know Mother was going to California. Of course it is too bad that they are not together, but we belong to that unfortunate group too elevated to be foot-loose and too low to be independent whether thru financial considerations or others. Marthe, I am sure, will love it in Santa Fé.

My own plans become less settled every day. I had hoped by now that Lull would be here & that I could go over the situation with him, but he's not turned up. I don't know where I stand. The staff appointment at Yale was offered me last Fall & I turned it down to come here. They appointed two men in my place—numerically flattering but rather unfortunate otherwise for both are satisfactory & now it is not considered the thing to turn them out for me just because I have (theoretically) been enjoying myself in Europe. The last plan I heard was to give me a Fellowship of the same value until a staff opening appeared. In one letter I was told, *unofficially*, that the fellowship was granted—a University one (Sterling) & not museum strictly—subject to my application & acceptance. In the next I was told that the delay in receipt of my application—a delay which did not exist as I applied the day after receiving a request to do so—had made it necessary to defer its consideration until the middle of February. That time is now passed, & all I have is another letter acknowledging the receipt of my application & nothing more. In other words I am kept wholly in the dark as to what is going on & I am unable to decide what weight to put on past assurances, definitely as they were given. I am quite piqued & would accept any other opening that would give me enough to keep us alive on & give me time for research on my own. The Am. Museum opening is far from hopeful as the last word is that they have a deficit of $150,000 for the current year & the staff must be *reduced*.

So I have no idea what will become of me. I have my influential friends—I know all the important vertebrate paleontologists in the world now—and I have done & am doing good work which is in the scientific public eye, so that I at least won't have to lay bricks. But my present appointment ends Oct. 15th, & something has to happen soon.

The Magnum Opus, the descriptive catalogue, & the work on Mesozoic mammals are the same. Its present tentative title is the simple one of "A Monograph of the Mesozoic Mammalia of Europe, Including a Few Extra-European Forms Preserved in European Museums, with a Descriptive Catalogue of Specimens in the British Museum (Natural History)." I am pounding out the ms. now, & it looks like about 300 or 400 printed quarto pages.

I do hope you come over this summer. England itself is interesting & at least three perhaps four months will be spent on the continent. My first consideration, of course, is getting my monograph done, & there is relatively little to be done in Germany & nothing in France, altho I hope to have the time to make related studies to be published after I return to America.

I did have a touch of the flu but it's over now altho I'm very tired. It takes all my energy just to get my daily task done at the museum. at [sic] least the mental fatigue which is more distressing has passed. It kept me comparatively sterile last summer & it nearly attained me again in December, but now it is gone and I am now so full of ideas & plans that I have to drive myself to do the polishing of old ideas now in hand & my mind is so active with new conceptions that I can hardly sleep. It keeps me burned out physically but is infinitely less depressing than the occasional mental blankness from which one sometimes suffers, when observed facts evoke no response of explanation & inspiration is dead. I have not had more than a week really withdrawn from my work since I undertook it in 1923 and I'm afraid I won't before 1928 if then—unless, as seems possible—I am favored to give it up for a time altogether.

With love,
George

I have extravagantly spent your $10 for a copy of Cunningham's "Human Anatomy" (a voluminous standard work which I got second-hand in new condition—a great bargain) & one of Duckworth's "Morphology & Anthropology"—also a bargain, a rare out-of-print book of greatest value to me. I'm spending time studying human evolution & I hope your gift will blossom some day. Thank you.

[London, 1927]
Feb. 27—

Dearest Marthe—

I've been gay & giddy this week, at least so my landlady (who is suspicious of all men under 40—or over 40) tells me. I've practically lived in dinner clothes—this dear old formal London has done that to the rough son of the West. Dinner at the Elliot Smiths—he's Professor of Anatomy at the Univ. of London, an Australian most of whose younger years were spent at Cairo. The other guests were the professor of anatomy from Hongkong and the former court physician of the Rajah of Sarawak, a pleasant lad named Le Gros Clark. And of course wives and an odd—very odd—young female of the species. Then it was dinner (at Martinez'—a spanish dinner) & the theatre (Blue Mazurka at Daly's—very amusing) with Hopwood & Dr. & Mrs. Matthew (my old friend with whom I was in Texas—he's on his way home from China, India, Java & other points east) and Professor & Mrs. Stensiö—he runs the paleontological business in Stockholm, one of the best. And then the Lulls, who are here now—he's on sabbatical leave. So it goes. And I'm booked (English for "dated") for the Matthews, the Le Gros Clarks & the Shellshears this week and the D. M. S. Watsons the next. And so I muddle along thru the lonely days & nights.

And an occasional evening with Sherborn.[4] No woman but the housekeeper has entered his rooms since he moved in, a young man of forty, thirty years ago, & they look it. We sit around & smoke our pipes & look at his treasures & he tells me about the illustrious dead. He knew Darwin, Huxley, & Owen & many others of the heroes who died before I was born. His rooms are a positive museum. The accumulation of no less than 800 years of Sherborns of taste and moderate means well expended. He's given me some treasures, an ms. & autograph letter of Owen's, a sheaf of notes in Clift's hand on the famous "Missourium"—a fossil mastodon found in America about 1825 or so & reconstructed into a marvelous sea-beast, etc.[5]

Evenings not spent elsewhere I study brains & bones or, if tired, indulge in Mr. Fletcher's ratiocinative mysteries.[6] Noons I lunch with Matthew & Hopwood, occasionally joined by Erik Stensiö. The rest of the time I clack upon the typewriter & add to the 250 ms. pages of monograph already completed or draw little things of great purity of line—"Twilight on the

4. Grafton Elliot Smith (1871–1947), Australian-born professor of anatomy at the University of London; known for major studies of fossil humans and their evolution.
Wilfrid Le Gros Clark (1895–1971), English physician and anatomist; early work in the East Indies, later professor at Oxford. Major work in human paleontology and evolution, including *The Fossil Evidence for Human Evolution*.
Erik Stensiö (1891–1984), vertebrate paleontologist at the Royal Natural History Museum of Stockholm; expert on early primitive fishes.
Joseph Lexden Shellsear (1885–1958), Australian-born professor of anatomy at the University of Hong Kong.
Charles Davies Sherborn (1861–1942), invertebrate paleontologist at the British Museum whose specialty was microscopic single-cell shelled protozoans; he compiled massive bibliographies of both microfossils and of British vertebrates.
5. Richard Owen (1804–1892), great British anatomist and paleontologist who was one of Darwin's severest scientific critics; he coined the word *dinosaur*.
William Clift (1775–1849), Owen's father-in-law; physician and first conservator of the Hunterian Museum; also dabbled in vertebrate paleontology.
"Missourium," the name given to a fossil elephantlike mastodon found near St. Louis, Missouri, in 1839 and acquired and displayed by the British Museum. The fossil's finder, Albert Koch, promoted the creature as the Leviathan described in the Book of Job. Its scientific name is *Mammut americanum*.
6. Joseph Smith Fletcher (1863–1935), English mystery writer who was a favorite of President Woodrow Wilson, who brought him to the attention of the American public. Simpson had already begun what was to be a lifelong recreational activity: reading mysteries.

Jaw of *Amblotherium soricinum*" or "General View of the Sixth Molars of Purbeckian Mammals Seen from the Tower of London" or "Pastel Study of the Head of a Young *Triconodon mordax*"—light bits, full of feeling.[7] There's talk of hanging me at the R. A. [Royal Academy], or at Newgate [prison].

Sunday mornings, like this I cultivate that virtuous glow by writing silly epistles to the dark parts of the earth, extolling London's murkiness to the sunshine dwellers in Holy Faith or elsewhere.

In eleven days we had six minutes of sunshine. Yesterday morning it was dazzling sunshine for an hour—that is, one had a sickly shadow for that length of time, but nature mercifully shut off this upsetting display & it has poured rain ever since. Everyone says "well, thank heaven, winter's about over—though of course March is our worst month." You can always forecast the day's weather in the morning. If the sky is overcast, it will rain, while on the other hand if, as happens once or twice a month, the sky is clear, why, it will rain. I'm reduced to talking about the weather, but that passes for clever chit-chat here. "Young man" says Dr. Colman, a true Dundee Scot, "therrre is no weatherrrr"—r's rolling like thunder in the hills—"in London, only samples. Forrr weatherrr go to Scotland." Which, like most advice, I welcome as a good thing not to follow.

Occasionally I go to the so-called "learned societies"—good example of that dry English humor. If they are learned I am a super-superman—and I wasn't, the last I heard. The zoological society, otherwise the zoo, to hear a red-faced old beef consumer talk solemnly about the female genitalia of certain Bornean insectivores. Or to the Linnean [Society] to sleep through a dissertation on the influence of calomel on *Boreanthis borealis*—he'd feed arsenic to little children (and a good thing, too, to London children) that chap who dopes defenseless flowers. Or to the British Association for the Advancement of Science—familiarly & not ineptly called the British Ass—for a drop of tea. Or to the Geological [Society] to hear some callow Oxonian argue for hours about whether the boundary between something silly & something sillier should be drawn above or below the zone of *Somethingia somethingensis*, as if anyone cared.[8] No one seems to do anything sensible, such as Mesozoic Mammalia, for instance! The only way in which these scientists (not Christian) exhibit the famous English common sense is by eating before and after each meeting, & telling one another scientific dirty stories the while.

I have only the foggiest notion—London particularly foggiest at that— as to what I'll do next year. I'm disgusted. I don't greatly care. I have to do research because I have that sort of mind & can't help it, & I have to earn

7. *Amblotherium soricinum* and *Triconodon mordax* are upper Jurassic primitive mammals; both genera were originally described and named by Owen. Purbeckian mammals are primitive forms found in England in Purbeck rock strata of late Jurassic age, of about the same age as the dinosaur- (and primitive mammal-)bearing beds of the Morrison formation in western North America.

8. Because fossil species originate and disappear in earth history, their presence often provides a temporal datum for layers of rock strata. Specialists may argue about a particular time boundary being drawn coincident with the first or the last appearance of a given fossil species. Any debate among experts about subtle distinctions quickly loses the interest of the nonspecialist.

money because I have myself & others to keep alive & can't help that, but it really doesn't matter. I'm in a fair way to be a "distinguished man of science"[9] but the samples so far exhibited don't make that seem the ultimate which life has to offer. Not *quite*, at least—I'd almost as soon be an artist!

It was nice to hear from you at last. I'm happy to hear that you are about to settle down & start making good at last. I can already see the fingers point & hear the whispered "That's her brother!" As *you* no doubt have happy anticipations of the 'arf bricks describing peaceful parabolae in your direction at the cry "That's his sister!"

Also glad that you find life serious. Anything so damned silly is, of course, very serious. Especially the fact that one has what the euphonious & tactful English phrase calls "guts & offal," beautifully adapted for the taking in of nourishment & frightfully annoyed if they don't get it. Also this delightful "central nervous system" which raises us so high above the apes, & makes us so much less happy, & this fascinating "reproductive system" to keep us in mind of our simian origin.

Well here's a splendid lot of blather. Of course I don't mean any of it. You serve as safety valve to keep me from telling everyone what jolly boobs they all are.

I'm glad you're in Santa Fé, it must be jolly there now. I can see the automobile dens and movie mansions posing as holy spanish missions. The palace, so many hundreds [of years] old, reminding one of the golf club which had had seven new heads & ten new shafts but was still the same old club, or is it "stick"? The silly addle-pated mexicans trying to make their dumbness seem mysterious. The Indians getting drunk & doing what they fondly hope may be ancestral dances. The artists hoping to be mistaken for Artists. The upper Miocene (oh yes! they *are* upper Miocene!) badlands[10] turning orange & blue in an effort to be really bad, & producing *bad* whisky in proof thereof. I love it all—it's so genuine in its total lack of reality.

I feel like writing on & on, but I have no beard now.

Yours, sweetly & innocently—

[11]

9. In a German scientific journal Simpson was referred to as *der ausgezeichnete Forscher* (distinguished man of science) and was obviously quite pleased with the praise for he mentions it several times in his letters.

10. The highly eroded variegated rocks north of Santa Fe are of the Santa Fe Formation, a series of rocks Simpson had studied in the summer of 1924 and reported in one of his first two scientific papers read at a national scholarly meeting. Their age was uncertain for a long time, although, as Simpson realized, some of the vertebrate fossils yielded a definite late Miocene age.

11. Simpson began using this Egyptian hieroglyph, representing a cobra and taken to designate approximately the sound "gee." "Gee" was the name that Martha, Anne, and a few intimate friends called Simpson. Others referred to him as George; once in a while he was also called "Gee-Gee," for his first two initials, but he didn't like the nickname because it seemed to him too easily confused with "Gigi."

Appendix!

Stensiö of Stockholm is here now, & one has the edifying sight of the three greatest vertebrate paleontologists in the world foregathered together—Stensiö for fishes, Watson (of University College here in London) for amphibians & reptiles, & Matthew for mammals. Hopwood & I are taking Professor & Mrs. Stensiö and Dr. & Mrs. Matthew to dinner & the theatre (Daly's—the Blue Mazurka) Wednesday. They are all the best sort possible.

I am also enjoying some association with G. Elliot Smith, the leading authority on the brain & one of the leaders in anthropology & especially in human evolution ("Studies on the Evolution of Man" & many other books). He's coming in this week to spend a day going over my Mesozoic mammal brain casts & also the cynodont reptile brain—the cynodonts were extinct reptiles ancestral to the mammals.

I occasionally spend an evening with Dr. C. D. Sherborn—a fine old lad nearly 70 whose rooms are a veritable museum and who knew Owen, Darwin, Huxley & others of the giants of the past personally. He has given me a hand-written manuscript of one of Owen's papers as also an autograph letter of Owen's and a sheaf of papers chiefly in Clift's hand (Clift was Owen's father-in-law & curator of the Royal College of Surgeons' museum at the beginning of the past century) relating to an early mastodon skeleton found in America—Koch's famous "Missourium."

I've also acquired a little ichthyosaur skull from Lyme Regis[12] here in England & a small mammoth tusk which Bassett Digby (author of "Mammoths & Mammoth History") brot back from north-east Siberia.

G.

Speaking of mammoths—Pfizenmayer who collected two of the most famous mammoths in Siberia sent me his book "Mammutleichen und Urwaldmenschen in Nordost-Sibirien" & I have a review of it in one of the current nos. of the Am. Jour. Science.[13]

March 6th
[1927]

Dearest Mother—

Letters have to go about a third of the way around the world to reach you now [in California]. I've not had your new address until this week, but I suppose Dad has forwarded a couple of letters sent to him recently [in Washington, D.C.]. I wrote the longest letter of recent years to Marthe last week & sent it to Santa Fé—I suppose it will long repose in the pseudo-pueblo opposite the cathedral and then be tenderly destroyed

12. Ichthyosaurs are Mesozoic marine reptiles, somewhat resembling porpoises; Lyme Regis in southwestern England on the Channel has provided exquisitely preserved marine fossils in Jurassic limestones and shales.

13. Eugen N. Pfizenmayer, turn-of-the-century German-born naturalist, geologist, and paleontologist; he was curator at the St. Petersburg Museum and a member of the expedition to recover the Bereskova mammoth preserved in the Siberian permafrost.

unread. So much for epistolary encouragement! I shall write to no address hereafter not vouched for before a notary public (or, as we English call it, a commissioner for oaths) on the spot.

I myself am very stable despite my remoteness. Except at Christmas I haven't been even a mile out of London and am still at my original pension (Angl.—"private hotel"). I'm breaking forth this week, however, for I'm giving Oxford a whirl, starting Thursday. Matthew is giving a couple of lectures there, &, as I have to go up anyway, I'm making my visit coincide with his. He's going back to N.Y. next week after a year's absence. Mrs. M. is quite keen to get back & see how her "pawned" children are getting on—one has become engaged in the meantime & one has decided to become an artist[14]—two tragedies due to absence of parental suppression!

I shall look about for Don Stauffer—I'm not sure whether he's still at Oxford or not—I keep putting off writing him. I do know six or seven fellows there, however, so shall have a good time. I have a few days work in the museum there.

Life looks up a little & I've almost lived in my dinner jacket the past few weeks. We have quite a colony of foreign vertebrate paleontologists now—Professor & Mrs. Stensiö from Stockholm, Baron von Huene[15] from Tübingen, Dr. & Mrs. Matthew from New York, Professor & Mrs. Lull & Dorothy Lull from New Haven,—oh yes! & Dr. D. N. Wadia[16] from Calcutta, he's a pestiferous Parsee who tries to be a paleontologist and can't. He's the author of learned treatises on the Geology of India but remains a horrible example of the effects of educating the native. Then there is W. E. Le Gros Clark, a young chap who was for some time court physician to the Rajah of Sarawak (Borneo)—I think he's going to be the great comparative anatomist of my generation. I'm dining with the Le Gros Clarks tomorrow evening. And of course there's a flock of English anatomists, comparative & human, who are all very nice to me. I still spend Sunday evenings smoking a pipe (& drinking port) in old Sherborn's fascinating rooms. So if I languish, as I occasionally do, it's my own fool fault.

The magnum opus progresses—the descriptive part is done, about ⅔ of the tome, leaving only distribution, geological & geographical, & ordinal relationships to do. For a week or two I've declared holiday from the Mesozoic mammals—after three years continuous work on them I thoroughly dislike them—& am making notes & drawings of European Tertiary didelphids, jargon for fossil opossums.

I have enough work planned or in hand to keep me hopping for about

14. Margaret Matthew became a skillful scientific illustrator and later married Edwin H. Colbert, a vertebrate paleontologist, who was Simpson's colleague at the museum and subsequently his successor as chairman of the department of geology and paleontology.

15. Friedrich von Huene (1875–1969), paleontologist at the University of Tübingen; he collected and studied mammals in Argentina, reptiles in South Africa, and dinosaurs in Brazil.

16. Darashaw Nosherwan Wadia (1883–1969), Indian geologist with the Geological Survey of India. He later became first director of the Indian Bureau of Mines and the first geologist appointed a National Professor by the Indian government. An authority on the structure and development of the western Himalayas, he wrote the *Geology of India*, which went into four editions.

fifteen years. Professor Lull just kindly offered to get together all the fossil mammalian brain casts in America for me to study, but I kindly but firmly turned him down. Life is too short, I can't do everything.

I'm not certain as to where I'll be next year, not being so keen to return to Yale as I was when I was there. I suppose I will go back—I have two fellowships to run simultaneously, a Sterling Senior Fellowship of $1500 & a Research Fellowship in the Peabody Museum of $1000. I can't manage on less than $2500 & they're not giving any fellowships that large, so gave me two. An old hen who scraped an acquaintance with Lydia has been meddling in my affairs and, while no one takes her seriously, I'm so disgusted that I'd willingly turn to fields afresh. I have several other leads. One would take me to southern or southeastern Asia, one to Egypt, another to New York, but none of them is sure to materialize. I really shouldn't stay away from the U.S. much longer if I hope to work there permanently, as of course I do. Out of sight, out of mind, & new youngsters keep coming up. I may very probably prolong my stay a couple of months to do some collecting in France, however, if the museum (American M.) will put up the money, as they are somewhat inclined to do.

I'm spending what little spare time I can snatch studying the evolution of man from the anatomical, paleontological, & psychological points of view. I don't think the battle in the Beknighted [sic] States is yet over, or indeed half begun.[17] I have nothing for those whose only approach is emotional & who will always believe what is pleasant in preference to what is true, but if anyone cares for facts I intend to know them.

I'll try to remember the Houbigant Soap, but please don't be hurt if I don't. You know I can hardly remember anything not paleontological for more than ten minutes, despite the fact that I'm not one of those terribly earnest young men. But I'll make a most determined effort.

By the way, ask Marthe what I'm to do with two bottles of perfume I got for [Aunt] Betty at Marthe's request. I can't be either smuggling or paying heavy duty on them every time I cross a frontier & I'd like to throw them in the sewer. They've already cost me a violent row and more money than Betty supplied. I don't mind in the least, as I did it for Marthe & not for Betty, but perhaps Marthe would be satisfied if I lost the bottles & returned Betty her money.

Here I am through my letter & with a whole page and three quarters of unused paper. My thrifty soul recoils, & having no more to say I shall embellish—

> With all my love,
> George

By the use of a complicated apparatus which you, dear reader, would not understand, we are now enabled to read the thots of the creatures whose bare bones are so shamelessly displayed in our great museums. We are thus able to present the following

17. The Scopes "monkey trial" had taken place a year and a half earlier (July 1925) and Scopes's conviction for violating the Tennessee law against the teaching of evolution was subsequently overturned by the state supreme court on a procedural technicality, not on the basis of the validity of Darwinian evolution.

ANCYLOPOD ANTHEM
from the Oligocene Epoch.[18]

When I was small (though rather tall)
My mother used to say
"Oh brush your hair with greatest care.
Guard teeth against decay;
Protect their lime 'gainst thieving time
And watch them day by day—

CHORUS: For you're going to be a fossil by and by.
A portion of your snoot
Will strike a scientist's boot
Or an upper left premolar strike his eye.
He'll put you in a case
With others of your race.
You're going to be a fossil by and by!"

Though past my prime, I laugh at time.
I scorn my fellow critter—
He lives to die, but not so I.
Unlike the common quitter
When life is past, I still shall last.
My fate's sublime, not bitter:

CHORUS: For I'm going to be a fossil by and by.
When you're dead, you're dead
But my time-hardened head
Will wake the eager scientist's happy cry.
He will study it and measure
With undisguised pleasure—
I'm going to be a fossil by and by.

And when I pass a comely lass
I never heave a sigh.
I have no taste for ladies chaste
(Or otherwise)—not I!
I pay no heed, I have no need,
I merely sniff and cry—

CHORUS: I am going to be a fossil by and by.
I'm quite enough for me,
For my posterity

18. Ancylopods are a primitive group of odd-toed hoofed herbivores that are distant relatives of tapirs, horses, and rhinoceroses which lived throughout the Tertiary Period. The Oligocene is one of the epochs within the Tertiary, dating from about thirty-seven to twenty-four million years ago.

Could never, never live so long as I.
Though years and years will pass
I'll *still* live (under glass).
For I'm going to be a fossil by and by.

At this point the apparatus broke, so that we are faced with the painful necessity of wasting paper after all.

LONDINIUM
XX.iii.MCMXXVII

Dearest Mother (and all to whom these presents come—):

On Thursday, March 10th, I went to Oxford, where the following program was religiously followed:—

10th: Arrival. Lunch with a friend from New Zealand. Successful search for a cheap room. Rugby Match between Greyhounds and Nomads, the former winning. Tea with Zoölogy Department. Lecture by W. D. Matthew. Dinner and theatre with a friend from New Brunswick (M.A. Chicago). Bed.

11th: Breakfast in bed. 2½ hrs. work on Mesozoic mammals. Noon lecture by Father William (more reverently, Dr. Matthew). Lunch with Prof. & Mrs. Goodrich,—Dr. & Mrs. Matthew & Prof. Poulton also present.[19] Walk with the Matthews and then see them off for London (hence America, before I shall see them again). Tea in St. John's College with two Canadians and an Englishman. Dinner in Hall at Merton College with Don Stauffer. Meeting of Bodleian Society—discussion of Birth, Copulation, and Death from the Literary Point of View! Oxonian! Bed.

12th: 3 hours work on M. m. [Mesozoic mammals]. Lunch in New College with an Australian and a Canadian. Sightseeing, Tea, Dinner, and Oxford by Moonlight with Don Stauffer. Bed.

13th: Bed all morning, reading the Constant Nymph. Dinner in town. Tea at University College with some Englishmen and an East Indian. Supper at college whose name slips me. Heated discussion in Australian's rooms in St. John's. Bed.

14th: 2 hours work. Lunch in other rooms in St. John's with an American, a Canadian, & an Englishman. Afternoon with Professor Sollas[20] in Geol. Dept. Dinner in town and cinema.

15th: Breakfast with Don Stauffer and roommate in Cowley Road. Geologizing and sightseeing in villages and quarries about Oxford by auto with young Baden-Powell and young Kindle (father a Canadian paleontolo-

---

19. Edwin S. Goodrich (1868–1946), English zoologist who worked on vertebrate origins and evolution.
Edward B. Poulton (1856–1943), English zoologist and vertebrate paleontologist at the University of Oxford.

20. William J. Sollas (1849–1936), geologist and invertebrate paleontologist at Oxford University. Among other subjects, he studied human fossils and ethnology, the structure of Pacific coral reefs, and the Silurian geology of Cardiff, Wales.

gist.)[21] Lunch at Stonesfield.[22] Tea in Oxford. Dinner and cinema in London.

16th: Bed till 11 at the Hall, Trebovir Road. Lunch and afternoon at Museum renewing acquaintances not seen since the 9th. Dinner at Cabnurus or something like that, Leicester Square, with Hopwood, then lecture "Man's Origin" by D. M. S. Watson, then oysters and stout at Scott's. Bed.

17th: Pretense at day's work, inserting Oxford material into manuscript. Dinner in Trebovir Road. Conversazione in Sherborn's rooms at Parson's Green.

18th: Still inserting Oxford into "European Mesozoic Mammalia." Dinner, Theatre, and supper with young Australian lad.

19th: Half day's work at Museum. Kew Gardens all afternoon. Early to bed.

20th: Late to rise. Wrote one page and one-third of letter. We are now at the present moment.

It's easy to see why I'm lonely here. The Lull's have left for Italy and the Matthews for New York. The Stensiös have returned to Stockholm. Baron von Huene is still here, but leaving soon. I still smoke a pipe with Sherborn once a week (the conversazione there was a more formal affair with several guests) and he still adds to my store of ancient coins and of nineteenth century anecdotes! This week I am going out to dinner one night, only, & one night to a smoking concert—delightful institution where you sit at tables and drink ale & smoke a pipe while a really good lot of male singers do old English ballads and such. I'm to week-end with Sir Arthur & Lady Woodward in Sussex soon and also with the Le Gros Clarks in Surrey. As soon as my next check comes (I'm now broke, of course) I'm going to Bath. They have some Triassic mammals & some Roman ruins there! I'm also going to York before long to see a jaw of *Amphilestes broderipii*,[23] the minster, & the city walls. So runs the world along.

Did I say that I might not go to Yale? I have accepted an appointment, but am looking about elsewhere, nevertheless. There's an opening at Berkeley but I have not applied for it. I think my friend Horace Elmer Wood II will probably get it. Chester Stock has left there to go to Cal. Tech., leaving a place open.[24] I may go to southern Asia for a year, or to Egypt, or stay another year or two in France, or go to New York. Or, of course, I may take the bit in my mouth after all and return to Yale forever.

21. Perhaps Arthur Robert Peter Baden-Powell (1913–1962), the then teenage son of the founder of the Boy Scouts.
Cecil H. Kindle, an invertebrate paleontologist, about Simpson's age, who was later professor at the City College (now University) of New York.
22. Stonesfield strata of mid-Jurassic age north of Oxford yielded many of the important primitive fossil mammals that Simpson was studying.
23. Another Jurassic primitive mammal.
24. Chester Stock (1892–1950), vertebrate paleontologist at Berkeley who moved to the California Institute of Technology in 1926; also associated with the Los Angeles Natural History Museum, he published the first extensive account of the fossils recovered from the nearby La Brea tar pits. Stock's move to Cal Tech opened a position at Berkeley that Matthew, rather than Wood, filled; Simpson then replaced Matthew at the American Museum.

Intimate to all whom you meet in the highways & byways that letters
are strictly de rigueur not to mention bienvenues.

Thy adoring but rather simple-minded

Son

Oh! Thou imbecile reptile *Diplodocus*![25]
Whoever created so odd a cuss?
    With a tail like a neck,
    And a neck like a tail—
    I wonder, by heck,
    If you ever do fail
To remember your ends,
    And when danger impends
Do stand still, which is bad, or, still more, run tail first,
Or indeed run both ways, which is rather the worst!

(The irregularity of the meter is nothing compared to the irregularity of
the critter himself—I ask you![)]

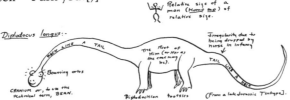

[London?]
About the 9th or 10th
[April, 1927]

My dear sister[*]—

I'm not going to write you in French because soon I'll have to speak it
all the time and now I would prefer to speak my native tongue—and I find
that the most beautiful of all, as a matter of fact. The French have, in fact,
something, I don't know what, of the spirit I love so well, which quickens
my pulse. When I speak French it seems a little like it is another world—a
world that I do not understand. There are too many worlds that I do not
understand without having that one too. [Continues in English.] So I burst
out into my more natural idiom. I sometimes sit with a Scot at dinner &
when I talk as I would in New York & he as he would in Glasglow we are
mutually incomprehensible—words pass, but no ideas. Not that *that* is
unusual. I remember about three—perhaps four—conversations (except
scientific ones) in which ideas have passed since I arrived in this decadent
kingdom.

Last week-end I spent with Sir Arthur & Lady Woodward at their place

25. *Diplodocus* was a quadrupedal dinosaur that measured some sixty to eighty feet
long and weighed thirty to fifty tons; it was among the largest land-living animals that
ever existed, along with related forms like *Brontosaurus* and *Brachiosaurus*.
  [*]Begins in French; translated by editor.

in Sussex. I was asked for an "informal week-end at our small country home." Luckily I scented (not to say smelled) a rat (*Mus rattus* Lin.) and, as I suspected, the "informality" included such details as dressing for supper Sunday evening and the "little country home" has about fifteen rooms and is ministered to by two gardeners & heaven knows how many house servants. My simple democratic soul laps it up however. I love having tea in bed when I first wake up; having my bath drawn for me at the specified temperature, my clothes laid out, & all the rest. Sir Thomas and Lady Holland were there & Sir T. drove us about Sussex between meals on Sunday (the idea that one need not necessarily have meals at Hill Place of course would not occur. We even faithfully returned there for tea and then resumed our wanderings.)

Today—which is Sunday—tomorrow, [and] Tuesday I am gracing the Anatomical Congress with my presence. Wednesday I am going to Bath where I shall stay over Easter. They have some mammal teeth there which I must see. Sir Thomas will be there & he is going to drive me about to the various geological localities which I want to examine. I am somewhat amused at the prospect of having as chauffeur the former director of the Indian Geological Survey, later Minister of Munitions there during the war & still later Member [Minister?] for Transportation and Natural Resources, now head of the Imperial College of Science here in London! He's a most decent, genial soul & about as near to being a back-slapper as the English ever get.

The English are delightful people anyway. I have made more good friends in a few months here than in four years in New Haven. They are neither cold nor reserved except with a reserve which is altogether admirable and which most Americans very badly need. I do loathe "good mixers"—when I think how upset Mother & Dad have been because I could not be one. I weep for them!

I have only encountered one snob—a professor at Oxford. He was positively fishy and so annoyed that I did not present excellent letters of introduction[,] which I had but let him take me or leave me at my face value. Half way through my stay he suddenly thawed and became downright affable. So much, thought I, for my winning personality! Alas for human vanity! The true cause came out! Dr. Matthew had arrived and had made flattering but embarrassing remarks about me in one of his lectures and was even so unguarded as to let our intimacy be apparent. Here, as in America, Father William is a great man—hence the honorable professors [sic] sudden thawing—a nasty brute!

I love France and rather dislike the French. I love the English and rather dislike England. If only the English lived in France—but then, alas, 'twould no longer be France. I neither love America (as a whole) nor Americans (as a race) but it is my country and unlike yourself I realize that I could not long be contented anywhere else. As to Germany, one waits to see. At present I feel kindly towards Germany as Baron von Huene, who has just left, is a most decent sort and the Neues Jahrbuch has just referred to me as "der ausgezeichnete Forscher" [the distinguished researcher]—how far a little judicious flattery goes! I hope the shock of finding that the distinguished research worker is a callow youngster does not mar their Teutonic placidity. My magnum opus runs now to some 500 odd ms. pages and is practically finished save for mopping up the odds and ends. It'll be out in a year or so. I'll also be bringing out a similar ponderous tome in America next winter, then, thank the lord, I can pass on to pastures fresh. The very thought of Mesozoic mammals, or mammifères Mesozoique or even of mesozoische Säugetiere makes me feel a trifle pale around the gills. I've belabored the filthy beasts for about four years and I hope to heaven no one finds any more during my lifetime—I've studied all there are in the world at present and I couldn't bear it if I had to study any more. I'm going to dinosaurs or something like that now, or so I hope.

On April 4th I received a note from Yale begging me to let them know by April 1st (sic!) whether I would be with them next year or not, but I shall wait a week or two before answering. I shall be, I suppose. I haven't the foggiest notion where Lydia will be next winter, or indeed where she is now. She has taken a notion to return to America but whether she has gone or not I am, humorously enough (if you care for that sort of humour), quite unaware. So it's almost impossible for me to lay any plans. I would be delighted indeed if we (you & I) could have a place together in New Haven for the winter, but of course I must wait to see what she is going to do.

I shall be most delighted to see Tante Françoise and Dorothée.[26] They'll have to wear little cabbages in their buttonholes or wear numbers or something. The old memory is a perfect blank—I have progressive amnesia, you know and shouldn't know them from Adam, unless they dress differently. I hope they won't mind, altho if they're likely to I can practise up. "Why of course, I'd know you anywhere. Those eyes! How could one forget?" But I'd rather not, if it's all the same to them. They'll know me—I'm the gloomy looking young man with a pink moustache & hair rather thin on top. Perhaps if I grew a beard it would help them.

Here's my tentative itinerary—I hope it coincides with theirs at some point—Shall be here and there, mostly here, in England until about June 1st, then I shall be dashing off to Paris for a fortnight, or probably until after the 16th, then I go to Lyons for a day or two, then Geneva, then Basel for a week, then Tübingen, about the end of June, for a couple of weeks, then the rest of July and August in Stuttgart, Munich, Frankfort, Berlin, Greifswald, then I hope and plan to spend a month in southern France or in Auvergne. Then I must once more turn my footsteps homeward and take up the white man's burden at dear old (or at least old) Yale once again. I wonder how chances are for Dad to come over and whether my itinerary would amuse him at all. I hope so. Hopwood is going to Paris with me—I think I shall enjoy him, he's not a bad stick.

I'm sorry that you're not contented, but who the devil is except a vegetable. I used to envy animals because they seem not to realize how discontented they are, but I have been reading animal (especially ape) psychology lately and have come to realize that a contented animal is as rare and as little admirable as a contented human being. Vegetables alone have no

26. Aunt Frances and her daughter Dorothy were Simpson's maternal aunt and cousin.

discontentment—Chunder Bose[27] of India calls himself a scientist and has much to say about the emotions of plants but everyone here considers kid Bose as an egregious charlatan and I daresay they are quite right. But of course even a vegetable isn't at all contented so that I am afraid nothing really is. You have to *know* that you are contented, or else you're not, but of course if you stop to wonder whether you are or not, why you're not, so there you are. Things are really managed very stupidly and I am sure that I could do much better, if given a free hand with the universe.

Apes are really very human in mentality and in emotions. They are, of course, less intelligent (in quantity, not quality) but the only really striking difference is that they have no inhibitions. If they feel like being filthy they *are*, properly. If they want to hit someone, they *do*, hard. No worry about what the neighbors will say or about whether it is proper and fit for an ape in their position to do such things. In spite of being imitative they follow a leader only because he amuses them or they happen to be going his way anyhow—if they decide otherwise, they stick his head in the slop barrel & go on about their own business. Oh—I'm not perfectly convinced that all progress from the apes has been upward!

Be that as it may, it is true I went out & had my tea (with which I consume pink-frosted cakes)—

Ton frère

What'll I do with bouncing Bettee's damned [perfume] Mimosa Bruno Court? Hop'll take it off my hands (he has a girl or two in the offing) and I'll send the money (fifty francs wasn't it?) to Bettina if you will permit. Yours—

———

20th
IVth
27th   Date ←
XXth
[London, April 20, 1927]

Dear Old Bean[28]—

When I call you a safety valve I wasn't swearing at you, I was only telling you, but I did not reckon on your abysmal ignorance, only equaled by my own. [I put that last in to make it sound less drastic—of course I really know practically everything worth knowing & several things not worth

27. Jagadis Chunder Bose (1858–1937), Indian botanist, founder and director of the Bose Research Institute in Calcutta. He wrote extensively on movement in plants.

28. A whimsical letter written on the empty spaces and reverse sides of Simpson's handwritten draft of his mammals manuscript. The letter winds in and around graphs and tables of data. The first paragraph refers to an earlier letter (3/27/1927) of Simpson's in which he likens Martha to the governor and safety valve on a steam engine, suggesting the role she played in his life, for he was able to confide in her and thus blow off steam.

knowing—such as the stuff on the other side of these sheets (I can't find any other paper so this'll have to do ⟨the data has all been incorporated in typed ms. now⟩)]. Go to the steam-engine, thou dullard! Observe the safety valve! Every other part is dull black or gray & pulls & grunts & sweats, while the gay little safety valve is shiny brass & spins & whirls & has a hell of a time. Then just to show who's boss it gives a jerk and there's a nice hissing noise and the whole thing slows down just because *it* says so. So if you can find anything prettier, gayer, & more useful to liken yourself to, why liken yourself to it and see how much sleep *I* lose!

I am just de retour from Ba-a-ahth. Everytime an Englishman says the name of that town he sounds as sheepish as he looks. Before I went I had been at the Royal Geographical with 3-initials Migeod,[29] Esq.—he calls it Mee-zhoh, but many's the unsuspecting lad who has addressed him as Mr. My God! This Migeod is a quiet dried up fellah that looks as if he might have a fairly sticky past tucked away somewhere—week-ends at Brighton (where all the beautiful & damned go with the not-even-handsome but equally-damned) or that sort of thing—but otherwise neither here nor there, as they say in Nigeria. Well, oddly enough Nigeria is to the point, for I will say that he knows his Benin as well as the Obi himself does. This little prune has tramped right across Africa on the equator (& that's a fair jaunt, mind you) & then not yet sufficiently fed up he tramped right back on the parallel 5° South, which is even worse. He thinks Lake Tchad is the height of civilization and he's made his tea out of water from the various sources of the Nile. The upper Congo rather bores him, but the Gold Coast is amusin' because there's some jolly good snipe-huntin' theyah. He's never had any trouble with lions or leopards—oh! of course he's had to kill one now & then that came bothering him, like you would a mosquito— but the fellows that get messed by those vermin go lookin' for them, & he hasn't lost any lions. Just what he has lost that he goes wandering through tropical Africa & can't wait to get back he doesn't know, himself. Well to return to our roast lamb (only usually it's the lamb's paternal grandpapa here [and their idea of roasting things is to place them in the same room with an oven for a minute or two ⟨or to give them a scorching look or a hot remark⟩]) to return, I repeat, to our moutons, this playful little fellah and I went one day to the R.G.S.[30] and there I examined some 15th-16th-17th Century maps & on them were laid out cities, only instead of a lot of crisscross lines & some black blobs, as on our own silly efforts, these old boys always drew the city as it would look to a one-eyed crow flying south of town at an elevation of 1,362 feet 9½ inches.

Well (well & er are my favorite interjections, but I prefer well, it is more liquid) well, when I got to Baath (quoting Dickens this time, the first was approximately original) when I got there I climbed Beechen Cliff and there

29. Frederick William Hugh Migeod (1872–1952), colonial civil servant in what was formerly the Gold Coast (today Ghana); he participated in various African expeditions, including two across the waist of the continent. In the 1920s he was the leader of several expeditions for the British Museum to excavate dinosaurs in East Africa.

30. The Royal Geographical Society, founded in London in 1830, encouraged and often sponsored research relating to the British Empire and global exploration, especially of the poles.

was Bath,[31] looking absolutely & in every last blasted detail like a city on one of those jolly old maps! Perfect! So I sat there for about three hours & was fairly happy, & then went & drank two pints (= one quart, as you won't remember) of Somerset cider (not this dishwater fresh cider either) and went to a street carnival and dashed about singing the Frothblower's anthem with a beggar on one side & a young lady (we will call her so) on the other and was fairly happy again, which was pretty good for one day, I think.

Well (again) well, I wrote a verse sequence entitled "Beechen Cliff" which started out with a sonnet on Bath as seen from the Cliff & went on with a ballade about Beau Nash, some Alexandrines about the Bathonian idea of town-planning (to which we shall return anon) dashed thru a silly interlude about the discovery of the springs, & finished up with a song about Pulteney Bridge, which is the nicest thing in Bath except the view from Beechen Cliff.

I got depressed one day (as we creative geniuses do, as you know) & gently dropped "Beechen Cliff" into the turbid Avon, whence it was probably fished by some water rat in Bristol & put to no good purpose. Characteristically enough the only part which memory has salvaged is the silly interlude which immortalizes (sic!) the accepted legend of the discovery of the waters. It here follows, so you may leaf over & begin where next the prose resumes:

PRINCE BLADUD

Prince Bladud, he
Had leprosy
About nine hundred years B.C.
His father great
Was heard to state
He'd have to give the prince the gate.

His money spent,
The poor prince went
From bad to worse, and not content
To starve and die
He thot he'd try
The post of keeper of the sty.

Prince Bladud, he
Gave leprosy
(Misplacèd generosity)
To all his swine—
Had they been mine
I shouldn't have thought Blad so fine

31. Bath, in southwestern England, in the Avon river valley. Legend has it that the city was founded by Prince Bladud, son of an ancient British king, who suffered from leprosy and went into exile in the region. He became a swineherd near Bath, and before long his pigs became infected with his affliction. One day, passing a steaming swamp, the pigs leapt in and were cured of their leprosy, so the Prince jumped in and he too was cured. Bladud returned to his father's court and became king, and in 863 B.C. established his court in Bath. Here King Bladud fathered King Lear.

They found one day
Along their way
A place where steaming waters play
And where they flow
The ground is low,
Black mud abounds & rushes grow.

O lovely mire!
O heart's desire!
It set the suilline soul on fire!
They wallowed thru,
The prince did too—
Those waters fixed them good as new!

Prince Bladud, he
And leprosy
At last had parted company.
The blessèd spring
Whose praise I sing
Had made him fit to be a king.

Which is every word of it as true as most of the things we were taught in school, & much more amusing. Well (I shall use this appropriate interjection as often as I please, so be resigned) well, the Romans came along and settled down for circa (as we say in Rome, & it doesn't mean a circus) for circa 400 years and built themselves a bathing establishment of utmost luxury—almost as luxurious as the average bathing establishment of a city Y.M.C.A. in the dear old Beknighted [sic] States—which one can still go for to see and to admire. And then the Saxons or somebody equally vulgar & rough & interesting came along & pitched the Romans out by the seat of their, well, togas, & busted things up just to see the splash and the place was a marsh, occupied only by teals & other critters. Then the Normans came along & made a city of Bath once more (Edgar was crowned there— oh, but my dates are mixed, I think probably Edgar was late pre-Norman— anyway he *was* crowned there, or perhaps it was someone else of the same name) but the city wagged along and got itself a Bishop & did fairly well in a puttering sort of way until the eighteenth century when Beau Nash & his boy friends ripped everything to pieces again & rebuilt it as it is today (for I don't know where you'll find another city as nearly true to period—all 18th century, except the Abbey, which is 16th). Then in the 19th century they began grubbing about & found the old Roman baths which had been buried & forgotten, and so wags the world along. The Pump Room, which was built for the benefit of suffering humanity, is chiefly devoted to thés dansants, real Charleston orgies which remind one that the waters are good for St. Vitus' dance, & the Assembly Rooms, where Nash arrayed the most brilliant balls in Europe, are devoted to the art of the Kinema. Gouty old gentlemen are wheeled about in chairs and ungouty young females are, on the contrary, wheeled about in chairs, & one pays twopence, please, to enter the Institution Gardens.

It's fairly amusing how antiquities are entwined, like chinese boxes which have smaller & smaller boxes inside as you go on opening. There are stones which were laid down along the shores of a Jurassic ocean (which is some time ago, I may add) and were later used by the Romans. Then the Normans used some of the Roman blocks, & the 16th century lads used them again & now, like as not, someone has sneaked off with one or two & used then to prop up the chicken house! To what period must we antiquarians assign such a stone? But then, how old is Ann? The answer, I believe, is as old as she looks, & the stones look very old indeed, so the bad old North Wind gave Benny back his Baa-lamb and, now, my little dears, Uncle Wiggly must kiss you all good night. Smack! There! Good-night all!

Ha! Ha! Ha! I bet I fooled you, for the letter isn't really more than about ½ (one half or fifty percent) done. I can tell, because it is still fifty minutes until dinner time, which means that we will have dinner in, roughly, one hour and a quarter. To return to Bath, (fooled you there, too, but you shan't get out of it until you're as bored with it as I was before *I* got out of it.) It's [sic] chief charm, from Beechen Cliff, is its regularity, & it's [sic] chief drawback, when you are in it, is its regularity. Some one with addled brains (if any) sat down and designed a house with a Doric ground floor, Ionic first floor (for they count floors here as they do in France) and Corinthian top floor. Inventive genius could go no farther, so they repeated it ad infinitum, in circles (as the famous circus) in crescents (of which there are about fifteen, each worse than the last) and in straight lines all over the shop.

> "Here at least
> England, so shy of order, sets a feast
> Of gracious symmetry . . . "

—So warbles the sappy poet (not me!). 'Steeth! 'Sbones! 'SMusculus ishio-pubi-femoralis! "Gracious symmetry." They make one perfectly nauseating façade, repeat it a thousand times, & then vomit about "gracious symmetry"! They also call it "city planning." And the same sheep-like dunderkopfs howl like Spink's lesser hyaena in the throes of a severe belly-ache when some enterprising modern builder throws up a row of red-brick cottages, all alike! If that's gracious symmetry so's your maternal great-aunt's sororal great-nephew, and he ain't. If that's symmetry (piling three Greek orders one on top the other & then repeating them like telegraph poles flying past the 20th century limited) they should tow it out to sea with the Albert Memorial[32] tied round its neck (good riddance *that* too) and sink it off the coast of Coromandel, except the Coromandelians would probably declare war, & quite right too.

They do the same thing all over the place. Sometimes it's all right, & usually they have sense enough to be ashamed when it isn't. Regent Street

32. The Albert Memorial, built to Albert (1819–1861), German-born prince consort to Queen Victoria, in Kensington Gardens, London. Completed in 1876, the Gothic structure consists of a statue of the prince seated under an elaborately sculptured canopy.

is that way, or soon will be, & it looks very well indeed. "Oh! But this terrible Yankee! Sometimes he damns it & sometimes he praises it and how *are* we to please him even if we wanted to which we'll tell the cockeyed world we don't."

We will now take some short excursions out of Bath with Sir Thomas &
Lady Holland, as announced in our last. First we went to Murdecombe (or something like that. I've never seen it written, but it's pronounced the way the French would Meudecoume, so its [sic] probably Murdecombe) where there's a really precious little unconformity of the lower Oölite on the Mountain Limestone—the latter having been upturned in the Permian & peneplaned, or very nearly, in the Triassic.[33] Also to Holwell, near Nunney, which is near Frome (which rhymes with "room" of course) whence came the little beasts I went to Bath to study. (I know it sounds as if I went there to drink the local beverages and vituperate the architecture, but I really went to work, & actually did so, a little). Also to Glastonbury, where they used to have an abbey—you can see where it was, & very nicely trimmed the grass is, too. One goes thru the Vale of Avalon, chiefly remarkable for a brick factory—"Avalon Bricks"—and its perquisites, piles of red bricks just the color of adult Britons, several smoke stacks, dead grass, & a lovely stink. Those unchristian souls who don't believe that there faerie tale about [King] Arthur being taken away in a motor launch to return at the end of the last quarter with the score 6–0 against the Knights and only one minute to play—those candy-snatchers say that Arthur is buried at Glastonbury. The spot was marked by a delapidated [sic] white cow chewing her cud (she does it fairly well, although an expert might criticize her corner work—she lifts her shoulders a bit just at the critical moment when the regurgitated titbit is shifted from the left buccal cavity to the right—but there, I mustn't talk shop) and if she hasn't moved she still marks the spot. I suspect an unfeeling paleontologist might find the bones of the cow's husband beneath that sod, rather than of that Pendragon whose wife became a nun after she was too old to enjoy her little escapades any more. But dear! dear! just see how suspicious I've become ever since an Englishman laughed loudly at one of my jokes even before I explained it and a Scotsman gave a girl ten bob on an eight bob dinner bill & didn't wait for change. Next I'll be thinking no Earl ever lived in Earl's Court or that no Parson would ever look at Parson's Green or that no hide is to be seen in Hyde Park—and *that*, at least, is a lie.

Well, we're back in Bath (don't squirm! I'm saving you the trouble of ever having to go there, or wanting to). On Sunday I took a middling long walk. (I have a classification [or taxonomic] system of walks—up to five miles = a short walk; five to ten = a middling walk; ten to fifteen = a middling long walk; over fifteen = a long walk). This was about fourteen and [Now we'll have to turn over and use the blank spaces on the other

33. Simpson is describing a geologic structure comprising a marine limestone of Carboniferous age that was uplifted in the Permian age and eroded more or less flat ("peneplaned") during the subsequent geologic age, the Triassic, and then another marine limestone, the lower Oolite, was deposited during the next geologic epoch, the Jurassic.

sides. I hope you can find your way—] *Don't* start *Volume Two* yet! was most enjoyable. It led up to Lyncombe hill, thru L. vale, up Entry Hill (thru fields) to Combe Downe (typically enough you have to climb for all you are worth to come up to Combe Downe) then by highways & byways, fields & ditches, past young couples doing things they didn't ought & older couples not but wishing, past dogs chasing sheep and boys chasing dogs and bulls chasing boys & men chasing bulls and sheep chasing men—past little girls crying & little boys bathing au naturel in streams just large enough to moisten them if they lay down in them one at a time, past country taverns—slips! again, not *past* them—past old men in [horse-drawn] rigs & young men on motorbikes, and past a good many other things besides, thru Midford & Hinton Charterhouse, & Wellow and Combe Hay and English Combe and I don't know where else besides, to return to Bath with wild violets in my buttonhole & banks of primroses in my memory. Then tea (to which I am by now passionately addicted) and then a good long walk in the Park, for a change. There little children feeding the fish (oh! you nasty, nasty thing they were feeding them cracker crumbs which their mummies had brought for the purpose in little paper bags—[most of the bags were white but I saw three yellow ones & I think I got a glimpse of a pink one, but it was gone when I got there, altho I ran as fast as I could]). They have a Yucca [plant] there—Alas! I tried to feel homesick for the great open spaces, but the climate had had its baneful effect on the yucca & had made it a lush green thing, not spiky in the least and I could only wish ruefully that it had been *this* one I sat upon that time in Texas—but never mind!

Well, all things must end, & so will my trip to Bath, give it time, so, even, will this letter, improbable as that may seem. But first we must return to London. The obliging policemen [—they *are* obliging, as everyone says, & give you a detailed & civil answer to any question, the catch, for of course there is a catch, being that the directions they give you are almost invariably wrong; I have been late to dinner parties three times on that account, & once—but that, too, is another story, two other stories in fact] well, the obliging policeman told me that if I stayed on that there platform & took a train at 2:06 I would be in Paddington (which is in London, dear reader) at 4:07. I did as bidden. I was reading Kingsley's Water Babies, which is the most fascinating & most caustic satire on my chosen science ever written, & it was only by luck that I noticed the wrongness of things before I reached Scotland (which is *not* in London, dear reader, altho all the Scots with brains are). After taking three other trains (I have quite a good collection now) I reached Paddington at approximately ten o'clock—& lucky, too. ‖ Here intermission for dinner. You are completely had, this time, you thot that that dinner would put an end to this nonsense, but it hasn't, you see.

And here I found a very flattering series of letters, neatly divisible into two (2, II) one half urging me to signify *immediately* that I would come to Yale, the other half urging me in anguished tones to come to the American Museum. Father William (that eminent paleontologist W. D. Matthew, Ph.D., Sc.D., F.R.S., etc., etc.) has been hinting for two years that he was going to get me some day and when we parted at Oxford not long since he seemed to feel that the day was at hand, there it is. But I haven't the slightest idea which I'll do. "How happy I could be with either," etc. So I suppose I'll find out which will pay the most or promise the most (for both keep their promises) for the future, & then choose whichever one I ruddy well please. You see how it goes to my head. Next week they'll both tell me that they've changed their minds & I'll get neither to the greatest Natural History Museum in the world nor to one of the greatest universities but will end up working in some 'dobe town in the southwest which, such is the spirit of restless youth, I should like quite as well.[34] I have only three reasons for going on with paleontology & none of them are noble. The first & predominant one is that I'm so built. I like it. I am curious in a super-apish way. I like finding out things. That (tell it not in Gath) is all that the "noble self-sacrificing devotion to truth" of 99 $^{44}/_{100}$ % of all scientists amounts to—simple curiosity. That is the spirit in which nearly all pro-ductive scientific research is carried on. Some scientists admit it (most do in their hearts) & claim that it is rather noble after all, but is it? I, for one, am not at all sure. The second reason is laziness. One *has* to live, after all, scientific research is not work at all but merely fun, although a bit sub-dued, I grant, & offers a living for practically no labor whatever. The third is vanity. By a peculiar stroke of luck (this is *not* modest, it has been luck) I have risen to a certain modest eminence very quickly and I meet on a common level men who would always remain my betters if I were in busi-ness or anything like that. I have a very good chance of becoming an emi-nent scientist—if I go into anything else chances are strongly against my having another such run of luck or of becoming eminent—so when I'm frank I know it's only curiosity, laziness, and vanity & I feel as if I'd be doing the decent thing to chuck it and go to work.

The fundamental trouble is that science is so beastly unsatisfying, as I may have intimated to you before, but haven't to anyone else. Beastly little grubbing after something called truth, whatever *that* may mean, picking up little bits & fragments & glorying in them like a child, and the most the biggest of them know (thank God I've learned it soon and need not waste time if I wishn't) is that what little they do know mostly isn't true and what is true isn't very important, at least not in comparison with what isn't known and isn't knowable. The scientists are such a hopeless lot. They run yapping after theories like a lot of little puppies and believe anything they themselves say if they say it three times. I know about three who don't. I might, I very much hope I would, be a fourth & that might be worth while, but after all—[Don't supply this

SEE VOLUME TWO

VOLUME TWO

(Don't read until you've read both sides of volume one.)—ammunition to the enemy. Scientists aren't one tenth, nor one hundredth of one percent as silly as the asinine theory-hawking befuddlists who attack them].

What else is there which touches basic things—only art & religion I

34. Ironically, Simpson did end his career in Tucson, Arizona, which from one point of view might be seen as "some 'dobe town in the southwest."

think. To be an artist you have to be not merely a lover of beauty—I hope I am that—but also to be able to produce it, and I haven't faith that I could do that. (by [sic] art of course I mean literature and so on as well as merely sculpture and plane-embellishing!). Besides, little as I know about art, I know that for me it would be vastly less satisfying than science for (please keep your seat) science progresses, it *does* find out new things that were absolutely never known before. Art does *not* progress and can't. There's not one fine thing or beautiful thing too many in the world & too many can never be produced, but after all, people produced things six thousand years ago which to our eyes (to mine at least) are as beautiful as any which are produced today or ever will be. Beauty, relative as it is, & personal as it is, is stationary & cannot be expanded. Scientific knowledge can be & is expanded. That is what I mean by art being (always from my own personal point of view, in ordering my own life)—being impossible of progress. When I find & describe a new pantothere I am making a positive & permanent advance over all that the human race ever knew before, & that is a fine thing, even tho it be inconsequential & futile, as fine things so often are.

Religion is equally hopeless, or more so. To live a life, as not too many people realize, one must live a life. I have a religion (so has everyone). I have a god or God (as you will)—so have most people. But even apart from the very real material difficulties, I can't live on so simple a fact as that. I don't criticize those who can, but simply confess that I can't. I have to get on with things. And my spirit is absolutely wrong for a professional religion-monger. I despise theologists, & I wouldn't for the world make a deliberate attempt to modify anyone else's religious views.

What's left? Why, only doing something that doesn't drag the mind about at all, not spurring on thru the ranks of the mind & the soul but running on thru life with as much simple enjoyment as possible in a land which appeals (and none I have ever known has as much continued appeal as the southwest of the United States). Which is a lovely lot of bosh, for of course curiosity, laziness, & vanity will win as they have since the world began.

How's that for a letter going serious on one? It must be the dinner—an Earl's Court boarding house dinner would induce serious heart-burnings about the inner (not to say gastronomic) meaning of things in a tailor's dummy, or even in a Utah congressman.

The longer I am over here the better I like it, and the more American I become. Americans are the worst buncum-artists, stump orators, mob-formers, theory-mongers and general damfools [sic] in the world. But they're better men than any of them, Gunga Din.[35] They're even better, by and large, than the English, which is the highest possible praise. The English are a lot of damfools too, & they're like to wreck their country— the chief difference being that we are large and still underpopulated and so

35. Recall Rudyard Kipling's lines:

> Though I've belted you an' flayed you,
> By the living Gawd that made you,
> You're a better man than I am, Gunga Din.

P₁ P₂ P₃ Pant M₁ M₂ M₃ M₄

as if I'd be doing the decent thing to chuck it
and go to work. ¶ the fundamental trouble is /that
science is so beastly unsat-

5.4 isfying) as I may have intimated to you before, but
5.2 haven't to anyone else. Beastly
5.0
4.8 little grubbing after something called
4.6 truth, whatever *that* may mean,
4.4 picking up little bits & fragments
4.2 & glorying in them like a child,
4.0 and the most the biggest of
3.8 them know (thank God I've
3.6 learned it soon and
3.4 need not waste time if I
3.2 wished) is that what
3.0 little they do know
2.8 mostly isn't true
2.6 and what is true
2.4 isn't very im-
2.2 portant, at
2.0 least not
1.8 in com-
1.6 parison
1.4 with
1.2 what
1.0

T. *ferox*
group

T. mordax
group
+T. b

) Phrantten

can stand an immense amount of lack of common sense, but England can't. And now a very large minority in England are firm advocates of the theory that those that don't work should be kept alive by those who do, those who have nothing should have something given to them & those who have things should have to give them. They actually pay people for *not* working (that is exactly what it comes to) already, & more & more is being demanded. I don't mean people who have a fortune and don't work—their capital, accumulated by themselves or by their ancestors, works, which is even better for the nation—but people who have nothing & do nothing, demand that they should be given to freely, & are getting it too. Not even war can wreck a nation as quickly as that idea, & England is actually tottering. Which is bosh, too, of course except that it is fairly true.

'Steeth, I *will* keep getting serious. If this goes on I must stop, which would be a great pity. By the way, read Mother as much of this letter as is, in your mature judgment, fit for young ears, as I haven't written her for a couple of weeks & all the news is here (really) if you plough thru & pick it out.

Your artistic soul would like this room, by the way. It contains what was a piano—every newcomer plays (euphemistic for "tries to play") one chord on it—a miscellany of chairs, one fairly comfortable, none attractive, a serviceable & fairly unobjectionable table or two—an open coal fireplace (only source of heat). The pictures are very fascinating. My favorite is an eighteenth century colored engraving of a very plump ripe young lady lean-ing coyly on a balcony, her bosom artfully & temptingly concealed yet not concealed, & this chaste motto below:

> Had I a heart to falsehood fram'd
> I could not injure you
> For though your lips no promise claim'd
> Your charms would keep me true.

Which again may pass for a great, great truth. "How much one learns in this wide world" sighed little Agatha as her aunt Hetty lifted her into her carriage at the end of a most instructive stroll through the Zoological Gar-dens (the same, my little dears, as is called the Zoo by bad little children who, I am sure, are not in the least like you, my little dears.) But Rupert, who will come to no good end, merely stuck his finger in his nurse's eye and went off on a binge on ginger beer (which his mama had told him was strictly non-alcoholic, with more good intention than strict truth, for as a matter of fact ginger beer is approximately as alcoholic as good malt & hop beer which bad gentlemen swill in the inns—and now you know how superior dear old England is to la belle France where they give little chil-dren light wines [notably less alcoholic than ginger beer] but [here is the sin, my little dears] *admit* in their wicked way that they are giving their children a beverage which, in much larger quantities, would intoxicate them. And so would candy, my little dears, but don't tell aunt Agatha [sic] that I told you that). And now you may all go out & play & don't get your fresh pinner [Ang.—"pinafores"] dirty.

—SELAH—

May 22, 1927
—MCMXXVII—
• LONDINIUM •

Ma Chère Mère:

Ere long I shall have departed from London, and the place hereof shall know me no more (for a while). Otherwise said "O! to be out of England now that Spring is here!" We had a warm day last week and ever since people have been arguing whether that was this year's summer or whether it was left over from last year.—But I'm afraid that's a stock joke.

Another, of which I am very proud and which I have memorized in order to dazzle the less fortunate with my intimate knowledge of London geography is about the taxi driver who told his mate that he was stopped in Piccadilly Circus by an American young lady who asked to be driven to Swan & Edgar's "And w'at did yer do?" asked the mate. "W'y, I took 'er, o' course. Now, as we was pahssing the Halbert Memorial . . . "—The joke (I hasten to explain) being that Swan & Edgar's is *in* Piccadilly Circus & that the Albert Memorial is in South Kensington several miles away.

I really do know London like a book, the best proof being that I am leaving at the earliest opportunity. Friday, May 27th, I leave D. V. & W. P. or not. Arthur Tindall Hopwood, B.A., M.Sc., F.L.S., F.G.S., F.Z.S.[36] and I are taking the night train for that sink of fleshly iniquity, Paris, or, as it is better known, LUTETIA PARISIENSIS,[37] or something like that. He's going to stay just long enough to visit all the music halls & low resorts & to try all the French intoxicants (as well as to study the PROBOSCI-DEA FOSSILES[38] in the Galérie d'Anatomie Comparé du Jardin des Plantes) and then he re-parts for Londres & I for Lyon & points east & north. Just think! A week from today I shall be drinking vermouth (dry) on the pavements of Paris! Joy! Joy! Like most things English, London is a dear old place & one likes to know it's there, but one can take it or leave it, preferably the latter.

I've just been at EBORACUM, which the modern savages call YORK, and a very pleasant place it is. I asked a taxi driver there to take me to the YORKSHIRE MUSEUM and he gave the lie to my poor joke, given above, by refusing & pointing out that I was within a few steps of that famous institution. I arrived there at tea time, & so, bien entendu, had to inhale three cups of tea & innumerable bisquits (which is English for crackers—sweet ones in this case) in the company of the curator DR. WALTER E. COLLINGE[39] and his bevy of handsome female assistants

36. Hopwood's honors, including a master of science degree and membership as a Fellow in the Linnean Society, the Geological Society, and the Zoological Society.

37. Roman name for the island in the Seine River—the île de la Cité—where the primitive tribe, the Parisi, first settled.

38. Hopwood's research included proboscidean fossils on display in the gallery of comparative anatomy in the Botanical Gardens in Paris as well as those at his home institution, the British Museum, in London.

39. Walter E. Collinge (d. 1947) of the Yorkshire Museum was an invertebrate zoologist whose specialty was insects and molluscs.

before I was allowed to do any work. They only had one specimen of interest to me and I finished it that same afternoon & then went off to explore & investigate a legend I had heard that York old ale is especially good for adenoids, rickets, or what have you. Quite true. After absorbing seven-and-six-worth (—rather stiff, over $1.75) of as good a dinner as exists I

wandered about beneath a flaming red sky until dark. Next morning (after a luxurious breakfast of porridge, kidneys, eggs, bacon, toast & marmalade, & tea—so English, you know—in bed at my very superior old hotel) I explored the cathedral in & out, up (to the top of the lantern tower) & down (into the crypt). The view from the top is magnifique, as we used to say when I was a haggis-hunter in Caledonia,—one sees the whole city and most of central Yorkshire. The building is not equal to most of the French cathedrals but is fascinating, quand même as the cyclopean pigmies of Mombasaland are wont to exclaim over their samoans-on-toast. There's a most amusing—or do I mean amazing?—arch (the central one of the main front) sculptured with most enticing little nude ladies—I'm sure the church waxed so fat because everyone who entered had grave mental venialities to confess. They all represent Eve, so of course it's all right, but I have my doubts as to the celibacy of the sculptor. Some of the glass is exquis (a word I picked up while studying with the Chief Sauce of the Krim Tartar) altho mostly later than the period I admire most—XII & XIII centuries which produced the glass of the Sainte Chapelle, much of that of Notre Dame de Paris & most of that of N. D. of Chartres. The most famous, & (strangely) most beautiful is the Five Sisters window. It is a glowing silver, flecked with spots of color. Idolator, as I am, of the windows of the French churches mentioned, with their harmonious blending of the most violent colors, I would never have believed that a window so nearly colorless as the five sisters could be so ravishingly beautiful.

After beer and skittles (only unfortunately I have learned that skittles are, or is, something to do & not something to eat as I had always supposed & as it should be) on a balcony overhanging the well named great gray green greasy Ouse (called "ooze") I went out & walked all the way around the city wall—for York retains its mediaeval fortifications almost in their pristine state.[40] I felt most middle-aged (historically, not personally) altho I was disappointed to find that the head of a disobedient Duke of York which used to hang above one of the gates has been taken down long since. If it had *only* been left there it would have revealed to the faultlessly logical archaeologists of the year 5000 after Lenin that the Yorkshiremen of our aera were head-hunters who derived all the elements of their culture from infrequent intercourse with Papua. For that is the way archaeologists work, you know, in unraveling the history of the most extreme past. And they would have measured the D. of Y.'s skull & decided that is was that of a dolichocephalic negroid, proving that the bushmen came from England—evidence exactly as strong has in sober reality been used to "prove" that the bushmen came from France, a view now widely

40. Skittles is an early English precursor to nine pins bowling. "Great grey green greasy" comes from Kipling's description of the Limpopo river; the Yorkshire Ouse runs through York and later empties into the Humber and thence the North Sea.

accepted.[41] However, I must not begin to upbraid my fellow seekers after what they are pleased to call truth.

I've about decided to go to N.Y. when I return, if Lydia is agreeable. Dr. Matthew is going to Berkeley, as I think I told you. If I do I shall have the resonant title of ASSISTANT CURATOR OF VERTEBRATE PALEONTOLOGY IN THE AMERICAN MUSEUM OF NATURAL HISTORY—with the salary, for the first year, of $33.33 ⅓ per letter of my official title. "Further increase in salary & position to be the reward of effort."

I'm shipping off a birthday present to Marthe. Please give it to her, with a chaste fraternal salute on the brow, on June 22nd—if that is her birthday. It's the most useless thing I could find—an alabaster unguent vase, Egyptian of the Ptolemaic Epoch—and was used by a lady of the time of Cleopatra to conceal the smell of Egyptian femininity by the method of the greater evil. The whorl about the top is unique—not being known on any but this and four (I believe) other vases of the same lot—there is none like it in the British Museum. I hope it arrives safely—I got it from a little buried shop far from the dealer's district which only a few of us real connoisseurs know. The Metropolitan Museum sometimes buys thru him but he's unknown to the vulgar collector. My better genius Sherborn sent me there (with a letter of introduction, for he (the dealer) doesn't like strangers!)—Everything he has is genuine & inexpensive as things go.

I also enclose a Roman coin from my collection. It's not clear enough for a formal series but the patina is very fine & may please whoever of you is successful in retaining it.

Work has lately absorbed me and, altho I am not sleeping well, I have only had one evening out in two weeks—Hopwood & I went to see the Terror, a delightful idyl with two murders on stage & several off, which we followed by a lobster apiece & a bottle of claret—strangely enough, I *did* sleep well that night!

> Your accomplished and devoted
> homme du monde,
> GEORGIUS VI, otherwise

I'll be in Paree c/o Am. Exp. jusq'au 12me ou 15me Juin after that my next address will have to be at Baron von Huene's, where I'll be until about July 5 perhaps—c/o Prof. Dr. F. v. Huene[,]

> Geol.-Paläontol. Institut d.
> Universität Tübingen,
> Württemberg, Germany

41. Dolichocephalic refers to a vertically elongate oval-shaped head, a key attribute of the "racially superior" Aryan as claimed by late nineteenth- and early twentieth-century social anthropologists of the Gobineau school.

Dearest Dad—

This is a true red-letter Sunday as it's the first on which I have written letters to two members of the family—one usually exhausts me! I hope they are passed around, as I'm afraid it's not been to you for some time. I have several letters and other things from you to acknowledge—two $5 bills for which I think I did thank you, altho I do so now, most sincerely, in any event, also a lurid clipping from some Hearst paper about Gidley's Florida skull and a Lit. Dig. with abstract of Gregory's article.[42]

Gidley's find is the same one he exhibited at New Haven at Christmas, 1925. I have examined it several times. The "scientific" reconstruction of the newspaper is absolutely ridiculous. Any photograph of a modern Florida indian would be a good "reconstruction"—it is thoroughly modern in type (altho undoubtedly old),—nothing of the cave-man about it. Gregory's article is, of course, about the beasts which he & I described together.

My magnum opus progresses famously. The manuscript is completed—just 450 sheets of typescript. It's name changes daily, almost. The Keeper [i.e., curator] now wants to call it "Descriptive Catalogue of European Mesozoic Mammalia, based chiefly on the Collections in the British Museum." I object, politely but strenuously, as I don't like that title, but that's what it will probably be. It's now being edited—two die-hard britishers are going thru ruthlessly eradicating all Americanisms. If there's anything puts the wind up your average red-faced Briton it's an Americanism intruding itself into what they fondly imagine is *their* language, and as this is an official publication of the government, or a branch of it, out they *must* come! I find some of my most innocent phrases are vile barbarisms to them. I'm afraid that when my American colleagues read some passages they will think that I have deserted the tongue of *my* fathers. I am happy to say, however, that the editing consists only of slight verbal changes and that only on due consultation with me, so that I am allowed to say exactly what I please even when, as it does in places, it absolutely refutes what English scientists have been repeating parrot-like for generations.

I've just received some reprints of a paper in the annals of the Carnegie Museum. I had no opportunity to revise it in proof & it seems to me a bit awkward in places as a result, but the facts are there. It upsets the apple-cart of some people too, for North American Oligocene edentates are not supposed to exist & it's been proven that they *cannot* exist, so that the fact that one *does* will not be palatable![43] I enclose a copy. Matthew was rather upset about it, when I told him, & rather declined to believe me (scientists

42. *Literary Digest*, founded in 1890 and continued until 1938, dealt with current affairs and personalities in the news; it was later absorbed by *Time* magazine.

43. The Carnegie Museum of Natural History in Pittsburgh, founded in 1896, had a particularly good dinosaur collection. In 1927 Simpson published in their annals a description of a previously unknown extinct thirty-five-million-year-old anteaterlike mammal. Almost a decade earlier Matthew had published on somewhat younger mammals, including anteater relatives.

never believe each other, you know) but I think the details in my paper must convince him.

Matthew's a deserter anyway. He's leaving the American Museum to go to Berkeley. It's very probable that I shall go to the Am. Mus. to help reinforce the sagging department (altho, of course, at the bottom not the top!)—they have urged me to do so as I've said before, I think, & have about decided to accept. Salary for next year would only be $2500—which is what I would get at Yale & barely enough to live on with a wife & two ravenous & destructive offspring—but in many ways the place would offer larger opportunities both present & future. It is especially attractive in one way—it would put me in the field, either in America or elsewhere, for several months every year, if I wished (as of course I do). As I think I told you, I'm mildly disgruntled because Yale wasn't able to offer me a teaching & research post, altho they offer me as much money for research alone. Still when you don't get exactly what you want there's a temptation not to take what they will give—when you can get as good or better a place elsewhere!

I'm leaving London soon—next Friday, May 27th, if I can get away. I'll have to come back here for a fortnight or two at the end of the summer, however, to go over proofs, plates, & clear up a few bits of work I haven't time for now. At present I am mostly supervising the photographing of my specimens for the plates—a tedious job. I'm also studying the cynodont reptiles, the group from which the mammals were derived, as I hope to publish on the origin of mammals & their early history when I can find time. I've just read proof of a very brief article on a cynodont which I'm publishing here in the English journal "Annals & Magazine of Natural History." I'll send you a reprint when it's out, in a month or so.

I am still hoping that you may decide to come over for a time. Of course it may not be worth trouble & expense to come for so short a stay, but it certainly would be fine if you could. We haven't really had any time together for years now,—I still look back on our excursions in the mountains with pangs of homesickness. My life seems to have changed as suddenly and completely that I have lost all grasp on the old life. I seem to myself to be an entirely new person!

It is less than two years since I published my first paper and yet it seems to me as if I had always been an accepted paleontologist. In a way that is the temptation & the great weakness of going to Yale again—a tendency to rest on the few laurels won. Dunbar urged me to come because I'm already accepted & successful there, the battle won, & I shouldn't have to work so hard.[44] If I go to the American Museum it will be largely to do over, for in spite of having a good body of research behind me, to them I would be new & untried & success would depend on still more hard work—so I think I'll go there! I have a lot of ideas seething which I long to get into print. I am really very lucky to be so young & yet to be able to turn out research, as it gives me so many more years to accomplish things

44. Carl O. Dunbar (1891–1979), professor of paleontology and stratigraphy at Yale from 1920 to 1960, when he retired. Dunbar had been appointed a few years before Simpson arrived at Yale.

in than most people have. But there's also a temptation to feel that I've already beaten the game, which of course I haven't, & to simply ride along or turn to something else. I was quite thrilled when a German periodical— the Neues Jahrbuch I believe—referred to me as "Der ausgezeichnete Forscher der Mesozoische Saugetiere"—"the distinguished student of Mesozoic Mammalia"!—the fact that I am thrilled at being called distin-guished revealing how young I still am! I really should grow a beard before going to Germany!

This is descending to the realm of blather, & besides, the supper gong has rung!

With love,
George

Till June 15th—
  c/o American Express Co.
  Paris
Till about July 5th—
  c/o Prof. Dr. F. von Huene
  Geol.-Pal. Institut der Universität
  Tübingen, Württemberg, Germany

[Lyon]
14me Juin, 1927

Chères Toutes!

In two words—Me voici! I have quitted the mirk, merk, or murk of London for the sunshine of France and so far have nothing for my pains but a cold (pas de danger! very light). The Irrevrend [sic] Hopwood & I spent 15 (or, as the French say— 75) days in Paris. We went to 2 music halls, one play by dressed actors and one cinema. We watched boats sailing in the octagonal ponds of the Gardens of the Tuileries & Luxembourg, respectively. We stayed up late & arose late. We ate a great deal and drank—well, somewhat. We worked—also somewhat. I will now elaborate.

The play by dressed actors was Cyrano de Bergerac at the Theatre de la Porte St. Martin. As usual the roles were recited rather than acted and the parts taken too rapidly but nothing could spoil that play, which is the greatest ever written, & Hopwood & I both had to blow our noses very violently. The cinema was a French one "Mauprat" & was as bad as might be. In the Tuileries we saw a most heart-rending sight. A little girl had a new tin boat which her father insisted on sailing way out into the pond where it promptly sank. The brute wouldn't even let her cry. Hopwood & I did it for her—into two (II-2) demi-blondes (this is not a person but a drink—quiet your fears & only almost non-alcoholic beer at that).

Most of our late sitting up consisted of watching the crowds along the boulevards. I was surprised to learn that Boulevard St. Michel—which I had always thought dull as ditchwater despite its reputation—does become pleasantly rowdy along about midnight.

Hopwood tendered me a birthday party in advance & we had the best meal I ever ate, also the best wine & *fine champagne* (which is not wine but brandy) I ever drank. We went home to bed afterwards because it was too good to follow by any lesser pleasure! That was the only night we were in bed early & then we lay awake reading (Gautier's Capitaine Fracasse for my part).

The little work we did was several days spent at the Galérie de Paléonto-logie au Jardin des Plantes where Professor Boule (a reputed ogre) upset calculations by being most extremely nice. We had two assistants run-ning[,] fetching us specimens to study. You will be interested, not to say intrigued to hear, that I studied chiefly the ptilodontid plagiaulacoids of the Cernaysian, *id est*, les multituberculés du Paléocène supérieur des envi-rons de Reims.[45]

All things end except circles. In short, we parted, very sadly (for we love each other like brothers)—he to return to Londres and the task of seeing my memoir thru the press & I to fields afresh.

Lyon is not a field afresh, for I got drunk here once (second & last time in my life I ever did so) but I didn't remember much about it, not unnatu-rally. I only arrived yesterday & leave tomorrow, but I've had two full days at the University & enough time to see the town, which is small.—How one's ideas change! I used to think Denver a teeming metropolis & was intensely surprised to learn that this little town has twice as many inhabi-tants! Comes of living in London.

Professor Depéret has been extremely kind to me here. Among other things, he gave me *carte blanche* with his entire Cernaysian collection, which he has never described.[46] I've worked hard & have it all in hand, with careful drawings of the important specimens which I shall publish when I get back to America. Tomorrow I go to Geneva. I do look forward to seeing high mountains again, even tho I shan't have time for any excur-sions. I already feel them, for the Rhône here is almost a mountain river, shallow & very swift & milky-white with rock-dust from the glaciers. After that I go to Bâle & then to Tübingen, where I shall get in touch with mail again. Also it will be a pleasant introduction to Germany & a break in hotel life for Baron von Huene has invited me to stay with them.

I'm not sure if I told you that I have finally decided what I shall do next winter & probably for many winters—In fact I'm sure I haven't for I didn't make the *final* decision until I got my mail in Paris. I am going to the American Museum in New York as Assistant Curator of Vertebrate Paleon-tology. The salary is far from princely—only $2500 to begin—and won't go far with a voracious family in New York, but after all this is the first real job I have had, & the salary should go up considerably in the course of a year or two. From every other point of view the job is a very good one. I'm getting $2500 this year, too, but what with traveling expenses & two sepa-rate ménages to keep up, so to speak, it's as little as the $1000 I had before

45. The multituberculates of the upper Paleocene from the neighborhood of Reims are extinct rodentlike mammals almost sixty million years old.

46. Charles Depéret (1854–1929), French vertebrate paleontologist and geologist at the University of Lyon, studied Tertiary-age mammals as well as the Cenozoic geol-ogy and stratigraphy of southern France.

& I am, as usual, broke all the time.[47] We should begin to recoup next year. Lydia's back in New Haven now, of course. She was going to stay over here, but decided not to. I haven't seen her or the nippers since Christmas & shan't until the end of October.

Marthe will be happy to hear that I'm letting my beard grow—but I'll shave it before returning to London!

No more news—

Love,

(George)

[Stuttgart]

[June 22, 1927]

[Seven lines of pseudo-German in pseudo-Gothic script]

—This pious thot occurs on my calendar for today "Juni 22, Mittwochs" [June 22, Wednesday] and I send it to you as a birthday gift, hoping that it will make a better girl of you—altho I doubt if it would, even if you read it. In four words—[more pseudo-Gothic]—the copy book would put it or [more pseudo-Gothic] as they actually write it. I have been in 1—Paris, 2—Lyon, 3—Genève [more pseudo-German], 4—[more pseudo-German] ou Bâle I am descended in a Hospiz "Für allein reisende Damen besonders zu empfehlen" [an inn especially recommended for traveling ladies] (as my guide naïvely remarks)—I am allein reisende [traveling alone] and if I am not eine Dame [a lady] at least I mean them no harm (no more would Don Juan once he saw them). I have kaffee und ein weichgerottenes Ei [coffee and a rotten soft-boiled egg) for breakfast, Suppe, Sauerkohl, Wurste [soup, sauerkraut, sausage], & a glass of beer 12 inches high (by measurement) for dinner (= noon meal, here), and Käse Bröttchen [sic] [cheese rolls] for supper. I exclaim "Ach" and "So!" or rather "Zoh!" In short, I am verdeutscht [germanized]. We have here:

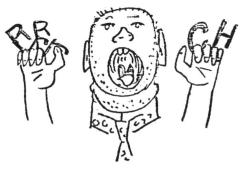

The German language.

In person.

Morgen [tomorrow], I visit 1—A art galery. Zwei [2]—Degerloch—geologically interesting Ort. I then go the following day for a day's ramble in

---

47. Recall that while Simpson was working in London Lydia chose to spend the year in southern France. By now she was six months pregnant with their third daughter, Joan, who was born in Boulder, Colorado, the following September.

the Swabian Alps—both these trips with Herr Dr. Direktor Berckheimer. After that I go to spend a week or so in the Zeppelinstrasse, Tübingen with the Baron & Baronness von Huene, not to mention the Whatever-Baron's-daughters-are-called. Berckheimer's one of the best. I have just returned from a long evening—6 to 11—with him, (& for part of the time with Frau B. & the two little B's—just the same age as mine). He has been in Amerika & fortunately speaks American. Even Frau B. knows a little, so we get on well. I've done some good licks in the Württemberische geognostische Sammlung, where they have some of my especial pets—the Mesozoic mammals—or did you realize that I am the world's leading student of the Mesozoic mammals? That is, I lead but who will follow? Fortunately no one knows enough about them to know what I don't know—and so forth & so forth.

Well, es ist spät [it is late] & I must to bed go—ĝin revidi[48] (as we say in Esperanto)—

PS—I've had no letters since Paris—I'll no doubt get some at Tübingen. In case there isn't one from you—thank you! now while I don't know.
PPS—I find I've used no French or Spanish, giving a meagre notion of my linguistic ability—sales Américains, la paloma—there!
PPPS—Ach ja! And Danish—Tændeskifte!
P.P.P.P.S.—Und Hawaiian—Poli anu-anu!
P.P.P.P.P.S.—Also Italian—Chianti blanco [sic]!

[Berlin]
July 13, 1927

Dear Dad—

Your letter of May 30th reached me today! My birthday this year has been a most prolonged one, for it started about June 10th when Hopwood gave me a birthday dinner in Paris (he had to return to London before the 16th) & continued until today, over a month later! Your birthday wishes gave me a great deal of happiness & your unreasonable pride in what little I have accomplished makes me determined eventually to justify it! Also your equally unreasonably large gift is very much appreciated and I thank you a thousand times. I hope you have not been thinking hard thoughts because I did not acknowledge it before—I couldn't very well! Gay's birthday was July 6th, & I shall consider part of your cheque as for her, & get her something here or in Paris, or in London.[49]

Your letter went to London, back to Chicago (why, I don't see), to Paris, to Tübingen, to Basel, & finally to Berlin—it should be quite broadened if traveling is all it's supposed to be!

48. Literally, "to see it again," when presumably Simpson meant to say "to see you again."
49. In fact, Patricia Gaylord's birthday was on July 10.

I left London a few days before June 1st. Hopwood & I had a little over a fortnight in Paris, when I went to Lyons on the 12th, thence to Geneva on the 15th, to Basel on the 17th, to Stuttgart on the 18th, to Tübingen on the 25th, to Munich on the 1st of July, & here yesterday, the 12th. I've been traveling for over 6 weeks & have covered a lot of ground in three countries. In Paris we did some work (afternoons!) in the Natural History Museum. At Lyons I put in two very busy days working on some unde-scribed material in Professor Depéret's collection—I shall publish some-thing about it this autumn or next year. At Geneva I spent several hours with a Dr. Revilliod going thru the museum, but mostly I simply wandered about seeing the city. It is a very small place, somewhat to my surprise, but in a beautiful situation. I even had a glimpse of Mont Blanc from the Quai Woodrow Wilson.

I had planned to stay a week in Basel but when I got there Professor Stehlin was away & his deputies, Dr's Helbing & Schaub, were *most* polite & *most disobliging, so I left forthwith, in something of a huff, for my pre-vious happy experiences had taught me to expect a great deal of profes-sional courtesy.[50] Dr. Stehlin wrote me immediately after, expressing his sorrow at being away & begging me to return, & I planned to do so, but Lo! the day before I was to go (I was then in Munich) he wrote me telling me that to avoid a misunderstanding he must let me know that it would only be possible for him to be there a single day & that unfortunately it would be impossible for me to see the collection in his absence! I at once wrote that unforeseen circumstances suddenly prevented my return to Basel (which was euphemistic but true) and spent the extra time in Munich.

At Stuttgart a young man named Berckheimer is in charge of the collec-tions. He was extremely nice & I was able to study everything of impor-tance to me. We also spent several evenings together & also he guided me to the geologically important places in the vicinity. He was in prison for *five* years during & after the war. The French took him off a neutral vessel (Dutch) as he was returning from America in 1914 & did not release him until a year after the Armistice—but he is optimistic & says if he hadn't been in prison he'd have been in the army & now be dead, so it's all for the best.

In Tübingen I had a really delightful time, staying with Baron von Huene & his very nice family. v. Huene himself is far the most brilliant paleontologist in Germany[,] one of the leading lights in the world—I would place here Matthew, Gregory, Watson (of whom I believe I have written much from London), Stensiö (of Stockholm, but whom I also met in London, you recall), & v. Huene (possibly Stehlin should also be included, but I'm not sure). Tübingen is a small place, the ideal German medieval town, surrounded by hills, with the broad Neckar valley on one side & across it a superb view of the Swabian Alb, with Hohenzollern at one end & the Teck at the other—the two being the ancestral homes respectively of the ex-Emperor of Germany & the present Queen of England.

50. Hans G. Stehlin (1870–1942), Swiss geologist and vertebrate paleontologist, at the Museum of Basel, who worked on Cenozoic mammals.

—I've written all about my wanderings thus far to some member of the family or other, but no one seems to get any of my letters any more—I'll try again.

In Munich I came to know three more paleontologists whose reputations are greater than those even of my private big five & who are all very pleasant—Baron Stromer von Reichenbach, Professor Broili, & Professor Schlosser. The latter is the successor of Zittel, who may be said to have inspired the whole of modern paleontology, altho he was a teacher & not a great original worker. Schlosser is retired & is a very old man—It was probably my last chance ever to meet him, alive, & I'm very happy to have done so. Broili is a red-faced, hairy brute, very profane but most obliging & kind beneath his rugged exterior. Stromer is equally nice, but very different, small & rather lady-like.

Here at Berlin Dr's Janensch & Dietrich are apparently going to be also exceptionally kind to me—with the single exception of Basel I have been treated *extremely* well everywhere, invited into my much more illustrious colleagues homes, & given every facility & assistance. I now know personally nearly all the real vertebrate paleontologists in the world. I do not know Baron v. Nopcsa (now in Vienna), or Professor Kiaer (Oslo), Wiman (Stockholm), Pilgrim (Calcutta) or the three South Africans Broom, Haughton, & van Hoepen—but all the others who amount to anything I do know. Of course there are a lot of people whom I don't know who dabble in it occasionally, but they don't count.[51]

I arrived here at midnight & went to a hotel near the Anhalter Bahn-

51. Ernst Stromer von Reichenbach (1871–1952), vertebrate paleontologist at the University of Munich, was an expert on fossil fish and mammals.

Max Schlosser (1854–1933), German stratigrapher and paleontologist, best known for his research on Miocene vertebrates of Germany and Tertiary vertebrates of China and Greece.

Karl A. von Zittel (1839–1904), German paleontologist, stratigrapher, and historian of geology at the University of Munich. Well-known too for his *Handbook of Paleontology* (1895) and *History of Geology and Paleontology* (1899).

Werner Janensch (1878–1969), German paleontologist at Friedrich Wilhelm University, Berlin, who worked on fossil proboscideans, armadillolike glyptodonts, and dinosaurs.

Wilhelm Otto Dietrich (1881–1964), German invertebrate paleontologist and stratigrapher in Berlin.

Ferenc Felso-Szilvás von Nopcsa (1877–1933), Hungarian paleontologist and stratigrapher; director of the Hungarian Geological Survey, he studied dinosaurs and other reptiles as well as vertebrate footprint fossils.

Johan A. Kiaer (1869–1931), Norwegian paleontologist at the University of Oslo; he researched fossil fishes and early invertebrates of Norway.

Carl J. J. F. Wiman (1867–1944), Swedish paleontologist and stratigrapher at the University of Uppsala. Wide-ranging specialist from primitive fossil protochordates to amphibians to ichthyosaurs (swimming reptiles) to pterodactyls (flying reptiles). Wiman collected fossils from Spitsbergen to Antarctica to China.

Henry Guy Ellcock Pilgrim (1875–1943), English-born geologist and paleontologist with the Geological Survey of India, then with the British Museum; he made the first studies of Bahrein and Oman in the Persian Gulf as well as of the rich mammal-bearing beds in the Salt Range of Pakistan.

Robert Broom (1866–1951), South African physician and paleontologist at the Transvaal Museum who did important work in human paleontology—he discovered a decade later the first robust australopithecine—and in South African mammal-like reptiles.

hof, where I arrived. Today I have taken a room—small but very pleasant
& only two marks a day including breakfast (which is cheap for any place
but extremely so for Berlin, which is an expensive city) and which opens
by one door into Dietrich's home & by the other into my real landlady's.
The Dietrich's insist on my taking dinner with them every evening & apol-
ogize because their apartment is so small that they could not have me as
their guest altogether. He is rather deaf, although not old, & understands
very little English, but we manage to exchange ideas. They also have a
young son, who is much awed by me!

I shall stay in Berlin 10 days, then go to Frankfurt, then to Paris, where
I must get some money & see my people (—Int. Educ. Board) then to Cer-
nays, a village near Rheims, then back to London about a month hence.[52]
I may go thru Brussels.

I am accumulating so many notes, without time to work them up into
complete papers, that I shall probably be publishing the results of my trip
over the next two or three years, possibly longer. It is being very productive & I have many new ideas & facts to bring forth. The Mesozoic Mammalia are now completely finished—there is only one specimen in the
world which I have not personally studied—but won't be completely published or even prepared for publication for some time. I am turning to several more or less related fields—the history of marsupials, the origin of
mammals &, in this connection, the structure of the cynodont reptiles,
the radiation of mammalian faunas—a general, rather philosophical problem of great difficulty—the Cernaysian fauna, which is the oldest *post*-
Mesozoic one in Europe, etc.

I think I wrote to you that I am going to the American Museum, or at
least you have heard it by now. It is incomparably the greatest center for
vertebrate paleontological research in the world & I feel that the opening
there is one I could not afford to refuse.

Everywhere I turn, France or Germany, American faces greet me in the
newspapers, magazines, & even the postcards—they make even more of
Lindberg [sic], Chamberlin [sic] & Byrd than if they had been French or
German, & it makes me very happy.[53] The longer I stay in Europe the better American I become. I appreciate & like Europe & realize how much
they have that we haven't, but it becomes more & more apparent that
today we are the one really great nation in the world, great both in position & in Character—altho of course we have our unpleasant individual
characters just as every country has. I could be happy to stay in Europe
indefinitely—even for many years—but only if I knew I would return to
America finally. I very much dislike our uninformed, self-centered dema-
gogues who call themselves patriots, but I certainly am becoming more &
more patriotic in, I hope, a more enlightened sense of the word.

52. Cernays is the type locality for the late Paleocene Cernaysian, almost sixty
million years ago, where a variety of fossil mammals are found in deltaic deposits of a
large Paleocene river.

53. Charles Lindbergh (1902–1974) had, of course, just flown solo across the
Atlantic two months earlier. Richard E. Byrd and Clarence D. Chamberlain had made
the first nonsolo transatlantic flight that summer, 1927.

I have not met Dr. G. C. Simpson, but we very frequently receive each other's mail in London![54] I publish all my scientific papers under my full three names, however, & we work in very different fields, so I trust that posterity will not confuse us!

By now you know that Lydia, Hélène, & Gay are back in America. At present they are in Kansas, & are going soon to Colorado for three months or so.

With much love & many thanks for your remembrance of my birth-day—

George

July 13th, 1927
Berlin

Dearest Mother—

This is a birthday letter, written a long time before the event—but my own experience has taught me that one must allow three weeks. Perhaps my thinking of it so long ahead will be added evidence of the fact that I *do* think of you constantly & that I do love you & appreciate all that you have done for me. I wish you many more birthdays, & all very happy ones! My growing up has had the usual, inevitable result of separating us widely in space and of giving each of us new interests, but I feel that it has not really separated us in spirit & that the coming years of fuller maturity for me will bring us closer together than when I was a little boy who knew nothing else.

Thank you, too, for your birthday letter to me. And please thank Peg for hers & tell her that I shall write her before long—you know how upset one is while traveling & how occupied I especially am with everything, so that it is very difficult to find much time for the many personal & profes-sional letters that I have to write, & I usually have to let Peg get her news from you or Marthe. Marthe's promised letter has not arrived—Dad's just reached me today, after six weeks of traveling!

I am very glad that Marthe has the opportunity to go to Hawaii & I hope that she enjoys her trip very much & I *did* mention it before, tho, didn't I? I think of things to write constantly, but find so few moments to put them down that I never know what I *have* written & what I only *meant* to write! And as I branch out into the world my correspondence assumes alarming proportions, altho of course my families come first.

I have not yet been in Berlin for 24 hours. I arrived at midnight last night after a 16-hour ride in the 4th class—a very enjoyable ride too, although most extremely fatiguing. I went to a hotel, but did not sleep well, so am all[-]in today. I now have a very pleasant room, & a very inex-pensive one, in the next apartment to one of my colleagues here, a Dr.

54. George C. Simpson, a generation older than Simpson, British meteorologist on Robert Scott's ill-fated South Polar expedition in 1910–11; he was later director of the British Meteorological Society and president of the Mathematics and Physics Sec-tion of the British Association for the Advancement of Science.

Dietrich. He is being most kind to me & I am sure I shall enjoy my stay very much. I shall stay ten days, then by degrees back to London, where I plan to arrive about a month hence. Much as I dislike traveling alone, almost everyone has been so kind & hospitable that I have found good friends in every city & it has not really been lonely at all. And I have been in Germany long enough now not to feel so foreign & awkward as one always does at first. They have their peculiarities, but I think the German people on the whole are very pleasant and attractive.

I sail for America on the [passenger ship] Arabic, Oct. 15th, & I start work at the American Museum on Nov. 1st. The more I think about it, the happier I am to be going there, & it really is an exceptionally good position for my first professional post, & with great opportunities for the future.

I shall be traveling a great deal, I am delighted to know, both in & out of America, so that even if you do settle down, as you have so long threatened to do, I shall see you frequently.

I am not tempting fate by sending anything from here, but I shall either send from England or bring to America some necessarily slight added evidence that I am constantly thinking of you.

With greatest love & birthday wishes,

George

Paris
Le 26 Juillet
[1927]

Ma chère Soeur—

Here I am once more in Paris after a circle thru Germany—Stuttgart (where I last wrote you if I remember rightly, & I *do* remember rightly, Miss Simms) Tübingen, München, Berlin, Frankfurt a/M. (which is not the time of day but the river on which this particular F. finds itself) hence to Paris.

As Cheops would so aptly put it—

—in case you haven't your Egyptian dictionary by you perhaps you will forgive me for translating freely "I heard a noise of thunder so that I thot it was a wave of the sea and I discovered that it was Paris." The rain has descended in torrents but with brilliant sunshine in between so that one is alternately soaked & roasted. I arrived yesterday and went in the morning to the Am. Exp. & to see my checked baggage thru the douane, & in the afternoon to the Van Dogens where I stayed until late in the evening. This morning I went to my people at the International Education Board & then after lunch to the Trocadéro, which as the Germans would say, is my liebling[s]-museum of all in Paris despite the fact that no one who hasn't been in Paris, & only the elect of those who have, ever heard of it & that it has only two or three originals in it. My passion for romance (Romance *sculpture*, I insist) & early gothic sculpture burns unabated (or however the devil one spells it), and is only approached by that for the ancient sculpture of southern Asia, pre-Pheidian Greek sculpture, Egyptian sculpture not too far either way from the Vth dynasty, & pre-Columbian American (esp. Mayan) sculpture.

No one ever painted a picture half as moving as the west façade at Chartres or wrote a symphony half as beautiful as the wood-carving of Hesire (unfortunately in Cairo—but I know many good photographs & have seen a cast.)[55]

What rubbish it is when our rising young artists insist that a work of art should not tell a story. Almost all of the great art of the world has either simply decorated objects intrinsically useful, or portrayed some person or god (which, of course, is also telling a story) or quite simply told a plain story. The incomparably gorgeous temples of southern Asia are not *sculptured* simply, but *storied*—and exactly the same is true of all non-decorative (not *purely* decorative, je veux dire) medieval art. The artist expected to be *understood*—he expressed something for everyone to understand. It is not necessary, & he did not expect, that the observer should be able himself to produce such works or even should know anything of technique or of the methods of production. The observer must know the *story*, & of course in addition his enjoyment will be increased in proportion as he is also able to appreciate the beauty of form & composition—but that is not necessarily dependent on a knowledge of pure *technique*. I haven't the foggiest notion how one goes about making a tympanum like that at Vézelay, & I am sure that it is extremely poor technically, but I know that it tells a story, I know something of what the story meant to the artists & to the people they worked for, & I know that the whole thing is marvelously beautiful. It would move me, to be sure, if I did not know the story, but *then* the artist could justly reproach me with not understanding him, & *then* I would fail in comprehension of it even *as a work of art*.

I find it absurd for an artist to paint something which means nothing, as so many of our bright young people try to do, & then (as they always do) to complain that they are not understood. A large percentage of modern painting looks to my benighted self like the work of people who don't know how to paint & have nothing to say. Most illustrators were not geniuses, but most geniuses in art were illustrators. I have just seen much of Holbein, Dürer, & the older Cranach—almost heart-breakingly beautiful engravings & paintings—& they never produced a simple picture which was not explicitly meant to tell a story or to portray a known individual (or place, or object, of course—one must give much latitude to the idea).

All of which means quite nothing. I must go eat.

—I have now had my excellent 4 course dinner, wine, service, & a cloth napkin for 25¢—I believe it's now only possible in one *clean* place in Paris. Who started the legend that *all* French restaurants are good?

55. Tomb of Hesira, Third Dynasty, in Saqqara, Egypt, ten miles south of Cairo.

I can now speak German. At first in Germany I had many amusing difficulties with the language. I had a little conversation-guide which enables
one to ask any question on earth, but unfortunately the inhabitants did
not use the same book, & usually replied by some brief reference to eco-
nomic conditions in Hindustan (so far as I could gather) followed by several pithy quotations from the minor poets. Consequently I could ask for
anything I liked but could not understand the answer. This is very sporting
and of course it is distinctly vulgar only to ask questions for utilitarian purposes, but, as a mutual friend of ours might say, still, at the same time—. I
was led to invent a conversation-manual which does not require one to
understand any replies. You may retain for ten days & then, if you are satisfied, remit by cheque or postal money order, otherwise return at our
expense, but please send cash (not stamps) $100 immediately to cover
expenses of packing only. *This offer is only good for one hundreds years so act
NOW!*

The following scenes (all rights reserved except those of publication,
translation, & presentation on the stage or in the cinema) take place on a
darkened stage, which should be well lighted by four military search-lights
and a pocket flashlight. The scene is a café in the Latin Quarter. A poet, a
stranger to Paris, has come here by mistake, believing it to be the Latin
Quarter. It is the solemn hour when tourists ascend the Eiffel Tower
(rhyme copyright in all countries, but none too good, at that) & only one
tourist is present. The waiter has been paid & does not appear until
called—

Scene One
As above

. . . .

Scene Two
The same. Enter 63 more tourists.

. . . .

Scene Three
Poet: O my sweet! In the eve I shall wait by the stream.
You will come! You will come! We shall rest & shall dream
By the swift-flowing water and hear in its strife
The sounds of the hills where it sprang into life—
Tourist: Gársson!

. . . .

Scene Four
Enter waiter.
Tourist: Parley vooz English? A glass of water!
Exit waiter

. . . .

Scene Five
Poet: Of the hills which remain, although life come & go
Of the stern virgin peaks with their mantles of snow.
Tourist: Gársson!

. . . .

Scene Six
Enter waiter
Tourist: A dish of ice cream!
Exit

....

Scene Seven
Poet: We shall flee from the world with its banter & vice,
And those moments shall be beyond praise, beyond price!*
Tourist: Gársson!

....

Scene Eight (the last, I'm glad to say.)
Enter waiter
Tourist: The bill!
Curtain.

   *This seems a rather C 3 poet, but, then, he wears spats (we forgot
to put this in the stage directions) &, then too, you can't expect
everything for your money.

In case you didn't like that (& we don't much care for it, personally) we
also present the following which, since it hasn't been composed at the
moment of writing these words, still has infinite possibilities. It is going to
be, I believe, a very modern quartette in free verse (which, of course, is so
called because it is not verse & not, if the poet can help it, free).

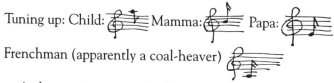

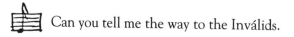

Andante mais pas too schnell.

Can you tell me the way to the Inválids.

Vous êtes Américain?

Oui! Yes! To the Inválids.

Voleur! [Thief!] Sale singe! [Dirty monkey!]

What did he say, mama?

Hush, dear! He's telling papa the way to the Inválids.

Tas d'ordure! [Heap of garbage!] Sac de porc! [Pork bag!]

Parly plu lentment, please.

Fils illégitime d'un chameau! [Camel bastard!]

♪ Thank you very much!

♪ What did he say, papa?

♪ He said to take a taxi.

♪ I don't see how people who don't speak French get along in Paris. Those conversation lessons have already been worth the sixty dollars.

The orchestra then plays the postscript or whatever it is that orchestras play in order to get people to leave. And that is all there is to that.

And finally, as a last wish from your antipodeal brother, I beg of you to take Iron Jelloids and be strong & well again.[56]

A kiss in the neck—

Budge says it [the hieroglyph] should be transcribed TCH; Erman usually calls it D, but sometimes Z; Maspiro is neutral; but *I* say it means *me*, & I am very gratified to see its frequent, and if I may modestly say so, prophetic occurrence in Egyptian literature.[57]

POSTSCRIPT—It is remarkable how much one learns in traveling. I am, I admit, a rather keen observor (or "observer", you may have your pick. Personally I prefer the latter, as it is correct, but you may have "observor" if you like—it looks rather well, don't you think?) & I've noticed a good many things one doesn't see in guide books. For instance, I find that all Germans bend their knees when they sit down. With a few exceptions, most of the Swisses or Swees or whatever the plural is were wearing what I can only describe as shoes. Apparently it has also escaped notice how very few Frenchmen walk on their hands. Another striking peculiarity, less national than those I have already mentioned, is that almost everyone over here has *five* fingers on each hand—there are exceptions, but I think I may safely make that rather interesting generalization. I was also told in Germany by a man whom I am not allowed to name but who is very high in political circles, that most Prussians prefer to lie down when they go to sleep. I have not been able to verify this by personal observation, however, so I can only advance it as a subject worthy of more intensive first-hand study. I could go on for hours, but you see how much one can pick up if one keeps one's eyes open & how impossible it is to really come to *know* a foreign nation except by seeing it, or reading about it, or hearing something about it, anyway.

In some ways, Europe is very backward. In medicine, I find that doctors

56. Lozenges of gelatin, containing iron; popular at the turn of the century.
57. Wallis Budge (1857–1934), Adolf Erman (1854–1937), and Gaston Maspero (1846–1916): distinguished British, German, and French Egyptologists, respectively.

are very slow to prescribe prussic acid for internal use. In music, I have, I swear, *not* seen a single orchestra which included a marimba. In art, I find that the use of buffalo-chips for house decoration can hardly be said to have taken root here at all. In industry, not a single drawn-steel combined toothbrush & postage-stamp moistener was produced in France last year. In short, they are really years behind us.

I shall now go to bed, as it is getting late & I am afraid if I went on I should, in my weariness descend from the lofty plane which I have so far succeeded in maintaining in this letter & possibly, if it is possible with a serious & crystal-clear mind like mine, become even silly.

George Gaylord Simpson (in case you didn't know)

THE SIMPSON METHOD
Learn all you need to know of French or German in TEN MINUTES!

Price: $10.00   Price: £2.2.0

Café: France: 1. Un demi blonde, vite!
                         2. Encore un biere.   3. may be repeated as
                         3. Encore.   often as necessary in
                         4. L'Addition   either language. As
        Germany: 1. Ein helles, schnell!   given, our method only
                         2. Noch ein Bier.   secures you three half-
                         3. Noch.   liters.
                         4. Zahlen.

Conversation: (Only for advanced students)
        France: 1. Quelle jolie fille!
                         2. Sale temps!
                         3. Oui!
        Germany: 1. Hübsches Mädel!
                         2. Schlechtes Wetter!
                         3. Ja!

Hotel: If you can afford to buy this method, you can afford a hotel where English is spoken.

Restaurant: See "hotel." Or else point to anything on the menu & say:
        France: Ça!
        Germany: Das!

Railway: Buy tickets from Cook's or the American Express Co.

Stores (for Americans)⎫ I. If you want to buy something point to it & offer
Shops (for English)   ⎭ a bill (American) or note (English).

When the change is given you, look the cashier in the eye & look doubtfully at the change, she (he) will then give you the rest of your change.

II. If you are just looking, ask:
        France: Combien?
        Germany: Wie viel?

After the clerk has replied, say

    France: C'est bien trop!

    Germany: Das ist zu teuer!

and walk away.

Streets: Reply to all volunteer guides, street vendors, & young ladies (if you are male) or young men (if you are female):

    France: Non!

    Germany: Nein!

All Emergencies, not elsewhere provided for in this comprehensive method—

    France: Je parle qu'Anglais.

    Germany: Ich spreche nur English   and then

explain your side of the affair in as rapid English or American as you can.

The Advantages of Our Method—

1. It can be learned in ten minutes.
2. It does not require you to understand anything that is said to you.
3. It provides for every possible contingency.

Copyright in all countries, including New York.

(This is rain,
in case you
don't recog-
nize it.)
[August?]
1927

Soeur Marthe—

I don't know why I write to you so often. It is stupid, but then I think that it probably soothes me, & I don't believe it does you much harm. I am glad to hear of your unspoiled arrival in the 100% American paradise of the northeastern south seas and I hope that you & the Sandwich [Hawaiian] Islands do each other good. You may write me hereafter in Hawaiian—or will you shatter another illusion by telling me that no one ever speaks Hawaiian anymore unless to show off? I imagine that is probably the case. Hawaiian has always struck me as being a language inadequate for the purposes of vertebrate paleontology, anyway, & hence useless.

I am just back from everywhere else. From Berlin I went to Frankfurt, from there to Paris, to Reims, where I fell illish, then to Paris where I luxuriated in bed reading (Tartarin sur les Alpes, le Compte Kostia, Les Trois Mousquetaires, Vingt Ans Après, etc.[).] A very preoccupied french doctor studied my case fifty francs worth & finally decided that I would either die or get well. After thinking the matter over, I have decided to do the latter. He gave me (to put it euphemistically)—gave me some medecine

[sic] which I poured down the drain. The following day the fish in the Seine turned belly up & expired—you know the fish I mean, the one they are all pretending to try to catch. Well, he's dead, I'm sorry to say, & after the solemnities in Notre Dame—the funereal pomps as we other french say—the National Club for the Pursuit of the Fish in the Seine (C.N. P.P.S., Soc. Anon.) purchased another fish and are slowly acclimating him in the hope of being able to introduce him into the river. He is placed every day for a slightly longer period in a compound of stale dishwater, vin ordinaire, garbage, & mud. He already complains of pains in the chest, but the lungs seem all right & he may be able to survive in the river. The fine for catching him has been raised from the former 5,000 francs to 10,000.

All of which reminds me of one of the most delicate lyrics of that neglected poet, Jean Sérien. I grant you that some of his work displays a certain ignorance. It is not, for example, sustained by the erudition of that other neglected but great artist Moy Jessé-Toux, but what he lacks in factual background he makes up by the purity of his emotion.[58] As you don't read french, I translate the little lyric to which I refer.

Le Poisson (The Fish)

O Poisson! (O Fish!) Murkily swimming
Thru you[r] native solution of Parisian ordure
Lazily fanning your not altogether inviting
Physiognomy with a languid nageoir[e] (fin)

O Poisson! (O Fish!)
What are you thinking of? Or do you think?
Of that long distant day when in your re-
Splendent coat of scales (now badly tarnished)
You emerged from the oeuf (egg)?

O Poisson! (O Fish!)
Do you give heed to the thousand pêcheurs (fishers
    or sinners, we haven't our dictionary by us)
Who pursue you carnally & relentlessly
Are you a poor victim asking nothing
But to have your own vie? (life?)

O Poisson! (O Fish!)
Or are you a merry playfellow who
Enters into the game with them?
Are you prey or companion? Goal or
Excuse-Qui sait (who knows?)

O Poisson! (O Fish!)

58. Simpson's word play here. "Jean Sérien" sounds like "je ne sais rien," which translates as "I don't know anything." "Moy Jessé-Toux" sounds like "Moi, je sais tout," which translates as "Me, I know everything."

To emphasize the sublimity of this lyric, I append an effort of my own which takes you from the sublime to the ridiculous:

Ballade of Ancient Beauty

Whence is the sweet endowment of the breeze,
And whence the clouds which blossom in the sky—
The softly singing whisper of the trees—
The lovely ripples as the stream flows by?
Were they not made as joys for the mortal eye?
Are they not ours? Ah no, they cannot be,
Or how could those more ancient beauties die
Which were ere there were any eyes to see?

Oh! Mourn the childhood of our ancient world
Whose million million years in passing by
Have known so many unseen flowers unfurled
Have seen so many haunting seasons fly—
So many suns to set & winds to sigh—
So many golden evenings on the sea!
Oh! Mourn all the gorgeous earthly panoply
Which was ere there were any eyes to see!

The sky was not less beautifully blue
The sun less bright nor his more pale ally,
The moon, less silv'ry, the pole star less true
The earth was not less sweet or bare of witchery
In that most distant age when we deny
That even life itself had come to be.
That fresher, newer beauty has passed by
Which was ere there were any eyes to see.

Envoi
O man! In might & hauteur do not cry
"The beauties of the earth were made for me!"
Lest to thy human pride that earth reply
"And those which ere there were eyes to see?"

That contains some very bad geology & some worse poetry. You see why I prefer Jean Sérien's work to my own.

Speaking of sculpture, as we were not, when I went to Van Dongen's last there was a quite lovely Khmer lady on her back in the garden—a cynic would say, well placed to acquire the marks of spurious age. As a geologist impassioned for sculpture, & who has seen much French & some Khmer sculpture, I was able to make the interesting observation (which I did not, in my selfishness, communicate to my host) that the East Indian statue was made out of French stone! What interesting vistas of a wholly unsuspected very early trade with the orient that invokes! Jean [Van Dongen], I tremble for your morals!

What is sad, the lady is rather better done than if the Khmers themselves had done her—just as a large dog at the atelier ("Mais il me semble qu'il y a beaucoup d'égyptien là dedans, Jean!" "Mais non, un peu, pas beaucoup—c'est tout à fait original")[59] avoids the little awkwardness of its egyptian prototype which, unfortunately, I had seen & which my love of egyptian sculpture had fixed firmly in mind.

À propos of nothing, an artist may be endlessly gifted technically but quite devoid of ideas. Indeed the egyptian artists (or Khmer artists) never had originality but merely copied, with slow modification, the processes of their forebears. Only now we are so widely endowed that we can copy, with modification, the work of peoples altogether foreign to us in time & place & the process loses its redeeming feature of natural evolution & becomes—copying! One must live—& please avoid the classical reply "I do not admit the necessity!"

Kees Van Dongen is now in the Luxembourg—a thoroughly *hideous* canvas.

I'm happy that you met the Gregorys, especially if you're to be neighbors. I'm taken aback, tho, to hear him called sweet! Of course he was the hero of all the young geologists at Yale, however. He shouldn't worry about paleontology demanding imagination—most of it demands far too much!

By the way, I have two Gregorys not to be confused. Yours is Herbert E.[60]—the other, with whom I have even more to do & who is also sweet, in the masculine equivalent, is my very much admired friend William K. who is at the American Museum.

Enfin, here I am again in London with a year's work to do in two months & not feeling quite up to it. I've lost your Hawaiian address—put it *in* a letter, not on the envelope only, as I usually lose envelopes.

> Your eternally devoted but
> thoroughly
> unworthy brother.

I'm sorry there's no postscript—I'll send it in the next.

> [Unsigned]

London
August, 1927

Ma chère soeur, folle, loitaine, belle, cherie;
pittoresque, précieuse, demi-ridicule, sage, et
encore mille choses mal entendus*—

I am a complete fool, because here I am writing you once more and the ink of my last letter isn't yet dry. But as a matter of fact I received your letter this week and I'm writing, I'm writing. It's a disease. Friends, work,

59. "It seems to me that there is a lot of Egyptian there, Jean." "No, a little, not much—it's completely original."

60. Herbert E. Gregory (1869–1952), Yale professor of geology and founder of the Connecticut Geological and Natural History Survey in 1903. Later he became director of the Bishop Museum in Honolulu, eventually retiring in Hawaii.

*In French; translated by editor.

sleep, nothing works any more. I do nothing but write you. Morning, noon, night, nothing but pen and ink (or sometimes the typewriter). And why? All for nothing. To make some jokes, to show my empty brain, to vaunt my talents which don't even exist, to give news which hasn't happened, to dirty white paper, to spoil you—in fact, to display myself as silly as a Camembert cheese.

In short, what's happening here? Nothing, young lady, nothing. It rains, or much more rarely, it doesn't rain, the days come, and the days go. I get up, I wash, I work on my discoveries. I eat, nothing changes. I sleep, then, as they say in Hollywood, "Comes the dawn" and there we are! Everything to be done all over again.

Mesozoic mammals! Little tyrants destroying my youth! Time-wasters! Who wander about in my dreams. Gray like a London day. Neither remarkable nor boring. A drama of emptiness! A life of emptiness! Days and years for nothing, nothing. And then one fine day, really—nothing more! A sad spectacle, this life, my dear sister. What can one hope for? Pretty girls?—their necks are dirty. Wine?—and then a headache—not worth it. Books?—They're all beastly. Money?—to do what? Amuse oneself?—There's no way to do that. Beauty?—if there isn't any within, there isn't any without. Friends?—what's a friend? People? They'll accept your hand only if there's a drink in it. Work?—fatigue, and nothing lasts. God?—I don't follow in His Name.
[Letter now continues in English]

In short, it is Sunday afternoon & I didn't sleep well. So damn the world!

[HERE INTERVENE JUST 168 HOURS]

I've had letters from all the family but you this week, but Dad made me the final recipient of three letters from you which had gone the rounds. One must be careful in writing to anyone in this family, as one then writes to all.

Hopwood has dragged his weary self through Capitaine Fracasse, but it took him from the middle of June to now, not because of the language, which he reads and speaks fluently, but because of the descriptions which bored him. Let it be said subrosa, tout bas, etc., they *bore me too*; I like them, I appreciate their beauty, their effectiveness. But I ask you, is there or is there not a lady in distress? Are there or are there not duels to be fought? Does one or does one not have an engagement (for what purpose we will not enquire) with a pretty soubrette? Then why must we wait while for pages on end quite relevant but unnecessary things are recreated by the gentleman's fine pen? And then too, one reads between the lines "What ho! I am Gautier. Others can describe stirring scenes, others have seen the old pictures of the Pont Neuf in the Musée Carnavalet, but only Gautier can write such descriptions." Well, quite true, only Gautier can, & they're quite perfect of their sort, but still, at the same time.

Now there's Dumas, there's a man! If he mentions a room, it's because someone's going to be killed or kissed in it. If he describes a street it's only so you can follow the hero as he dashed down it pursued by relentless enemies or pursuing still more relentless ones. Nothing is described except to help the action & there's a fight every three pages & a murder every six. To hell with your fine writing, this is the stuff for us he-men!

And Gautier's heroine! Gods! How I hated that female, her prurient modesty, her fine airs! Inhuman little wench! I absolutely longed for someone to do real violence to her, & I'm not vicious either & even admit that chastity is a very good characteristic in its place.

Which leads one by a single transition to Herr Lion Feuchtwanger's *Jew Süss* which I have just read in a very good English translation. Whatever else may ail his feminine characters, they are not bothered by chastity, except for one who appears very little & his quite high moral tone of his times is well hint[ed] by the Duchess' father—"Certainly his daughter had amused herself a little with an Englishman. Why should one not amuse oneself with an Englishman? . . . If he were a wife, he too would seek out an Englishman. There was no need to make such a noise, such a song about it." And so on. The scene is mostly in Stuttgart, which I know fairly well, & the book is really compelling, very powerful, & very well written. The English reviewers have been enthusiastic. It is thoughtful enough to appeal to the highbrows & salaciously exciting enough to appeal to the lowbrows & so is the book of the hour here. As I am a skillful blend of high- & lowbrow it pleased me very much.

I also went not long since to Mr. C. B. Cochrane's revue One Damn Thing After Another which, despite its title, is a good revue in the English manner. Not so tuneful as the American nor so undressed as the Parisian, but good stuff. I also went to a cinema & saw a picture about Lafitte the pirate, which amused me because I know his stamping ground. But unfortunately, inevitably, he became a thorough gentleman & polished dandy for the purposes of the fillums. That concludes the recital of my cultural experiences in recent days if you exclude a long visit to the zoo.

You know (of course you don't, but that's a bad habit I have, saying "you know"—English, it is, so should go down well among the lofty Art Hounds in U.S.A.—but a very bad habit) well, you know, I do get fed up with people who talk about about [sic] the degrading effect of the theory, or rather the fact, that man's descended from the apes. The great contribution of the theory to human thought is, quite unlike what is thought, that it shows man's infinite superiority to the lower animals. Everyone knows that what we earn is more precious & less likely to be squandered than what is given to us. Our *humanity*, our character of being human beings, has been earned by the handicaps & battles of a hundred thousand generations. It wasn't given us as a toy by a gentleman in a long beard. We've fought for it & it's up to us to keep it. We aren't poor silly weaklings who couldn't even keep Yahweh from foreclosing his mortgage on our garden, we're Men who've made ourselves such & have raised ourselves above the brutes. We're not on the way down, but on the way up. We didn't inherit our wealth, we earned it by the sweat of our brows. Because we were once apes is the more reason for not acting like apes now that we are men.

It makes me ill, too, in view of all this to see humans lavishing human affection on lower animals. They are degrading themselves by not clinging to the vast gulf which we have put between ourselves & those unsuccessful competitors. Every day an old lady, usually of the male sex, writes to the papers (a favorite indoor sport here) demanding that anyone who hits an animal shall himself be flogged. Disgusting! As if any amount of animal suffering was an excuse for making a Human Being suffer. I like animals. Treat them well as animals, but never forget the barrier which our ancestors have raised between us & them. The same old ladies never write to complain of the thousands of cases of visible *human* suffering here & to demand that someone should be punished for *that*. They think babies are disgusting little objects & lavish affection on their Pekinese—those filthy caricatures which should be mercifully put out of their deformed lives.

The human race is a pretty nasty sight on the whole, but it's better than it's ever been before & it may get better yet—Heaven knows it could!—and it *is* human. The most imbecile of us knows depths not even comparable, in degree, to those known by other animals. The best of us may attain to heights which are sublime, in the proper unspoiled sense of the word.

I don't, myself. I only know they are there, which is something, after all. I only dimly glimpse what might be. I haven't the core, the simplicity, the steadfastness to climb up, to do more than glimpse the truth of living & dying. I am not a great man in the non-popular sense. I am a shell which knows in the end that it has no insides to fall back on. I live on my periphery, giving & taking from the stream of things outside & appalled by the knowledge that there is nothing to draw on inside, yet unable to grasp the obvious advantages of such a situation & to forget the inside & follow the stream. I do silly things, excentric [sic] things, then suffer the tortures of the damned because I can't help seeing them as silly & excentric [sic] my own self. I'm neither fish nor flesh nor good red herring. I don't fit anywhere. I know wonderful things to do, & I don't do them. I keep on doing the things I long to do, going where I want to go, accomplishing almost everything I try to accomplish, & never achieving anything that's any good when I have it. I'm never happy, & seldom profoundly unhappy. In short I'm an idiot, as I think I have shown.

61

From a traitorously forwarded letter to Peg I learn that you are still in bad ways. (1) Despising the army, the hope & glory, etcetera (as you would say[)]. (2) Lending books—when will you learn? Never lend a book to anyone, even to me (especially not to me) if you ever want to see it again. (3) Being done out of money, which is a sin, although, like almost all sins, quite unavoidable. (4) Not charging enough for your [puppet] shows. 15¢ & 25¢ is ridiculous. Charge 50¢ & $1.00 & people will think it *must* be good & flock to see it, & even if you only got ¼ as much you'd make as much. Be chic, make the lousy aristocrats come & the soldiers with their girls—they don't dare spend less than $1.00 on an entertainment. Only let children—poor ones—in for less, or, better, give special free performances for them & soak the bloated at other performances. People want to think they get a lot for their money but they really want to spend a lot of money

61. Egyptian hieroglyphs for "Life, Prosperity, and Health."

for things. Take it from one who doesn't know anything about it. Also you do without ink because you don't want to buy till you move—what are fountain pens, Post Offices, & banks provided for by a beneficient providence if not to keep us in ink?

I'm glad you heard Paderewski. I heard him in N. Haven just before leaving there, but I went to sleep—not because I didn't enjoy his music but because I was sleepy, & when I am sleepy I generally go sleep. I shall now give a practical demonstration of that physiological phenomenon.

Eternally—

AMERICAN MUSEUM OF NATURAL HISTORY 1927-1930

W HEN SIMPSON returned from London in the fall of
1927 he took the position at the American Museum of
Natural History as assistant curator of fossil mammals at the modest salary
of $2,500 per year. Within several months he was advanced to associate
curator, which is roughly equivalent to associate professor with tenure at a
university. Simpson thus began an association with the museum which was
to last for thirty-two years, except for two years of military service in World
War II, until he resigned and went to the Museum of Comparative Zoology
at Harvard University.

In these early years at the museum Simpson engaged essentially in pure
research in vertebrate paleontology; for the most part he was deciding what
he wanted to study next. Having completed two comprehensive monographs on the earliest, Mesozoic, mammals, he turned to their descendants
of the early Tertiary Period, some fossils of which were already in the
museum's rich collections, others of which he collected himself in the San
Juan Basin of northern New Mexico in the summer of 1929. What particularly intrigued Simpson about these mammals is that they record an enormous expansion of species radiating out into a wide variety of terrestrial
environments following the widespread mass extinctions of dinosaurs
toward the end of the Mesozoic Period. From half a dozen surviving lin-
eages of Mesozoic mammals, some two dozen orders of mammals evolved
within about twenty million years—the first third of the Cenozoic Era—
forming most of the major categories of all mammals extant today.

The museum also had a large collection of Pleistocene-age mammals of
much more recent vintage from Florida; these had been donated by a
wealthy retired businessman and amateur fossil collector, Walter Holmes.
In the early winter of 1929 Simpson did fieldwork in Florida to determine
the geologic setting for these fossils as well as to explore for still older ones.
The following year he returned to Florida and other places in the southeastern United States to examine other sites, check related specimens at
nearby institutions, and see what else he might find. To honor his benefactor for the Florida fossils, Simpson named an extinct giant armadillo after
him: *Holmesina septentrionalis*.

Lydia had returned to the U.S. from France some months earlier than   115

Simpson and had gone on to Colorado, where their third daughter, Joan, was born in September 1927. After Simpson's return, the family was briefly reunited, living together in New York City. But Lydia soon grew restless and took off once more, this time for California, where she remained for

116

AMERICAN
MUSEUM OF
NATURAL
HISTORY
1927–1930 more than a year. Elizabeth, the fourth daughter, was conceived shortly before Lydia left and she was born in December 1928 in California.

Anne Roe, who had married impulsively while Simpson was in London, was working on her doctorate at Columbia. For a time she went with her husband to Princeton, where he was teaching, then followed him to Stanford, where he completed his doctorate in psychology. While in California, Anne had a serious attack of what was later diagnosed as brucellosis, or undulant fever; she was in bed for nine months. Anne later returned to Princeton, where her husband had returned, but she went up weekly to Columbia in New York City to finish work on her doctorate. Anne and Simpson met yet again in New York and they started seeing each other, first as friends, soon as lovers.

Martha too had returned from France in 1927 and later in the year proceeded to Hawaii, where she continued her artwork and also making puppets and giving puppet shows. About a year later she moved to Washington State, where she taught art in a private girls' school, the Annie Wright Seminary.

Despite his topsy-turvy private life, Simpson continued publishing the results of his research. Major publications included additional articles on Mesozoic mammals, descriptions of newly discovered primitive mammals from central Asia, publications on Florida and other southern fossils as well as the first of what would be a long series of articles on the early Tertiary mammals of the western United States.

In re The World, which is too much, etc.
[AMNH letterhead]   [New York City]
                                     About Dec. 26, 1927

Miss Marthe or Martha Ssimpsen:
Honolulu, Pacific Ocean
Dear Miss Scihmpsan:
In reply to your kind letter of unknown date, I am sorry to say that your manuscript has no scientific value. You have obviously been reading much, if not wisely, and have been thinking for yourself, as far as your mental equipment will permit. This is all very commendable, even laudable, but the fact remains that in order to do really important scientific work it is necessary to have gone to Yale, and from your letter I am sure that you have not even been to Harvard. Acquiring sufficient knowledge to come in out of the rain, scientifically speaking, is a long and expensive proceeding and after considering your case for forty days and thirty-nine nights I am sure it will not pay you, as one has to take a pitcher to the well if one is to come back with any water, if you have brains enough to follow my meaning. I hope however that you will continue reading and "thinking" about paleontology, although please don't bring the results to me again as I have

a wife and three children to support. I especially recommend that you read all my own books, which cannot be obtained from any good bookseller and on which I receive royalties. Further please consult "Fossils I have known, and How!" by Miss Dryas Bohn and "The Secrets of the Devilish Devonian" by Dr. Senile de Kay.

117

AMERICAN
MUSEUM OF
NATURAL
HISTORY
1927–1930

As for the specimens which you send for identification, your dinosaur egg is pronounced by a committe [sic] of authorities on dinosaur eggs and whatnot to be a quartzite pebble.[1] And as regards your bones of a prehistoric monster, do you miss a cow around the farm, and shall I bury them or will you?

I shall take up your suggestions regarding hobbies when and if I get around to it. I fear it is impractical, however, for me to take up vertebrate paleontology as a hobby. It is true that hobbies make one young, but I already have so many that I am in my second childhood. The following are my chief hobbies, in order of relative unimportance, from left to right, except on alternate thursdays, when one must change at Bridgeport or Saugatuck:[2]

    #. Egypt, its language and antiquities.
    7. France, its language, literature, cathedrals, music halss, and wines.
    $. Crying babies.
    ¾. Sanskrit.
    &. Animiles, living and dead or worse.
    @. The art of the Kinema.
    68. Operating subway turnstiles with wooden nickles.
    ?. Mountains, their history, topography, pestology, and uses.
    %. Travel—it is *so* broadening.
  105. Eating.
Not ½! Poetry, its imitations and limitations.
    ". New York crowds, or isn't war justified?
    ¢. Prehistoric man, or have we ascended or descended?
Non¢¢. Nonsense, living and extinct.
    (. Sculturs, penting, and moosic.
  100°. Sisters and why.

The last brings us to boiling point on Mr. Centigrades famous thermometres, so we will now boil:

    o o o o o o o o o
        o

I am sorry to hear that you are now wasting time on hand paintingoils, personally I use my oils for boilingin purposes, and anyway modern repro-

1. In the 1920s and early 1930s the American Museum mounted several expeditions to central Asia on which were found many new, interesting fossils, including primitive mammals and caches of dinosaur eggs, some of the latter with embryonic skeletons. These and other discoveries, as well as the adventures of the expedition itself in a remote and robber-infested terrain, were well described in the popular press and magazines. Rounded pebbles of quartz-rich rock are to dinosaur eggs what iron pyrite is to gold.

2. Local stops on the Connecticut portion of the old New York, New Haven and Hartford railroad.

ductions of the stag at dawn and beauty and the beast (rather clever, that, a girl and a bull dog, you know? Do you get it? Beauty, you see—that's the girl, and the dog is the beast. Think it over, you'll get it after a while) are so natural you would almost eat the apples and thats what I mean, art, although I really don't know much about art but I know what I like. And what's the use of these beauties descending the cellar stares to see is a bootlegger or only a burglar or what when you can go to the vanities and see nudity as is nude, and no prunes and prisms about it until the police come. Well, what I mean, its pretty but is it art? Or is it pretty? or what?

118

AMERICAN
MUSEUM OF
NATURAL
HISTORY
1927–1930

I will now guess who wrote which poestry, neatly written on one side of the page and addressed to the pizzle editor, no contributions returned and prizes given only to members of the committees [sic] families: The soft guitare (whatever that may) be which has a bell and twinkles like a star, and mixed is right and the author aint the only one shivering and shuddering, is Mr. V. Hugo strained through Miss Symmpsnne, and badly strained and virgin moon who is destined to be raped and stolen to boot is Miss slimmepsonne with no one else to blaim. And I hope I am wrong, but Ime afraid not.

As the great grey poet so ineptly puts it:

> who wold be a Mer-
> Maid bold sitting a
> Lone singing alone under.
> The sea with a Crown;
> of Gold on a
> Thrown.

And I must, in honesty reply that I, for one, would not consider it for a minute. Ide surely catch cold. My favourite passage, however, is in Longfellow's In Memoriam[3] where he says, among other things—

> "Tears of the widower when he sees"
> "The cattle huddled on the lea"
> "And flash at once, my friend to thee"
> "And falling idly broke the piece."

Which all goes to prove what I have always said, that the multituberculate scapula is of normal eutherian type resembling that of the monotremes in no essential respect and showing that the views of Dr. Abel, for whom I have nothing but respect and admiration, are pure poppycock and that he is a damn fool.[4] I will now go out and kill a few jews and get me acquitted on the sole ground of companionate insanity.

Fortunately the presence of two (II, 2, //. B,ΥΊ,Ր,Ὶ,β,=,ϟ,ʔ,Ͳ,)

3. As part of the coy tone of this letter, Simpson apparently intentionally credits *In Memoriam* to Longfellow instead of Tennyson; he also rearranges and misquotes lines from that poem.

4. The upper bone of the shoulder, the scapula, in a group of primitive rodentlike mammals, the multituberculates, is like that of other placental mammals and not like that of the egg-laying, or monotreme, mammals. In this opinion Simpson disagreed with Othenio Abel (1875–1946), German paleontologist and director of the paleontological institute at Göttingen.

prevented a Bronx Love Nest Horror Murder, the Horror Killer's Own Exclusive Story, or something snappy in the way of uxoricides. For the first one to emerge started a knockout scrap as to who was to get to wear it and the emergence of a second gave me one for myself! A thousand (*1000*/— no we won't bring that up again) thanks for evertin all of which are much appreciated and in constant use and employment, save the paper dolls which are too nice for my rough nekks but are hung on the wall for them and us all to see and admire. Oh, by the way, in the beginning of this paragraph i was Talking about the coolie jackets, we havent a coolie and prefer airedales anyway so we wear them on our own selfs.

My teeth are calling me, au revoir and let who will be clever—Hoping to expect the favour of the kind granting of an early order of this choice line of fall models in the latest fabric, goods, and cloths,

Sincerely yours,
Tischbaum, Finkelstein,
and Reilly
G. G. Simpson, the sole agent
for Siberia and the Hawaiahan
Islands, including the
Scandinavian

119

AMERICAN
MUSEUM OF
NATURAL
HISTORY
1927–1930

[New York City, Summer,
1928?]

My dear sister—

You're a hell of a correspondant & it runs in the family. I haven't heard from you since Pele[5] was a pup and I suppose you've gone high-hat since entering Annie-Oakley's or whatever. The last I knew you were panicking the Cross-roads of the Pacific (adv.).[6] Mother slips me the dirt occasionally & I suppose she keeps you informed of my formless existence too. I have had a letter to you in a cubbyhole above the plane of my endeavours for exactly four months & seven days, embellished with sketches of Faith, Hope, & Charity & festoons of baloney, but it never achieved a 2¢ stamp & it's mildly antique now, like us (I guess you're 30 whether you admit it or not—many happy returns).

The real reason for my reformation is having been out late last night & being afflicted with the leaping melancholia which makes me remember your sins. The theme of the play which kept me up, however, was not sufficient moral to enforce this radical revision of character, although it was designed to make one think of relatives—all about the noble Sir Basil, V.C., K.C.B., K.C.G.M., & his three little bastards—"The Bachelor Father." Deshed Diverting, even from where I sat in the second row (of the second balcony).

Like yourself I dissipate by the bell. Dissipation regularly on Wednesday,

5. Pele is one of the Hawaiian fire goddesses.
6. The Hawaiiian Islands were then as now heavily advertised for tourism, one attraction being their central position in the Pacific Ocean.

6 P.M. to 1 A.M. Companion & evil influence, my only friend in this heartless city, Creighton Peet.[7] A cynical devil whose girl married a bond salesman & who goes out for Bitterness in a Big Way. Stout fella.

I always did say, it's a nice place to visit but I wouldn't like to live there. Niggers & Jews, L, sub, wheels, the same & then the same—am I getting ga-ga modernistic?[8] Or what?

Anyway the history of art must be fun. I'd like to do it myself. Venus-de-Milo, Leonardo-da-Vinci, Rosa Bonheur, & Lorado Taft.[9] At least they may find out what they like, even if it is pretty.

Well this is all rather sick & brief, but you have to take it easy after such a long lay-off.

Vox Populi

120

AMERICAN
MUSEUM OF
NATURAL
HISTORY
1927–1930

[New York City,
Summer, 1928?]

Ma liebe sister—

[Three days pass]

I'm sorry I didn't acknowledge birthday present if and when any, but I never do. I know it aint cultured or seminole like your girls, but I just don't, I mean don't. Excuse it please.

[Forty years elapse. Clarice is now broken in spirit & mind.]

I may have acknowledged Christmas present—I really don't recall. But if so, after your snooty implications, I take it back. However I will back down: I'm sorry I mentioned Rosa Bonheur & Lorado Taft as great artists. I only did because I couldn't remember who painted "Beauty & the Beast."

[Two hours intervene]

Your questions are pretty hot, that I give you. Without looking up things & without ever hearing your nodoubtpriceless lectures (I mean it too) I'd bat about 47.83%, or thereabouts. I know exactly 27.2% (no fooling) of the 18 parts of the Doric order. I know stupa, & Akhenaton, & the Shive Dragon,[10] & I suspect that by first two periods of Greek art you mean Minoan I & II, although if so I suspect you of giving Argos a dirty deal.

[The Roman Empire falls.]

The rest I pass up, sorry.

7. Creighton B. Peet (1899–1977), journalist and author who was one of several longtime friends of Simpson's. Peet, and later his wife, Bertha Ann, and their son, Creighton H., are frequently mentioned hereafter in the letters. Peet and Simpson had met at the University of Colorado and worked together on *Dodo*.

8. Simpson is describing New York City with reference to its ethnic and racial diversity, but crudely put and uncharacteristic of Simpson; its elevated ("L") and underground ("sub") trains; and its car traffic.

9. Lorado Taft (1860–1936), American sculptor and writer on art and aesthetics who taught at The Art Institute of Chicago where Martha had studied. Rosa Bonheur (1822–1899), French painter of animals; her most famous work, *The Horse Fair* (1853), is in the Metropolitan Museum of Art in New York.

10. *Stupa* is a Sanskrit word for a domelike tower containing a Buddhist shrine. Akhenaton was an Egyptian king of the fourteenth century B.C.; also known as Amenophis IV. Shive Dragon apparently refers to a demon associated with the Hindu god Shiva.

But I seriously take you to task for implying that the flagellations of my involuted sneering is [sic] casual.

> [The Roman Empire opens its eyes, finds that
> Al Smith[11] was elected, & drops dead again.]

I enclose elegant pictures drawn by me. Two artists in one family! It is too much, or quite enough at the very least.

> [The same, the next afternoon. Richard de Vere
> is discovered upstage, right.]

Speaking of the Drama, I've recently been to Elmer Gantry—Night Hostess—Bachelor Father—Gary [sp?] War—Goin' Home—etc. All characterized by language that should get their mouths washed out with Fels Naphtha Laundry Soap—Otherwise (or even thiswise) swell, mostly. I'm taking up the Drama in a big way, at 50¢ per perch at Gray's.

But now, then, (I can't make up my mind which) I really didn't mean to hurt your feelings. I know that it must be hell (that would just slip in [I hate parentheses]) to tinkle about by the numbers & have dumbells to instruct, but at that it doesn't sound like a bad job. (I *just* can't seem to be sympathetic without going it-might-be-lots-worse all over).

> [Ten years pass. The 8th Avenue subway is finished.]

Let me see—where was I? Oh yes, on the corner of 42nd & Broadway, the Main & Center of New York. I am now dilloing in an armerdillo's armour.[12] It's more fun than poison ivy. You take thousands & millions of separate little bones of his armor plated hide and you try to piece them together. If you get two together the joke is on you and everyone laughs heartily & you play tail the donkey next. The dillos come from a rather unsanitary stream deposit in the Sunny South, Down where I Want to Be, on That Tamiama Trail with Ma-dra-ah-mee, in F-L-O-R-I-D-A, that spells home to me.

> [A few minutes later.
> The banker's body has disappeared]

I'm going there soon, if my angel is angelic—the bird who shells out the sinews of war, I mean.[13]

ARMODILLO, believe it or not.

In a way, I suppose that this letter is not one of those deeply significant flashes which one values throughout life. In fact, after taking the matter up

121

AMERICAN
MUSEUM OF
NATURAL
HISTORY
1927–1930

---

11. Alfred E. Smith (1873–1944), American politician and governor of New York State. His Catholicism and opposition to Prohibition are reasons usually given for his unsuccessful bid for the presidency in 1928; he ran as a Democrat against Herbert Hoover, a Republican.

12. An armadillo is a toothless mammal covered with bony plates; found in South and Central America and in the southern U.S. Also common as fossils because of their easily preserved bony armor.

13. The museum benefactor and amateur fossil collector, Walter Holmes.

thoroughly with the Board of Trustees, I am inclined to question it. It's more or less a matter of poisonal opinion, of course. As for me, I like it with a dash of soda.

Yes, yes, I'll be serious—a true picture of my mind is given overleaf—

Georges de Sacré Son

[Weeks and weeks elapse—just wait & see]

[New York City, 1929?]

Hermana mía [Sister mine]—

Mother can't imagine anyone owing me a letter—but you owe me two (2). Postcards don't count. However I'm both persistent, broad-minded, & forgiving—three of the most obnoxious virtues in the catalogue. Good idea for an epitaph: "He had all the obnoxious virtues and delightful vices." At the moment I'm taking a vacation & hard put to it for things to do anyway.

Last week was hectic—Monday I went for a spree in Greenwich Village & Time's Square Triangle. Wednesday I went to the Circus—although I imagine the ladies are the very same I saw twenty years ago they didn't seem as beautiful somehow, but it all smells the same. Thursday I lectured at Yale & then went out for a spree (Spreeing is here)—I was supposed to put up, thanks to a dignified friend (my *one* d.f.), at a snooty club but I arrived there in a frayed condition at three A.M. & found it all closed & sealed so went off to a dormitory with some of the boys. The dig. fr. probably believes the worst. Friday I motored to N.Y. with an undig. fr. & his girl at an average speed of 60 m.p.h. in a downpour, & then went off to Intercollegiate Fencing Finals and ball. Saturday I talked over radio (having written talk at 2 A.M. after coming back from Intercollegiate binge), & then went to Columbia Club & some new movies in evening. Sunday had a string of friends in, Monday had a very minor operation at St. Lukes, & here I am in bed but fairly fine & restive. I am to reform, but can't make up my mind whether to cut out work or to cut out going places & doing things.

Did I mail letter about you going to S.A. with me? If I did that's three (3). If I didn't—the upshot was, sorry but it isn't practical. In the first place, I may not go myself. There are lots of other places. To be brief, it just can't be done. I'm sorry.

Proof of bookplate arrived & I like it *very* much.[14] It really is swell & I thank you greatly, or at least will even more when all arrives for I know expression based only on proof must be inadequate.

Be as good as is compatible—

Your loving brother

14. Martha was designing a bookplate for her brother. When the bookplate was completed, Simpson placed it in all of his books thereafter.

St. Petersburg Field Branch
March 10, 1929

Dear Mother & Dad—

123

AMERICAN
MUSEUM OF
NATURAL
HISTORY
1927–1930

The fact is that a postcard is as difficult as a letter. I am well & happy, & I will be back in New York sometime next week, exact date & time not yet to be specified. Tomorrow I leave here for Lakeland. At noon I meet the State Geologist again & then spend several days rambling about, after which I will go to Tallahassee or into Mississippi or back to New York or somewhere, depending on how much money I have & whether the Museum succeeds in reaching me with instructions.

I have had a wonderful time here in St. Petersburg. The Holmeses are charming people & their home here is luxurious & beautiful but not ostentatious. It fronts on Tampa Bay & has large grounds & an orange grove, although it is in the city. The cook is a gem & my room is cool and very comfortable. [—I didn't realize until I had written that a cool room might not sound enticing to you at the moment. The weather here has been very warm ever since I came. I haven't had even my light coat on once & a vest is usually uncomfortable.] I have been learning how to cast a plug (for bait fishing & lawn casting) & driving about among the palms, oleanders, live oaks, alligators, & other tropical vegetables & insects. Also finding lots of fossils. In short, this is a great improvement over my former conceptions of heaven. Oh yes, strawberries are 2 quarts for 25¢, stone crabs are the most luscious of sea foods & favorite in this household, piles of oranges kept replenished in my room, mangoes, papayas, & all sorts of good things. Music from New York, as clearly as on our own radio there. Movies in the living room. New York, Connecticut, & Florida papers.

Last week we went down & explored Peace River, which is well named—a lovely spot. Saw a wild turkey, alligators, lots of ducks, cranes, herons, pelicans & other birds exotic to a New Yorker. Went part way by boat & motored up & down to various points along the river. Spent one night in the open, very comfortable & millions of stars.

Last week I spent two nights in Sarasota. The Leonhausers live about two miles out of town, & I should have never found them save for the coincidence that they live almost directly across the road from the man I went there to see—J. E. Moore. Once they gathered who I was, they become very cordial & had me out to dinner my second night there.

I hope you both are well, & gather from your letter (just received after trying to find me on the Peace River) that you are finding lots to do. I am very anxious to see you again, but I do hate to think of leaving here & going back to New York.

With much love—

George

My dear sister—

It is in very trowth with small bigheartiment that you give at me these ropreaches on the tongue english. I preach and I inscribe her at little near so good as you make. Had I but awaited at young fillies seminoles how you teach, so had I also those liquencies of repression like how you were, if not. This I corrate not in bittleness or corrasion but singlesome in expense of my one dibility and honorty, anyhow canbe.

These bookplate was gorgeful, the most gratuitous of you, sure. Them stupids is hearkworthy and at little near inhearable. The papers of science that I make all follow for to make names at these creamals so new! Them pterosaurien qui a des plumes! Ne sais-tu que les pterosauriens étaient des reptiles? [Don't you know that pterosaurs were reptiles?] Veritudinous crawlers with the skin and escales [scales], whenever *not* with the hares or plumes, positively not point. No. But, when even, is it most meek and ill tempered of you to think at me and I sinecurely depreciate it.

Now I does not loves the loti and the ailements of poesie or what have I? Both and the pterodactyle who holds scrawls in his bee is not all satisfulling at me. The authres is much nice. I am so concerted to like meportayal, but I am wish not that they know how concerted I be, so how to can to make? I think so. But I loves these tint-pot and those animals who comes thereout.

[Letter continues in French*]

I've thought a lot about what kind of design to put in my books. (The devil with English, it's too beastly—how many people there are who only speak with uncivilized and savage words!) I want only a single one for my whole life. Thus, it's something very serious, you know. Therefore, I ask myself what is needed? First, it has to be me. My name, yes, but also the design. Those designs that can belong easily to anyone at all are pretty perhaps, but not distinctive enough. Thus, it has to be me. (I've said it twice, so perhaps it's true.) Eight hours a day I am a scientist, geologist, a student of ancient beasts—some beasts then. I write a great deal, so perhaps a pen—it's all how you yourself see me. But it must be a design for all my books. There are art books, Egyptian books, books on travel, on languages, some novels, poetry. I myself have thought of a crazy design that embraces the whole world, all of existence. I think that's enough. It's even too much. I don't like it. (The Egyptian means "I shall make you love literature [or writing]. I will make their beauty penetrate into you"). Alright, I'll take it back. It isn't beautiful. You'll do what you want. In art, I love most the cathedrals of the 13th century, the sculpture of the 12th. Ancient Greek art, Khmer art, ancient Egyptian art (especially the 5th Dynasty, isn't that so), the paintings and prints of Holbein, Dürer, Rembrandt. As for animals, those included are some of the very beautiful ones. Before def-

*Translated by editor.

initely making the book plate, perhaps it would be good if you sent it to me? But it's all as you wish. I know that you know what will please me.

It's exactly three months since I've written you a long letter. I expressed my joy in having the female bust you gave me. I expressed my highest sentiments, and my theories about all that. I did not send it on—I don't know the reason why I didn't. Laziness, no doubt. But really, I am very grateful. The beauty always lies on the table next to my bed. I look at it very often, and I think of you, my dear sister.

Believe me, I am very happy to hear the news that you are coming here soon. I remain—

Your servant who kisses your hands,
Jean Marie Dieudonne
Itchattuchmee de Choucroute

[Letter ends in English]
Weret but here would can talk of these importances more detailly.

125

AMERICAN
MUSEUM OF
NATURAL
HISTORY
1927–1930

Plesiosaur

Stegosaurus (a dinosaur)

shull of Triceratops

Triceratops (another dinosaur)

Quelques uns de mes amis
de la Floride

Mastodon

(still another dinosaur)

UINTATHERIUM

Cynognathus

Imperial Mammoth

Dimetrodon

Camarasaurus   (still another dinosaur)

Cynognathus   (an ancestor of ours)

Dimetrodon   (not a dinosaur but just a kind of good reptile).

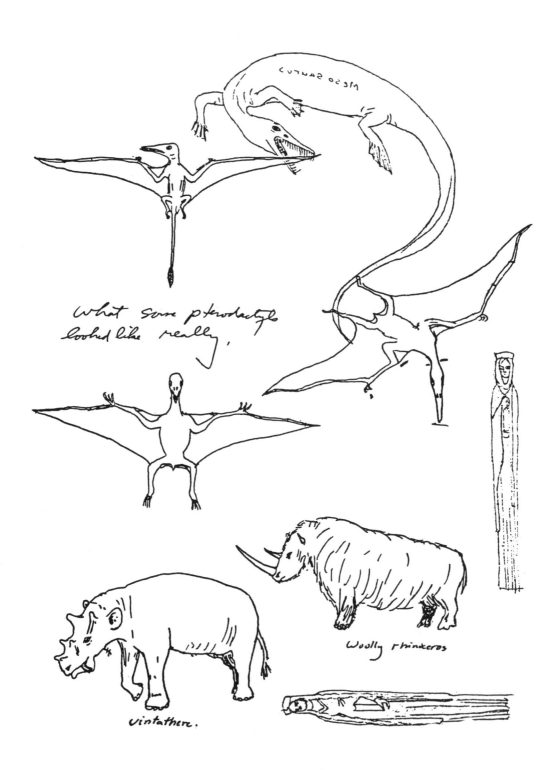

MESO SAURUS

what some pterodactyls
looked like really.

Uintathere.

Woolly rhinkeros

Meso Saurus

What some pterodactyls
looked like, really.

Woolly rhinoceros

Uintathere.

132

AMERICAN
MUSEUM OF
NATURAL
HISTORY
1927–1930

[Farmington, New Mexico]
About June 10th, if
I haven't lost count.
[1929]

Hermana querida [Sister dear]—

If the Annie Wright Seminary isn't in Tacoma or can't be moved there, you won't get this.[15] If you shouldn't get it, write to Tacoma & have it forwarded to wherever the ~~hell~~—dear! dear! this virile Southwest—to wherever you are.

I have been traveling for ten days, & have just arrived at the starting point of my expedition. Traveling alone is boresome, but I always talk to everyone who understands English & quite a few who don't. In this way I acquire much misinformation & narrow my mind, which was tending to broaden with too long residence in N.Y. I don't hesitate to say that I am one of the best misinformed & most narrow-minded men in the country. I attribute my success to an ability to avoid hard work, to Chesterfield cigarets, & to Simmons Spring Mattresses. I have recently inquired into the points of view of a house-painter who has been out of work for thirty years (I'm so big-hearted & such a dam'fool that I bought him a T-bone steak), a waitress with a glass eye (very pretty it is, too), a bath-house proprietor, a fisherman & several other liars, four hotel proprietors, three mexicans, an Indian who knows three words of English (none of which can be sent through the mails), a prep-school boy from Boulder, Colo., a young lady of no visible means of support, an old man who doesn't like Indians because one shot him once, a young man who eagerly explained my own profession (about which he knows nothing whatever) to me in a very patronizing way, & others too numerous to mention.

I am thoroughly sunburned & peeling, I have already been through the worst sand storm I ever saw (but said by the natives to be practically continuous here & not as bad as in the region where I'll be working), my vehicle which I jokingly call an automobile has broken down five several times, each time with unerring instinct for picking the most desolate spot in a God-forsaken region, I have involved myself in a damage suit, I have scraped most of the skin & much of the flesh off both hands, and I am thoroughly happy & unspeakably delighted to be back in a region where you have to be pretty tough to survive.

I have engaged a young cowpuncher as cook & handy man. He's not much on pies or biscuits, but says he's hell—there I go again—on beans & coffee. The water here tastes like a strong dose of epsom salts with a good shot of alum thrown in & made me violently sick when I first drank it, but they say it's much better than the water we'll get in the field, when we *do* get water. We leave the city (pop. estimated at 800 by a local enthusiast) day after tomorrow.

15. The Annie Wright Seminary in Tacoma, Washington, was an Episcopalian school founded in 1884 whose teachers were "college women of experience." The American author Mary McCarthy was a student at Annie Wright when Martha taught there and McCarthy describes Martha in her autobiography, *How I Grew*, as "amusing, unconventional, . . . who taught us some innocently naughty French songs . . ."

Well, whatever, if you should have a birthday (30th [31st?], although I suppose you still stick to your story that it's only 29th) soon & if I shouldn't be able to get a message to you by Navajo runners or carrier pigeon, consider yourself smacked & all. If you could use a Navajo blanket, N. jewelry, pottery, or basket or what not, hint in words of one syllable for I'm more likely than not to get something for an unbirthday present sometime & if you don't say yes or no you'll have to like it. Your loving brother.

[unsigned]

133

AMERICAN
MUSEUM OF
NATURAL
HISTORY
1927–1930

Farmington, N.M.
June 12 [1929]

Dear Mother & Dad—

Tomorrow I plunge into the wilderness, so don't be alarmed if I don't write again for a while. Mail carriers don't penetrate to the city of Kimbetoh (Population 1 [one]) which will be our only contact with civilization—if you care to call Shorty Widdows civilized, an extremely dubious point. I drove out there yesterday with an old timer to make a rapid reconnaissance of the country & decide on our first camp site. If you don't like traffic, come here. We drove 120 miles, passed two cars, one a truck freighting supplies & one a fliver driven by a scared Indian, & saw three white people, the truck driver & two traders. This is a swell place. The roads are unspeakable, mere sandy tracks through the sagebrush, often quite frequently washed out, & frequently there is no road of any description, just compass directions & luck. The place is full of Indians, but otherwise almost deserted. I'm going to like it fine. I have a young cowpuncher to cook, & bought an enormous sack of beans, a side of bacon, & some coffee (a few other things, too, of course, but Jeff says he's hell on beans & coffee but weak in other branches of the culinary arts.)

In short, I am well & happy, & hope you are. Has Dad got his license yet? This Dodge truck is just like learning to drive all over again, it shifts differently & the whole feel is different. But I've driven it about 700 miles already & am used to it now.

Much love,
George

Dick Simpson[16] has sold out & lives here in town now. They say he's been drunk for 20 years & is ready to retire.

[In field, San Juan Basin, N.M.]
June, 18, 1929

Dear Mother & Dad—

Just a note to surprise you. I'm going into town tomorrow & so can get mail off, & it may be two or three weeks before we get in again.

This is a rough, tough country, but I love it. Our camp has been blown away twice. We chew sand, sleep in it, get stuck in it. It permeates us. It

16. Presumably a relative, although I am not able to verify this.

fills locked trunks & sifts through two layers of canvas. We have to haul our water, which is not good, several miles. We freeze at night & broil in the daytime. We are all blistered, but shiver all night. Everyone is well & happy, however.

134

AMERICAN
MUSEUM OF
NATURAL
HISTORY
1927–1930

Fossils are far & few, but we are slowly getting some.

At present our party is quite respectable, including a Jesuit priest, who is a very good egg.[17] We all like him.

Tired & blistered, to bed now, as we'll be up by 5 (if you can believe it)—

Much love,
George

[New York, early 1930?]

Thank you:

A. For beautiful book. It is surely true that Maillol is very fine and I am grateful that my ignorance of him does not endure.

B. For postcards loaned to me. I shall try to keep them safe. I already have postcard reproductions of the sculpture so that I can perhaps be persuaded to part with some.

C. For kind sentiments postcardly expressed.

NEWS Since your departure my life has not been exclusively one long lingering joy. As someone has so well remarked:

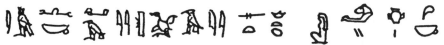

For over six thousand years wise men have known the major unpleasantness of life. It is a sad reflection on our human species that in this six thou- 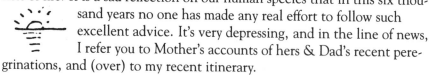 sand years no one has made any real effort to follow such excellent advice. It's very depressing, and in the line of news, I refer you to Mother's accounts of hers & Dad's recent peregrinations, and (over) to my recent itinerary.

ART In New Orleans I saw some really very nice things by a young Mexican whose name I forget at the moment. Etching & Sculpture. He is color-blind and the most curious thing was a painting, the only one he has attempted, which was really very impelling & attractive, with the strangest possible coloring but not unpleasant.

THE DRAMA My recent attendance on the drama has been sketchy but omnivorous. Eva La [sic] Gallienne Sea Gull, beautifully done but

17. A Catholic priest who visited the field camp and said Mass; one of the camp helpers acted as altar boy.

futile almost to the point of boredom. Various movies, the most interesting a well managed Russian film: Fragment of an Empire. One or two plays I can't even remember offhand.

FOOD My trip through the south was a partial success gastronomically as I dote on the very very black Louisiana coffee, on Oysters Napoleon, various Creole sauces and fish dishes. I had a grand time with an archaeologist in New Orleans who admired local food, & possessed good liquor.[18]

PALAEONTOLOGY News in this department is always plentiful, but, Alas! uninteresting to you. We have some lovely new bones and teeth and I am having a good time with them. Two skeleton's [sic] of last year's collecting are being mounted now.

LITERATURE I am reading or have not long since read "The Purple Land" by Hudson. "A Naturalist on the Amazon" by Bates. A volume of Hakluyt's Voyages.[19] All of these are swell. The Purple Land is a curious book that you would like. You would also enjoy Hakluyt by skipping about somewhat. A chapter or two of Bates would satisfy you, as not being a naturalist you would like only the general impression of the jungle, & not the repetition and details.

135

AMERICAN
MUSEUM OF
NATURAL
HISTORY
1927–1930

WEATHER Usually rotten, but I don't mind.

LANGUAGES I am rather languidly studying Spanish, making some progress. It's easy to pronounce and to spell, but full of tricks nevertheless. I am about far enough to try to read a novel in it, which I shall soon attempt.

INCIDENTALS I no longer have to worry about binding Elie Faure[20] or other unbound art books as I no longer have any of them, except the Maillol you gave me.

I have now published two books and 53 shorter papers and all of them are lousy. I owe apologies to posterity.

Fried chicken and grits with the State Geologist of Fla. and his mousey but tyrannical wife.[21]

German dialect recitations by the chief engineer of a steamboat in New Orleans Harbour.

A night in an enormous four-poster in a room 15 feet high in an old plantation mansion in the wilds of Louisiana, and the charming and extremely voluble New Orleans French hostess, and more fried chicken

18. The Volstead Act, commonly known as Prohibition, was still in force and would not be repealed for several years.

19. Henry Walker Bates (1825–1892), English naturalist and explorer. He accompanied Alfred Russel Wallace, codiscoverer with Darwin of the principle of natural selection, on the exploration of the upper Amazon River. Bates later developed a theory explaining mimicry in butterflies. Richard Hakluyt (1552?–1616), an English geographer who published narratives of his many voyages and travels. Those describing the New World are considered especially interesting.

20. Elie Faure (1873–1937), French art historian and essayist who wrote the multivolume *Histoire de l'Art* (1909–1921).

21. Herman Gunter (1885–1972), state geologist of Florida from 1919 to 1958; he is best known for his studies on the geology, mineral resources, and groundwater of the state.

and grits, and scrumbling about stream banks with an incomprehensible negro.

Driving furiously at night along the invisible Mississippi, with dense corporate clouds of fog oozing over the levee and flowing singly off over the dark marshes.

136
AMERICAN
MUSEUM OF
NATURAL
HISTORY
1927–1930

Lunch in the open court of an old house in the Vieux Carré, & tea on the balcony of another on the cathedral square.

A very wild party all night in the room next to mine in an otherwise very good hotel.

A stay in a town where I was supplied with a house and a servant and literally could not spend any money, all the storekeepers informing me that they were ordered not to let me pay for anything.

A day spent weirdly and unsupported by the solace of tobacco in the depths of a coal mine, looking for the footprints of animals dead since 250,000,000 B.C.

A long drive with a high-school boy for chauffeur who explained the personal attractions and morals of every young lady in town and nearly landed us in the ditch several times.

A very suspicious Professor of Geology in a tiny college who thought me a city slicker come to steal his lousy specimens.

A very jovial oil geologist who put (literally) a barrel of whiskey in his car and then drove me unsteadily but widely through the surrounding country, at his company's expense.

A heavy snowstorm in Alabama & freezing cold, a man dropping dead just outside my door.

and many others.

And so to—a meeting of the Section of Geology and Mineralogy of the New York Academy of Sciences.

Honi soit qui mal y pense—

SOUTH AMERICA
1930-1931

IN THE fall of 1930 Simpson led an American Museum
expedition to South America. Committed to a full
understanding of mammalian evolutionary history, Simpson was eager to
do fieldwork in South America, which had been isolated from the rest of
the world as an island-continent—much the way Australia is today—during most of the Cenozoic era, the time when mammals were increasing
greatly in diversity and abundance. The expedition allowed Simpson to
examine existing collections in the museums of Argentina as well as
uncover additional specimens in the fossil-rich strata of Patagonia. A similar expedition had been planned in 1929 to be jointly undertaken by the
twenty-seven-year-old Simpson of the American Museum and the distinguished vertebrate paleontologist Friedrich von Huene of the University of
Tübingen, who had collected fossils in Brazil several years earlier. But
when this arrangement collapsed, Simpson decided to proceed on his own.
Given the expense of such a venture, Simpson solicited additional funds
from wealthy businessman Horace Scarritt, who agreed to bankroll the
project. Scarritt's generosity continued over the years, providing support
for a second expedition to Argentina in 1933–34 as well as summer field
excursions to Montana in 1932 and 1935. Thus during the worst years of
the depression Simpson was able to generate crucial research support at a
time when the American Museum had to cut back severely on its
programs.

A second motivation for Simpson's departure from New York may have
been Lydia's removal with the children to California and Simpson's
increasing involvement with Anne Roe, who was also married. Simpson
moved in with his parents, who had come to New York City about this
time. (During his stay in South America his father was hired as a lawyer
with the Federal Trade Commission in Washington, D.C., and his parents
moved to the capital.) Simpson may have needed both emotional and
physical distance to determine how to resolve his marital situation. By the
time the expedition was over, Simpson had decided to end his marriage
either by divorce or legal separation and seek custody of the children.

Accompanying Simpson to South America was Coleman S. Williams, also associated with the museum and about Simpson's age. They sailed in August 1930 for South America along with Williams's new wife, Dora, who accompanied them on the first leg of their journey. Williams returned after the field season, but Simpson stayed on for another six months to work on the fossil collections in museums in Buenos Aires and La Plata. The field camps were in remote corners of the already remote Patagonia. Despite the hardships of living in a cold, windy, deserted land, Simpson obviously enjoyed his stay. He kept a detailed journal that became the basis for his first book, *Attending Marvels: A Patagonian Journal*, published several years later (1934) and reissued first by the TIME Reading Program (1965) and then by the University of Chicago Press (1982). It remains in print today.

During this time Lydia was moving about the country with all four children in tow: Helen, 6½; Gay, 4; Joan, 3; and Elizabeth, 1½. Helen later started school at a private boarding school in Connecticut. Simpson's sister Martha was in France once again, painting and giving shows of her work. She was to remain there until the fall of 1932. Anne was finishing her doctoral dissertation at Columbia. She completed it in 1931, but the degree was not awarded until she could afford to publish the thesis in 1933 (at that time Columbia required formal publication for the Ph.D. degree to be awarded). She took a research job in Philadelphia—she was still living with her husband in Princeton—with a team of clinical psychologists studying persons who had lost the ability to speak or to understand words as a result of brain damage from strokes or other injuries. One of the researchers, Katharine McBride, became a close friend of Anne's; she later became president of Bryn Mawr College in Pennsylvania.

Simpson's chief publications at this time were additional reports on his work with early Cenozoic mammals from the West and the much younger ones from the South. He also published a thirty-four page bulletin through the American Museum which outlined a major rethinking of the classification of mammals. His research on the Mesozoic origins and Cenozoic radiations of mammals gave him new insight into the overall evolutionary relationships of mammals, which, he argued, ought to be reflected in their formal classification. When Simpson joined the musuem in 1927 he had been given responsibility for maintaining a systematic catalog for the collections. The published classification therefore had its beginnings as a simple updated list of orders and families of mammals for organizing the many fossil mammal specimens housed in the museum. This published outline led to a much more fully developed scheme of classification, its theoretical justification being completed a decade later, just before Simpson entered the army in 1942. That work, *The Principles of Classification and a Classification of Mammals* (1945), became a "citation classic," cited 565 times in the scientific literature between 1955 (when the count began) and 1984. Simpson was chagrined by this success, for the work for which he wished to be chiefly remembered, *Tempo and Mode in Evolution*, was cited "only" 215 times during the same period.

S.S. Western World
Aug. 9, 1930

Dearest Marthe—

Now you can say that you are related to the w.k. [well known] Playboy of the Western World in person. At the moment I lie luxuriously on my bed (a proper bed & no berth) in my large cabin, complete with dresser, lounge, closet, private bath, & other conveniences, & sip a tall glass of distinct alcoholic content. On the lounge sits my assistant likewise writing—his wife is busy being very sick in his own cabin. This is their honeymoon—I'm glad to be on one at last. She returns to U.S. from Buenos Aires.[1]

Tomorrow we reach Bermuda, where this will be mailed. Thence to Pernambuco, Rio de Janeiro, Santos, Montevideo, Buenos Aires, trans-ship there & to Comodoro Rivadavia. Oh! Magic names! I can't remember when they didn't evoke delightful pictures & stir desperate longings. Now I am in a daze, it seems so unreal that I am really on my way, well supplied with funds, leader of my own expedition, to see the places & do the things I want most in the world to see and do. I feel a bit overwhelmed at the softness of life, that seems so inexorable but that yields so fully after long continued pressure. It makes me almost optimistic.

Your letter caught me by the skin of its teeth. It arrived actually after I had left for the boat, but one of the men at the Museum brought it to the pier. Thank you for it, & for the enclosed Beaudelaire [sic]-Quincy. I think an occasionally one franc fifty book would be a very sound idea & thank you for the thought.

You should not have feared for my morals, even from your narrow Catholic viewpoint, since I have been daily & continuously for some weeks under the eye & influence of a Jesuit priest—the same that was in the field with me last summer. He says that I am not a sybarite (his *very* words), so you see! Furthermore he argued dogma with me by the hour, unsuccessfully. He knows he's right, because he has faith, & I know I am because I have facts. Now you can begin to worry, as I am out of his influence again & on my way to wicked Latin America. But a little sinning is good for the soul perhaps—if I could only decide!

By the way—waste NO time in being acquainted with *Yerba Maté*, a South American drink akin to tea, non-alcoholic but stimulating. They must have it in Paris. There are so many South Americans there. Drink it five times before you decide, since no one likes it the first time. I am very keen about it. And—if you should care for it I could perhaps somehow get you a good outfit in Argentina [Like opium it requires a special outfit to prepare].

The proper point is for you to either (a) to stay in Paris until I'm through in S.A., or (b) to return there then and in either case (a or b) to

1. Coley and Dora Williams spent their honeymoon on the first leg of the voyage to South America.

insist on my returning to N.Y. *via* Paris. That might just possibly be done, & would be downright swell if so.

Anyway, hermana mia, hasta la vista—

Don Jorge

S.S. Western World

Aug. 9, 1930

Dear Dad—

Second night out, and last chance to write for a while as we stop at Bermuda tomorrow about noon. I'm disappointed that there's no chance to go ashore there, but they do take letters.

Everything is going well. I had a touch of seasickness, but ate a hearty meal after & walked about deck till it settled & so feel all right again. The sea is fairly calm, although we're having rain & lightening [sic].

The ship is steady, relatively, & seems comfortable enough. The food is good. I haven't encountered any interesting passengers, but Coleman & Dora are company. At the moment Dora is very seasick, but doubtless will recover & be about tomorrow.

I'm very happy about getting on this trip & know I shall enjoy it. Even now it's hard to believe that I am really on my way to new countries & a new continent. It seems a bit unreal.

My cabin is very comfortable & cool & I slept soundly.

By the way, we do stop at Pernambuco, although this boat doesn't ordinarily do so. It is this time, because it took some Voltaire passengers for there.

Thank you so much for helping me get off & especially for the suitcase, which I like more & more.

I'm writing to Mother too.

With much love,

George

[At sea, off South America]

Sept. 27, 1930

Dearest Marthe—

Perhaps this will reach you sometime, somewhere. It is written from what must be approaching zero as a place to be. Namely: an empty oil tanker bucking a very strong head wind & enormous seas in the Golfo de San Jorge off the coast of Patagonia, going on four days out of Buenos Aires, which in itself is some slight distance from home and mother. This triply damned tanker is standing on its head, & then shaking like a dog and recovering with a combined mighty pitch & roll that are like nothing on earth. The oil tanks boom with each wave that slaps us, & the whole boat groans & shrieks as if it couldn't last another minute. And yet, believe it or not, I am *not* sea sick. Oh miracle! But I don't like it. It's not *quite* fun.

Speaking of fun, just real good clean fun, you should have seen our revolution. You wouldn't believe the things I saw & did while it was going on, so I won't elaborate. But for the first time in my life I was *really* convinced that I was going to die, quickly and horribly, and that, dear audience, is a unique & unforgettable feeling.[2]

I was probably closer to death at other times during the revolution as when the Escuadrón de Seguridad charged us and killed the man next to me, or when I walked down the Avenida de Mayo under machine gun fire, but on such occasions you feel, as I felt, "of course it won't be I!" But when two hard cases, looters and murderers, stuck rifle barrels hard into the pit of my stomach and trembled on the triggers, it most obviously *was* I. For once in my life I talked Spanish, talked it fast & correctly, then while they hesitated[,] turned and walked away—lovely feeling in the small of the back! My nerves lasted till I got home, then called for whiskey straight.

Well, that's only one incident of many, but I'm not writing a history. It was fun, really, & since I wasn't killed I wouldn't have missed it for anything.

Aside from the revolution, B.A. seems dull by contrast, although we had no idle moments. It's a little like Paris, but deserves more distinction than merely being like something else. It isn't as gay as supposed to be—even London is gayer. In spite of being Latin there is a well known Argentine melancholy and the gaiety never seems spontaneous & is usually vicious. But we were lucky enough to meet many nice people & to get about. The ambassador fed us, as did other people at the embassy, & we visited some of the better clubs in & around the city. The grandest club, of course, is the Jockey Club, with its famous (but mostly lousy) objets d'art, its grand staircase, & what not. The wine cellar *was* a grand sight—I could become lyric about it. The only objets d'art really attractive were two Goyas.

Speaking of painting, we also were taken to the Argentine equivalent of the Royal Academy show or The Salon. It was not impressive. Practically all of it was either just tricks to cover absence of technique, or else painfully in the style of some French artist or other. The following are the leading Argentine artists. Do you know any?

Caraffer, Ripasmonte, Ayllon, Botti, Mathis, Lynch, Sol, de Quiros, Pelaez, Pirig, Cordiviola.

Not all of these are *criollos*, natives, but they all paint here. Leonie Mathis is French & some of the others are Spanish.

As for the rest, I am now past page 125 in my journal. Incidentally, this is the first time I ever succeeded in keeping a journal for so long, but I hope to make a book of it, so it's not labor lost.[3] But in the sight of such

2. About a week after their arrival in Buenos Aires, there was an uprising that eventually unseated the president of Argentina. This minirevolution is described in the opening pages of *Attending Marvels*.

3. Simpson is referring here to the journal on which *Attending Marvels* was based.

riches, I am unable to select and simply throw up my hands, fold my fountain pen, and steal silently away.

Con todo amor de tu hermano que besa tus manos,

Jorge

[I'm working at my Spanish, & find that I can't spell English any more,
and probably never could].

Nov. 8th to 15th or perhaps
earlier, anyway the lads say this
is Sunday, & I remember that
it's 1930, & it must be
November because the weather
is just beginning to get warm &
spring flowers are coming out.

South end of Lake Coli-
Huapi (or Colhué-Huapí, if you
prefer it so). [Patagonia]

My dearest sister—

Your touching letter from Paris XIV [fourteenth arrondissement] and
postmarked 15 septembre was received in Buenos Aires on Oct. 6 &
reached me in the wilderness about Nov. 1, which is rather good time.
This one going back will do better, because I have to go to town for more
garlic tomorrow and can mail it at once there—Colonia Sarmiento, a
place which looks upon Comodoro Rivadavia as the Big City, & Comodoro is the most awful little hole you ever saw.

Well, I can't tell you much about this spot because you wouldn't believe
it, so why bother? For instance you are too bright to think that I had an
ostrich egg roasted with sugar for dinner last night & an armadillo on the
half-shell for breakfast this morning, but I did. Nor would you believe that
one can walk upright on a slope like this

here, but one can (if it faces west—puzzle on that a while).⁴ Nor that drag-
onflies are the worst pest in our desert camp & parrots the worst while we
work. Nor that we go swimming often here, yet haul water from about 15
miles away to drink. Nor that the four people in camp speak French, Spanish, Portugese, Italian, English, German, & Lithuanian & that we sometimes encounter people none of us can understand. Nor that a young guanaco [wild llama] makes a noise like the whinny of a horse that was left in
the rain & rusted. Nor that the ants roar. Nor that we eat a lamb in one
day & an old sheep in two.

As you know your Cabell, you remember Dom Manuel. At the moment
he sits across the tent from me, scratching his whiskers & sucking maté.

4. The winds of Patagonia are notoriously strong, blowing often with gale force out
of the west.

Tastefully attired in rope-soled slippers, long pants buttoned at the ankle, woolly underwear, & a black sash. Others present at the mate party included Justino Hernández, the popular son of the Vuelta-del-Senguer Hernándezes, & Coleman S. Williams of Charlottesville, New York, & Saugatuck, who wore a gown of light cotton without sleeves, cut low at the neck & with a double shirt ending above the knees and carried as his bouquet an old pipe dug up from an armadillo hole where his companions had buried it. The remaining guests consisted of several insects of species yet undetermined although their host is actively seeking them. The table decorations were a large butcher knife, a plate of tortas made of ostrich egg, and a ten kilo sack of yerba maté, with the carcass of tomorrow's armadillo forming a charming centerpiece. Music was provided by Sr. Hernández Hijo on the guitar, Mr. Williams on the accordion, & by Dr. Simpson's refraining from singing. Favors, in the shape of large hairy spiders dropping off the tent, were lavishly distributed.

I'm sorry you haven't met yerba yet. It's all that makes life worth while after all. Perhaps it is at its best in Patagonia, where it helps to keep me going when one wants often to stop, but it's so good here that it should be some good anywhere. We didn't do right by it in New York, that was a mere pallid imitation. Fill the maté (gourd) at least half full of yerba. Suck out the first *hot* gourdful & spit it out (sorry, but that's the thing to do), then use warm water the second time, then hotter & hotter until about the fifth time it's almost boiling hot. Swell! I just had one. When done you may say "Gracias" when you hand it back to the pourer, & he says "Buen provecho" [enjoy!], & then you sigh & say "¡'Sta bien!" [that's good] and he says "¡'Sta lindo!" [that's lovely] and then you just sit a while.

Someone obviously made Spanish up for fun in his spare time. I now speak it rather fluently, or at least I speak it in the sense that an Arizona cowboy speaks English. I completely unlearned my scanty book knowledge & relearned the idioma nacional here by ear. Such swell words! "Burlap" is a poem—arpillera, & "canteen" a symphony—caramallola (or perhaps caramayola—it comes to the same sound here). Incidentally, my vocabulary is rather specialized as you see. Tent, knapsack, canvas, dirty weather in the west, chuck out that damned bug, blow through the little gasoline pipe, give me a lot more roast goose—all ideas I can express perfectly, but of how little use in the more frequented regions! Then they have such nice ways of expressing size. Our fossils come grandote, grande, regular, chico, chiquito, & chiquitito, in descending order, ranging from widespread arms to thumb & forefinger pressed carefully together for the pedacito chiquitito de animalito.

The maldita [damned] tent is about to blow away again—farewell—
—Don Jorge—

—It did, & is put back at great expense, but we're going to eat now, our evening armadillo. Our cook (Don Manuel) has a one-track mind. For a while we ate nothing but mutton, then nothing but ostrich egg, then an almost exclusive diet of agutarda (a very tough sort of wild goose), & now we're well started an armadillo (pichi, this sort is called), having eaten several & having one in the pot, one ripening, & one staked out alive by a string as future food. I like several things better, but heaven knows what his next streak will be—parrots probably. Oh yes, we had a fish period too (there are lots of fish in this desert, we often pick them up alive in the road—but then, you won't believe that simple truth either).

[Patagonia]
Nov. 16, 1930

Dear Mother & Dad:

The only noteworthy news is that our cook left us—servant problem entering Patagonia. We discovered that we really didn't care much for each other and by a delicate adjustment of tempers he resigned and was fired simultaneously, although we later parted with expressions of good will (completely insincere on both sides). So we dashed up to Colonia Sarmiento and got a new cook, one Ricardo Balina, who seems much better.

Aside from that, all goes on as usual. The collection is steadily growing. My journal for the last week consists almost exclusively of natural history, some of which may amuse you.

The so-called ostrich (avestruz), for instance, is very amusing. Their domestic economy is poor, to be sure. Each male has three or four wives, but all they do for posterity is to lay ten to fifteen eggs each in a nest provided by the male, and then go on about their business. The male sets [sic] on the eggs and looks after the young, which are called charitas until they change their feathers, then charas until they are nearly adult, then avestruses. The other day we came on a male with a large family of 25 or 30 just hatched and swarming about him. We chased them and caught one— he legged it, but soon came back and began clucking to them, ridiculously like an old hen, while they wobbled about whistling. Our captive whistled for his father all day (imagine a baby one day old crying for its father and not even knowing who its mother is)—a mournful sound slurring down the scale and ending with a pathetic low quaver. We still have him in camp and he is now quite tame and doesn't miss his family, although he whistles madly for one of us if he finds himself alone. He just now came and tried to climb into my pocket, which he loves because he keeps nice and warm there. Last night he slept with Williams—somehow the idea of sleeping with an ostrich strikes me as slightly bizarre, but almost everything here is strange! The most ludicrous sight is his trying to scratch himself. He props himself up on his ungainly long legs with great care, then lifts one foot to scratch and promptly falls plop! on his nose.

We also had a couple of armadillos, or one rather, as one just escaped, and have had seven or eight but ate most of them. They are extremely weird in appearance, but very uninteresting as pets.

A bird more admirable in its home life is the agutarda, a sort of wild goose. They are absolutely monogamous. The female lines the nest with down from her breast, and does all the house work and child rearing, laying two clutches of 9 and 4 eggs respectively. Unfortunately for them, the hen agutardas are excellent food and we eat two or three a week (along with a sheep, a few armadillos, a duck or two, and as many partridges as we can get). Every time we shoot one, the male stays around until the hen is definitely gone, and can even be seen near there several days later, still inconsolable. It seems a shame, but we must have meat.

The best game bird for eating is the martineta, which I surely mentioned in a previous letter, a sort of partridge-like bird. It also has the unnatural custom of having the male look after the young. He usually has two wives, but they spend their evenings at the club while he minds the babies.

The guanacos [wild llamas] are having their young now, chulangos, the young are called. They never have twins. The chulangos are hunted mercilessly, for their hides, and in due time the guanacos will surely disappear, but they are still very common now. Those common, however, are almost all old ones which are not hunted, and it is not unlikely that in a few years the people will find that guanacos have become scarce, apparently without warning. They live a long time. There is one in the Sierra San Bernardo which is marked peculiarly and hence is known to all the hunters and was about three years old when first recorded twenty-five years ago. It is still strong and hearty. There is, or was until recently, another one south of here with a white forehead which became very famous. The La Plata Zoo offered $500 for it alive, but it hasn't been taken. Guanacos fit beautifully into this landscape and I would feel quite lonely without one yammering at me from a nearby hill. The only thing that really annoys one is to toil laboriously up an almost vertical cliff several hundred feet high, finally reach the top, to then see a guanaco run up it as fast as he can go, and then dash right back down again, without effort and apparently just for the good clean fun of it. They also have trails all over the place, including one very good one which is an easy route from the Villa Hermoso down to the lake and can be followed for miles, but apparently it isn't sporting in guanaco circles to use the trail except for serious travel and they prefer a good shale slope at an angle of seventy-five or eighty degrees or a rock ledge three or four inches wide above a wide abyss. They'll dash up to an inaccessible apex, then stand with their silly tails curled like jug handles and make derisive noises at you, something like a susceptible fat man during a rough channel crossing. Their scorn for our sluggardly species is deep and inclusive. We want to get a chulango for a pet, but even when only a few hours old they are too fast for us.

My Spanish progresses well, and I can now call anyone a son of a big seven or swear by my mother-in-law of the big flute with great abandon. My appetite also improves, and I count it a poor lunch that doesn't begin with an ostrich egg (equal to at least a dozen hen's eggs) and go on through meat (i.e., mutton), two or three kinds of game, and wind up with cheese and membrillo (a hard plum jam eaten with cheese). Vegetables, I am happy to say, no longer play any part whatever in our diet, and thanks to twenty or twenty-five large gourds of maté every day we feel all the better for it. We do eat potatoes in our puchero, to be sure.[5]

5. A *puchero* is a cooking, or stewing, pot.

Probably will add more to this before mailing. If not, much love—
George

At last I have a chance to mail this, a couple of weeks later. It will barely reach you for Christmas—but it carries much love and warmest holiday greetings. It is impractical to try to obtain or send gifts from our wilderness, as you will understand, but I shall be thinking of you and missing you both very much at Christmas time.

G

Camp, a long, long way from anywhere else.
Jan. 1, 1931

Dearest Peg[6]—

Perhaps your heart will stand this shock. Of course I would have written daily if I hadn't known that Mother & Dad keep you informed, but this nice rainy day is so perfect for writing to long lost sisters that I can't resist it.

One bad thing about writing now is that I've been in Patagonia so long that it no longer seems strange to me & I can't think of writing anything interesting. Here are a few commonplaces.

Ostriches have no white meat & no meat at all on the breast. The best way to cook them is to cut off the neck & legs, remove the bones, then sew the body up into a steam-proof bag with hot stones inside. The result (on which I am munching as I write) is not bad, but a little tough. Armadillo, roasted in the shell, is on the contrary very tender, but a little greasy.

We are camping at the only water hole for some miles about, consequently have to share it with the local guanacos. They come in troops & families every morning, as many as twenty at once, complete with babies. Incidentally a baby guanaco is not a guanaco but a chulango. They muddy up our water so that the cook dug four water holes, one for us, one for small birds & mammals, one for ostriches, & one for guanaco. They don't mind very well however & we're thinking of putting up signs, also labeling one side of each puddle "Señores" & the other "Señoras, Señoritas, y Niños" for morality's sake. This is one of the plans of the This-Is-Your-Patagonia-Keep-It-Neat Society, whose activities also include spraying the badlands with Flit & dusting them off, delousing baby tucutucus, & other similar worthy objects.[7] Checks should be made payable to me.

Patagonia is said to be the windiest place on earth. Not long ago a fast airplane took off at Comodoro, flew hard for four hours, & landed exactly where it started, the pilot receiving numerous congratulations because he hadn't lost any ground. I've seen wild geese actually going backwards when they were flying hard to go against the wind. Fossils we are working on are sometimes blown away, & often to cross the top of a hill we have to get

6. Simpson's other sister, Margaret Anna, was seven years older than he and had left home at sixteen to travel and soon marry. They were never especially close.

7. Tucutucu is a native name that imitates the sound of the animal in its burrow, a ratlike rodent, *Ctenomys*, found in South America.

down on hands & knees and pull ourselves along by holding onto bushes or grass, if any. Perhaps the prize effort was the wind blowing a hammer away as it actually did the other day. And this not just in sudden gusts but for days at a time.

The only unpleasant things Patagonia does not have are traffic jams and snakes. Wind, heat, cold, rain, snow, hail, scorpions, poisonous spiders & lizards, lions (or pumas rather), bandits, sand storms, indescribably bad roads when any, millions & millions of flies, no trees, mile after mile of naked plains broken only by piles of black, infernal, jagged blocks of lava—It's as near to livid hell as the earth produces. I like it very much.

There used to be a tree a few leagues from here, planted by some optimist, but it blew away while yet a sapling. In sheltered spots, as here, there are thorn bushes as tall as a man (incidentally these were used last Spring, in October, as a hiding place by a band of Chilean robbers). Otherwise such vegetation as there is, is small & has the universal characteristic of being extremely prickly.

But I fear that if I keep on in this vein you won't come to visit me. You must really. We'll roast you a whole lamb & save the eyes (the favorite morsel) for you.

We have some swell fossils, and a notebook full of pretty colored geologic profiles[8] which I will publish someday and set on the parlor table so I can stroke my long red beard & say "Well, when I was a boy in dear old Patagonia["]—Except that there are no boys. Everyone is old when born.

Hoping that you are not the same & with much love,

Jorge

———

Jan. 14, 1931
Cow Canyon, Camp 3
Patagonia

Dear Martha—

Heaven, I hope, knows where you are. The last news I had was that you were en route [from France] for the United States (of North America), then a last word says not, that you are so encouraged at being hung that you're staying on to see if you can't be sold too.[9] It sounds mysterious and rather nasty, but I trust you and shan't come dashing up to that hemisphere for a day or two.

The only bon bon mottoes yet found in Patagonia I discovered at Christmas dinner—a forlorn little group of all the English & all the Americans (2—count them—2, including Williams & me) in Chubut. In my pocket when I woke up next afternoon I found a mouth organ, one of my host's spoons, a hair net, three olive pits, and a bon-bon motto in Japa-

8. An important part of fossil-collecting is the accurate sketching of the rock outcrops from which the specimens are taken. The nature of the strata is carefully noted as well as the precise location of the fossils.

9. Presumably Martha's several shows in France delayed her return to the U.S. as she hoped to sell some of her displayed work.

nese. I am used to being addressed in Russian, Ba-Ntu [Bantu], Pushtu,[10] and an obscure dialect of the Scottish Highlands, but this is the one occasion on which I was taken for a Japanese.

My Christmas mail, or what I childishly dream is the first trickle of the flood, arrived on January 8th, two weeks being just nothing here. Heaven knows when I can send off this & other letters I should write but probably won't, for Patagonia considered *en masse* is practically nowhere at all, & we are encamped near the geographic center of Patagonia. Wild beasts drink out of the water hole just in front of my tent every day (no kidding) and it's miles & miles to the nearest human face—and what a face! Furthermore the last truck that tried to drive in on the road we follow to town had its driver & his companion killed by bandits. Patagonia is so exhilerating. I love it.

What I mean is, it's wild, if you follow me. One can write letters between intervals of scratching and what not, but getting them or mailing them present technical difficulties of the highest order. However, two weeks ago a Russian who is a neighbor, camped within fifty miles of us now, said he might visit us tomorrow, & he knows a man who sometimes goes to town, so perhaps I could mail a letter by him.

Let's see, just why was I writing this? Oh yes! To thank you for the three books just received a week ago. They look swell. I haven't much time to read, but what I have is theirs. The more welcome as when Spanish comes in the door, French flies out the window. Someone addressed me in French the other day & after stammering a while I finally gasped "Oui, Señor, je parle francés perfectamente bien, como un criollo [native]. ¿No ve?" [Isn't that so?] I need to read a little so I can hold my head up the next time I visit Briand & Poincaré & the boys. As for leaving books on a mountain, I'll have to use a little discretion. The only local mountain, Cerro Salpú by name, is a mass of jumbled blocks of black lava and is inhabited by large families of pumas, the local lions, which reach a length of twelve feet and don't like books. I faithfully promised my grey-haired grandmother never to leave books, or even to read them, on jagged black mountains inhabited by twelve-foot pumas. Please cable permission to leave books, if at all, under a nice shady calafata or in a quilimbay.[11]

We had a puma who played about where we are working. We never saw him, or her as it proved, but she used to come about nights & left us mementoes in the shape of half-eaten sheep. But alas some neighbors of ours (the near neighbors less than 15 miles away) killed her after an all day battle with dogs—I have the skin.

I love animals anyway & I never saw so many before. We have two armadillos now as pets—and amazing creatures they are, too. The following (from that famous work "A Patagonian Menagerie") celebrates the armadillo, of which the local species is called a pichi:

10. Ba-Ntu is an obsolete spelling of the African language Bantu; Pushtu, more often spelled Pashtoo, is a language spoken in Afghanistan and West Pakistan.

11. Calafata is a local tree; quilimbay is possibly a native word whose meaning I have not been able to determine.

The pichi is an animile
That's very seldom seen to smile
And almost never heard to sing
Though he can dig like anything.

His sense of humor's very slight
And some would say he's not quite bright.
Like other people one could name,
His brain is dense, his skin's the same.

A most secluded life he leads
And very simple are his needs:
A hole, a mate, a bit to eat—
Just something plain like putrid meat.

Just think how easy life would be
If human-kind like you & me
Were so contented with our lot
As pichis are:
    —Thank God we're not.

There's more to the menagerie, but we rest assured that the sample will suffice.

I do get a grand thrill out of the various beasts, living & extinct. It's like the earth before man.

Life is routine more or less now. We work & sleep & every few weeks go to town and get drunk & provisions. It's a good enough life. I'm in no hurry to leave.

Hell (as my good friend the Archbishop so often remarks), you don't rate another sheet until you send me another postcard.

                  Amor y besos—
                  George

                  Somewhere near Paso Niemann
                  Rio Chico de Chubut,
                  Patagonia
                  March 5, 1931

Dearest Marthe:

At last I have finished the three books that you sent me. Of the three, I like *Confession de Minuit* the least. It is very clever, which both praises and damns it. If I knew this bird Salavin, I would shun him like the plague, because to me he is the complete and quintessential bore. The ability to present him so fully and in such nicely turned epigrams is quite as clever as training a seal to play "America" on a French horn, but I do not prefer the performing seal to a philharmonic concert. A bore is a sort of spiritual typhoid-carrier, one cannot even approach him in writing with perfect safety.

La Maison de Claudine, on the contrary, accomplished a miracle. It provoked my sympathies and interested and diverted me deeply—and that is a miracle because I have almost completely given up reading books by or primarily about women or children. That's a psychological complex, of course, and silly to boot, but I have it.

But I really place my money on La Brière! In my really humble and wholly inexpert opinion, that is a book that deserves to be a classic. It isn't clever and it isn't sentimental. It is written with a workmanship completely unobtrusive but almost flawless. Its construction seems inevitable, its sequence almost majestic. And its scene and its characters are interesting; strange as they are, and foreign, one feels that they are very real, that they do exist, that they would act just so and look just so. In short, I like the book, and do not intend to leave it under a Patagonian bush—I must reread it sometime, as it is difficult French for me here where I have no dictionary and have forgotten some of the idioms.

The unusual gift of enough spare time to finish La Maison de Claudine and to read La Brière through was presented to me by a short illness from which I have now practically recovered. Being ill in camp, in Patagonia, is a curious thing. The bodily difficulties are nothing, but mentally it detaches one from life and from the world in a way that is extremely curious and terrifying. Here where one has no roots, one must either die or get well without benefit of anything familiar and reassuring. There is nothing to gloss over the starkness of space and time. By fighting, one can retain the essential fact that one is oneself, but everything else becomes a strange, flat panorama, a thin surface hiding nothing. And the only thing that one comes to fear or to believe in is Nothing.—I do hope this doesn't sound like a complaint. One of my companions so often feels sorry for himself (I don't care to judge whether with reason or not) that I dislike even saying that I dislike anything. I don't, in fact, really dislike anything, in a certain sense, even though I use the word. I am not sorry for such troubles as have visited me of late, really. They are as essential a part of life as its pleasures.—I've had too much time to think, lately, you see!

I think you like people—I will present you with some who interest me:

Irigoyen, a young Argentine of Basque ancestry, is a pilot for the French Aeroposta which operates a line from Buenos Aires through most of the length of Patagonia. The day before I arrived in Patagonia, he took off for Deseado, to the south of Comodoro. They start at four A.M. because the wind often dies down a little about that time, but that morning it did not. He flew for nearly four hours without getting out of sight of the field, in constant danger of being blown out to sea, praying for a calm moment in order to land his passengers safely. The wind kept on increasing, and he had to land. With consummate skill he did land, so well that a woman passenger, quite unaware that she had been in the jaws of death, thought that she was at Deseado. After all were safe and the pilot also out of the plane, the wind turned it over, mortally injuring one of the soldiers who was helping to hold it in place. Irigoyen, who had been unnaturally calm up to that moment, suddenly broke out crying like a baby and had to be led off the field. Since then he has become one of my best Argentine friends. He is bursting with nervous energy, interested in everything, excitable as a child, emotional as a woman, but very much a man.

Señora Nollman, the mistress of a boliche (small country store, bar, and inn), was born in Denmark, grew up on a German Baltic island belonging to Holstein, and finally came to Patagonia where she married an Argentine of German descent. She looks as if she were made out of rubber and blown up. Fat, enormously fat, at the very acme of fat! But not flabby; solid, distended, bloated. She looks as if one of those innocently obscene Paleolithic goddesses of fertility had come to life and put on clothes. And true to her destiny, she too is innocently obscene. Herself barren, she broods over, watches, and abets all that is fertile about her. She assists the local human babies into the world. She rears pigs, cows, chickens, dogs, and even wrests fruits from the sterile Patagonian soils. Unconsciously, she follows the oldest cult of man, lustily luxuriating in every manifestation of fecundity.

El viejo galense has no other name than the "Welsh Old One." Forty years ago he came from North Wales. He was then a man grown, but for years now he has been forty-seven, for since he was forty-seven all the years are one to him and time is continuous and indivisible. He speaks English as if he were painfully recalling a silly rhyme learned in childhood, and with a tremendous Welsh burr. He speaks Spanish as if he were apologetically constructing the words from old lumber. He has no home and no occupation, and never uses money. Somewhere he acquired a couple of horses, but in almost blasphemous defiance of local custom he always goes about on foot. He is invariably accompanied by three enormous, gaunt greyhounds and both they and he live on what they can catch, hares, ostriches, guanacos. He gives things to everyone in his neighborhood: ostrich meat or feathers, arrow-heads. He does them small services, such as retrieving strayed animals. In return they give him wine and tobacco and occasionally some old clothes. He worries over the affairs of everyone, all confused in his fuddled old head, and is humble and apologetic because he is only a feeble old man.

Whiskey-proof Jones has the appearance and manners of an English gentleman and the habits of a beachcomber. He has done almost everything, acted on the stage, played a piano in a brothel, sold Fords in Punta Arenas, but almost always the characteristic which earned him his soubriquet also earns him his congee. Although middle-aged in years, he is always the youngest in every party, the last to go to bed every night in the year. He never refuses a drink of any sort, and acts exactly the same whether sobre or too drunk to stand up. In Patagonia no one is really quite ordinary or thoroughly respectable, but the other English people cannot understand his strange way of being extraordinary, his subtle model of departure from respectability. They tolerate him, even associate with him, but do not cultivate him.

Antonio de Kock is a Boer and was born and grew to manhood in South Africa. During the Boer war he fought stubbornly and unimaginatively under his idolized leaders until he was wounded, captured, and sent as prisoner to Saint Helena. There he passed the time calmly, carving curious objects out of cedar wood with a pen-knife. The war over, he put his tongue in his cheek and signed an oath of allegiance to Edward VII, carefully noting that it did not include allegiance to any future English sovereigns. When the Argentine government offered land to Boer immigrants, he married and came to Patagonia where he begat three daughters and saw his wife die of consumption. He respects the English because he fought them and knows that he gave them a good fight. He loves South Africa with the fierce, inarticulate love of an exile, and hates Patagonia as an unworthy antagonist and a dreary prison. His daughters are grown and are pretty, but so secluded that they are like wild animals in the presence of any stranger. He has taught them a few words of English, but not a single word of Spanish. He would kill any of them that married outside of the Boer race. One is successfully married to a good Afrikaner, Boer, but there seem to be no possibilities here for the other two. He longs to take them back to Africa where they can fulfill their female destiny without dishonoring his blood.

Ingeniero Piátnitzky was a provincial baron under the old regime in Russia. He went to the university and became a mining engineer, then and at that time a gentleman's profession, with all the actual labor done by assistants and servants. He married and settled down in a mining community, a quiet bourgeois sort of small aristocrat. When the revolution against the bourgeoisie and the aristocracy came, he opposed it on both counts. When it succeeded, and his army faded out, he slipped into Jugoslavia, took temporary employment, and waited for his wife to escape and join him. He had arranged to go to the United States, but a year elapsed before his wife could reach him, and then the United States would no longer admit him permanently. Eventually he heard of an opportunity in Argentina and came here. The government stationed him permanently in Patagonia. His wife came here for a time, but her health could not stand the terrible climate and she now lives in Buenos Aires. Once every two years he is able to pass one month with her, and with his single child. Although a mild, grey man, he rules his assistants like the small tsar that he is, keeping them completely in their places but considering himself their father, and they adore him. Summer and winter he walks and rides over the Patagonian pampas. His eyes are giving out from the constant wind. He talks and thinks very slowly, very simply. He smiles often but quietly.

George Cunningham's father raised sheep in Oregon. George raised sheep in Montana until he had a small capital, then came to Patagonia to run that up into a competency in a few years, then to go back and enjoy life in his native land. That was thirty years ago, and he will never return. As civilization began to reach the fringes of Patagonia, he fled from it until now he lives in the heart of the southern Andes. He has an Argentine wife and many children, none of whom speak English. The law does not reach his large and distant ranch, and there he literally exercises the right of life and death, but is himself ruled by his family. Once in a while he comes to

Comodoro on business. His business done, he is afraid to go back to his Argentine wife and his Argentine children. He stays on, sometimes for weeks, coming down from his room at 6:30 every morning and starting at once to drink rye whiskey straight. At midnight he is still drinking rye whiskey straight. He seldom says anything. If anyone remonstrates with him, he says, "We might as well be drunk as the way we are." He will look you straight in the eye for minutes at a time, and then burst out into loud, mirthless laughter. He thinks he is insane. He is probably right.

George Gaylord Simpson is—but we already have one American in this list of strange Patagonian characters.

Patagonia is like that.

Hoping you are the same,
Tu hermano que besa tu mano.

Here begins one of the world's longest postscripts.

———

—I just get you written to for another year, and along comes your very welcome letter full of things I want to talk about.

In the first place, although rumors reach Patagonia only some time after they are kitchen gossip in Tibet, I had heard even here that you are rapidly becoming famous. Need I say that I am delighted? Perhaps I do need say that I am delighted. I am delighted. I think perhaps it is a selfish emotion to like to have the ability of one's family recognized. It will be very pleasant to be the brother of a famous artist.

And then all this about science and religion—interminable and rather futile discussion, but of course one that interests me tremendously. I don't feel bashful about entering into it, for I am a scientist, not a great one but a good one as they go, and scientists certainly know more about God than the theologians do, for scientists study the works of God and theologians only study what has been felt and said and written about God.

Regarding B, that a miracle was necessary for the creation of life from inorganic matter, in two words, no. The scientific attitude—and that only means the attitude of studying facts and using common sense in interpreting them—is never to assume the supernatural if a natural explanation is conceivable; even if we cannot at the moment explain the details of a thing, it is fair to assume that there is a natural explanation if such is reasonably conceivable and fits in with what we do know and can explain now. Anything else is the attitude of savage superstition. The savage hears thunder, cannot explain it naturally, and so assumes that it is supernatural, the voice of the gods. The first scientists could not explain it either, but they felt it possible to assume that it had a natural cause, and eventually they discovered that cause. We do not yet know how life arose. But we do know from the study of matter and its properties that it is quite conceivable that it arose from inorganic matter according to the laws of nature, and therefore we assume that it did so arise. Much study has been and is being devoted to it. It is now possible (even on a commercial scale) to make compounds which were formerly supposed to require life for their manufacture. We will probably know the whole secret someday, and in any event it seems in the meantime proper to assume that the simplest forms of life arose from the lifeless without a miracle, and from those simplest forms to the highest there is no further real difficulty.

A, the initial formation of the universe, is in a different category altogether! Given the universe, even a universe devoid of matter as such but provided with the actual laws of nature, everything that exists could, and I firmly believe did, develop from and in this without outside, divine, supernatural interference. But that universe, with its laws! For one thing there is nothing inevitable about the laws. It is a fact that masses attract each other, that gravity exists, but a universe in which masses repelled each other is also conceivable, and in it nothing could possibly be like what does exist, and so with all the laws of nature. They are necessary to the existence of things as they are, but from the abstract point of view there seems no real reason for them. The primordial reality of the universe, the great Why of everything, is a great unified natural Law, of which the various laws discovered by scientists are small interrelated parts. Science does not explain where that law came from, or why. A real explanation is (to me at least) inconceivable. It will never, never be explained. It is the one great and true goal in the search for knowledge, and it can never be reached.

Call that great Unknowable by any name you wish, call it X, or Yahweh, or God, or say that God created it. Applying the letters "g", "o", and "d" to it or what created it is no explanation and no consolation. It is a common failing, even more among scientists than among laymen, to think that naming a thing explains it, or that we know a thing because we can put a name to it. But to say that God created the universe means nothing whatever. What then is God? Why, he is whatever created the universe— and so on in as many circles as one likes. In any event, God in this sense is certainly not the being one worships in church. The heathen who worships a tree is quite as near to it (perhaps nearer?) as the Christian who worships an invisible anthropomorph with a son.

I want desperately to believe in something more comforting, but how can one? Inner faith?—a gift which cannot be sought, and probably a false gift since it is common to all forms of religion and since faith often has been placed in things which are obviously false. Miracles?—none has ever been authenticated for which a natural explanation was not conceivable. Divine consolation?—Many people are consoled by delusions and errors. The intricacy, the marvelous beauty of nature and its works, of man and his?—The intricacy is completely comprehensible as the result of natural law, and the beauty is subjective and also the result of the laws which gave rise to man's mind.

I respect religion, any religion in which men believe honestly. I would never try to turn anyone from it. The truth is not good for everybody, perhaps not for anybody, and anything that makes men contented with their sad lot is worth while so long as it does them or their neighbors no harm. My only quarrel is with militant and prohibitive religion and with the insincerity and dishonesty that so often accompanies [sic] it.

—This is turning out badly, almost a credo, and a polemic one at that. Basta.

You probably are acquainted with Poe's critical essays. If not, they should console you after Keat's belief that poetry (Art) should come as leaves to a tree. Poe was also a true poet, and as I remember his argument he insists on the artist as a craftsman who labors to make perfect.

The Simpson method of work should be copyrighted.

$\left.\begin{matrix} \text{You} \\ \text{I} \end{matrix}\right\}$ begin a $\left\{\begin{matrix} \text{picture} \\ \text{piece of research} \end{matrix}\right\}$ from a sense of duty and [become] rather bored about it. We get extremely keen about it when well into it, and then deeply disgusted with the result when it is finished. In some cases I can reread a thing I did, after a year or two has elapsed, and think that it wasn't bad after all. Perhaps you also can admire your own pictures when you have forgotten exactly how you painted them, or perhaps that is only my personal conceit cropping out. But age does not stale or custom wither (or vice versa?) the badness of some of the work that I've done, and I'm never frightfully surprised when, as sometimes has happened, someone rises up and tells the world that I am no good. Nor do I get discouraged, because I am good part of the time, and that's about as well as anyone but a genius can hope to do. The only thing that really annoys me about criticism (I imagine this applies perfectly well to your work to) is that people will criticise the wrong thing. They will take a paper which I myself know to be full of faults, and will completely overlook the faults and criticise the parts that are all right. At least I have one great advantage, my critics almost always are colleagues and do know something about the subject.

Don't become anglicized. I endeavor to speak English in England, just as I try to speak French in France. But American English is quite as cultured as Oxford English, rather clearer and more forceful, and happens to be my native language. One has much the same thing here. Argentine Spanish is a perfectly good language, rather clearer and more forceful than Castilian Spanish, and I am rather unpleasantly impressed when I encounter an Argentine who tries to talk Castilian (or rather, true Castilian, for the Argentines also call their language castellano, although it is very different in many ways)—who tries to talk Castilian in Argentina. Incidentally, I am still improving slowly in the tongue. I understand everything perfectly and talk fluently, but still make many errors and use awkward constructions. Of course one difficulty is that the Argentine dialect is probably nearly incomprehensible to, say, a Mexican, but having a good Spanish base, no doubt one could pick up the other dialects very rapidly. Williams is very clever at getting along with a few nouns and one tense each of the common verbs, but I am amazed that anyone can be so completely deaf to the real rhythm and structure of a language after hearing it every day for months and having to speak it often. He is quite proud that he can make himself understood without making any effort to speak it correctly.

Our first cook was a greasy Portugee and left by request. Our second was very good but wanted to kill Williams, so I reluctantly let him go. Our present one is a lad of twenty, hard working and clean, but a rather heavy handed cook. This because the child of nature, Olegario whose father was a Garcia and whose mother was a Fanjul, has just placed a juicy estofado[12] before me and this letter, this tome, this library, comes to a close.

Comodoro Rivadavia,
metropolis of Patagonia
God help it
[1931]

Dear sister Martha[13]

I accuse receipt of a letter from you a week or two weeks or a month or two or so ago. I don't know why I write you so often you dont deserve it. and you seldom reply. But I do owe you one now. I gues I will write to you. I haven't anything to say. Except the swell, the grand, the unbelieveable news. I am leaving Patagonia, forrever. Well, maybe not. I am leaving though. All through for the nonce—and I hope a nonce is a good long time. I really dont know how long a nonce is. I must look it up sometime. But thats the trouble with us cerebral types. We always look things up and read them instead of going out and living them. But how am I going to know how to live a nonce unless I find out what it is. Oh well, I have probably lived lots of them and called them something else. Call them imgimps or rtyuipses if you like. Perhaps thats this etaoinshrdlu I read so much about in the papers. Anyway I wish id looked patagonia up in a book instead of going out and living it. Patagonia should be heard and not seen. Its a hell—oops sorry—well it is a hell of aplace. Anyway im leaving. Heh, heh to you patagonia, theres one you didn't get. The population con-sists one tenth of people who were born here against their wills and nine tenths who came here for a month and cant get away. If you drink mate or cross the Pampa Castillo (which is called a pampa—plain—because its a high mountain or plateau) or eat calafate berries (which arent very good anyway) you never leave. I did them all lots of times and Im going. I lived right on top the pampa castillo. What a place. A little narrow flat ridge a hundred yards or a mile or so wide and several hundred miles long no trees no water nothing but wind and if you dont hold on itll blow you off in the sea a feww thousand feet lower. And Ive crossed it lots of times. Lots of people have been killed up there. Well patagonia may be alright to die in. It doenst serve to live in much. The trick is that you have to cross the Pampa Castillo to getanywhere, it forms a barrier between the coast and evrywhere else. And when youre across, where are you? Nowhere.

Well, I like Patagonia anyway even if I am glad to go, or im not even sure I am glad to go. My god its getting me too. Someday id like to write a

12. An *estofado* is a stew.

13. This letter was typed with many errors in spelling and syntax, apparently intentionally so and therefore not corrected here.

poem about patagonia. a real savage brute of a poem with jerky swift lines that beat on one and wear one down, with wind, wind, merciless and eternal wind for a refrain. Why is it fascinating. empty as the cellars of hell just miles and miles and miles of space with the wind of the wide earth blowing through it. No smiling land. No seed roots here. The husbandman has no crop, the woodcutter no trees. Savage distainful bitter land. No one loves it. everyone hates it with a deep bitter resentful hatred and scorns it and maligns it and longs to leave it to its desolation and stays and stays and would not really leave. From its scanty peoples comes up a torrent of hate and a storm of groans. And the land resentfully sends its wind to beat upon them and blights their fruits and their hopes and their lives. Savage land where the guanaco stand on peaks against the sunset and yammer their hate and lust and the spirits of all the guanaco of a million years wander in foolish immortality. Land of ostriches and armadillos and pumas and all manner of wild and hateful things. Flaming sunsets that hurt the eyes already inflamed by the wind and the sand. Blazing white cliffs and red and orange and green and blue, and the bitter cold firmament at night with the southern cross wheeling eternally across it. Scant thorny bushes, jagged black peaks, flat endless plains covered with mysterious pebbles that came from no man knows where or how. Heartbreaking work for eleven months and then a month of irresponsible drunkenness. Land of slow life and sudden death. Remote, secretive, unresponsive, fascinating land. Jealous barren old maid of a land who refuses to coquette or to make herself attractive and then wails its neglect and returns it with scorn and bitterness. Even its long coast from one end to the other offers not one shelter and haven but only a mighty surge and terrible breakers and a tremendous tide deceitfully hiding the miles of jagged reefs. God turned his face away from it and man can only curse it and fight it, but in their hearts they know they cannot leave it.

Well I am going at last. Our boxes [filled with fossils] are on the beach waiting to be loaded on the steamer—for in this busiest port of Patagonia the boats must anchor far out and their cargoes be brought out perilously in little boats into the beating open sea.

Tonight I shall eat dinner with a proudnosed haughty secretly timid Italian and his pretty fading little wife. He will feed me good wine and brandy and call me a gringo and envy me and strut before me and try to pick my brains and tell me his secrets. His trouble is that he is not sure he is a good scientist. He thinks he is—and in fact he is—but he is not sure of himself. He should either be sure, as I am, or not think so much about it, as I also try. Last night I ate dinner with a German who was genial and argued fluently in English and German and Spanish and French and who is sure he is a good scientist and who is not a good scientist. I dont like scientist very much. That is why I have to dine with them now before I go. I have been neglecting them for Whisky-proof Jones, and little Ryan from Las Heras, and that sly bird Jack Davies and Lareta the only Argentine who always drinks straight whisky and Irigoyen and Luro the pilots who are going to be killed soon by the wind of Patagonia. I shall probably never see any of them again. And I like them. They are all insane, but they are all very good fellows. That is the one unpleasant thing about traveling. People pass through ones life too quickly. One doesn't have time to savour them fully and to see all their meannesses and their greatness.

Alexander cocktails are a little too sweet (I am answering your letter now).[14] I haven't the palate for champagne, it tastes exactly like sparkling cider to me. I prefer still wines, or if one wants sparkling wine for a change one of the red Italian ones. I dont know Italian rum but they have a South American rum that is rather like Bacardi and is good, also another that is much too sweet. There are only three usual apertifs here—gin and vermouth (Italian), same half Italian and half French, and vermouth and bitters—almost all the Argentines drink vermouth and bitters. These go by a dozen different names, but the waiter can pour, say, a San Martin and a Solano from the same cocktail shaker. In B.A. they have the Cubano seco. That is swell. No I have not quite become a sot. Drinking is the only even semi-decent form of amusement here and I get more social as I grow old in wisdom and realize how much more interesting to me other people are than I am.

I the voracious the continuous the rapid reader have read one book in the last two months, and barely managed that. El Camino de las Llamas by Hugo Wast. Its nice but wouldnt stand up except that there is almost no Argentine literature and it has no comparison to suffer by. But Wast does write well, and I can read him more easily than most Spanish authors because he writes the language I know, Argentine.

Your life sounds so full and pleasant. It makes me realize how Patagonia has sucked me dry.

We had four tame armadillos and I was holding a contest to see which of them was most worthy of being presented with the books you sent me. Then one day as we were driving from the desolate town of Pico Truncado to the still more desolate seaport (so called because it has no port) of Caleta Olivia their box turned over and a heavy jack fell on them and we scraped them up with a knife and threw them away. Now I have to take the books to Buenos Aires and leave them carelessly about because there a knowledge of French is necessary if one is to be considered socially. The real beaux and belles talk an amazing language concocted out of equal parts of French and Spanish. I'll be in B.A. all winter—thats summer to you of course. Maybe longer. I have a job offered me there, a good job, and I might take it but not permanently. Only that scares me, because everyone I meet here started by taking a temporary job and never got away again. Anyway I toy with the idea—perhaps only because it somehow flatters me to go to a strange land and forthwith be offered the best job of my sort in the country. But in spite of the airs they put on, South America is still a small puddle in the scientific world and I would rather be in a large puddle even if I cant be the biggest frog in it. In fact I dont think I want much to be a big frog. I want to go places and see things and indulge my itch for scribbling and study interesting problems and if it lands me somewhere all right, and if it doesnt thats all right too. But there the insidious

14. Alexander cocktails are made with gin, creme de cacao, sweet cream, and ice.

thing is—it is a temptation. I cant quite say no and I dont want to say yes. And I am an American after all. I even become almost rabidly so after listening for a time to a deamericanized american blaspheming the country of his birth.

What a drool all this is. Stream of consciousness sort of thing and as usual not a very limpid stream. Que te vaya bien [Fare thee well].

Don Jorge

May 10, 1931
Maipú, Prov. de B.A.

Dearest Mother & Dad—

If anyone ever passes your window saying "¡Marcanudo, che!" you will know you are in Argentina. It means more or less "Gee, swell!" & is purest Argentine without any relationship to Spanish. Someone just passed my window (high & shuttered of course & opening on the promenade of the beaux around the plaza) saying that. I can't quite echo his feelings, but here we are, the band is playing, the Autumn night is clear & only pleasantly cools & the beds look clean & comfortable.

Now I see why no one in Comodoro ever heard of such a wild idea as driving to Buenos Aires! So far we have come something over 1800 kilometers (about 1000 miles) & this is our tenth night on the road. One night at Boliche Nollman, (see sketch map especially for Dad's benefit [since lost—Ed.]), one at the Estancia Lochiel, two at Trelew, one at San Antonio, one at Conesa, one at Anzoategui, one at Bahía Blanca, one at Juárez, & now here at Maipú. With average luck we'll reach Buenos Aires day after tomorrow.

Everything that could happen, nearly, has. We've been badly stuck, mildly stuck, and almost continuously on the verge of being stuck. We've had puncture after puncture until we change tires to the numbers in record time. Our lights have burned out after dark far from shelter. Our steering gear broke while we were going fast. I'll never make an egg-nogg [sic] again, I know how the poor thing feels. Our normal position is half-way from seat to roof, rapidly moving either up or down. We've been lost, & benighted in little inns where cockroaches run over one's face at night.

Nevertheless, here we are, & lucky too. We keep progressing a little every day, & on all of our accidents & getting stuck we've been fairly fortunate in not being killed & in being able to fix things up again within a few hours. Here we intercept the road from Buenos Aires to Mar del Plata, which is said to be the best road in the Argentine—which could be true & the road still terrible. Too many times I've been told the road was fine, & an hour later been stuck in the middle of a sea of mud. However, I'm still hopeful & it hasn't rained here lately.

Now we are in the Argentine of the geography books, perfectly flat limitless pampa, not the high barren pampa of Patagonia but low wet extremely fertile plains with prosperous farms & thousands of cattle. Leaving Bahía Blanca we first passed through a great wheat district, then through a zone of rocky hills around Tandil, then down onto the plains. Everyday the country has been getting richer, more civilized, and, I must admit it, less interesting although more attractive. I find it hard to remember that I am in a foreign country, & have to stop & look for differences.

Of course there are many, but mostly I'm so used to them now I don't notice them. For instance, for a hotel to be of one story with all the rooms opening on a patio is just the way hotels are built & not noteworthy. And Spanish is just the language everyone but Williams & I speaks and sounds more natural to be talked than English would. And for a man in European dress to walk down the main street with a neatly folded blanket (or rather, usually, wool poncho) over his shoulder is just a wise provision against a sudden change in weather.

It has been very interesting watching the slow change from the typical Patagonian pampa near Comodoro to the provincial pampa here. The abrupt change does seem to correspond very closely to the usually accepted limit of Patagonia, the Rio Colorado. South of it, barren dry land covered with thorn bushes & pebbles & without trees. North of it, trees begin to appear & cultivated fields, even outside the river valleys. Around Bahía Blanca are passes through the wheat district, becoming more & more civilized, & Bahía Blanca itself burst on us as a revelation. A real city with handsome buildings, trees, grass, flowers, (palms even), pavements, street cars, running hot & cold water in hotel rooms. I was so overcome that I had my beard removed, & now feel & (it seems to me) look very strange & bare, almost indecent.

I like the Argentine people, even here in the provinces. To them we must seem quite insane & outlandish, but they are uniformly courteous & helpful. When asked about roads, for instance, they explain in great detail & even go with us to point out the right turns to take through towns, & when thanked assure us that they have done nothing & are ours to command. Only, they are hopeless optimists or awful liars for they always say the roads are good, & they are almost always terrible. A great deal of the heavy hauling here is still done with chatas, great carts with wheels up to ten feet in diameter & drawn by ten to twenty horses.

[Letter not finished, although apparently mailed—Ed.]

B.A.
June 13, 1931

Dearest Mother & Dad—

Mother's birthday letter arrived on June 11, five days to spare (but nevertheless the last mail that would have made it I believe). It was very welcome. It must be genius to be able to remember birthdays at all, let alone so long ahead of time.

There's no news now. I wrote you only a few days ago, I believe, & told you the collections were all cleared—extremely good luck. How blue I was when Williams sailed last Saturday—more (to be honest) because I wasn't going than because I was losing him. But lots of people have kept me from getting too lonely, in fact I have had a hard time getting enough sleep.

Of course I'm working hard at the museum, trying to get done, & it's coming fairly well. But it's a tremendous job, & don't count too positively on my getting back in August. At present it looks more like September, with luck. I don't see how I can possibly finish here before August at the *earliest*, & it takes three weeks to get back after I'm through. Paciencia [patience], as we Argentines say.

I am well, as happy as one could be so far from anywhere that one really wants to be in, & no shooting has occurred on the streets for several weeks.

Much love,
George

Buenos Aires
June [1931]

Querida hermana mía:

I suppose that you are now back in the United States (of North America, not of Brazil) and will therefore blend, unite, expand, and contract two previous abortive efforts to write to you. I will also thus kill several birds with one stone, as you can tell the other members of the family that I am alive and well, although not, of course, as happy as I deserve to be or as they would like such a close relative of theirs to be.

In the older of the two abortive efforts, written I don't remember when, there are several recipes for cocktails, but of course those are useless in North America, and they were not intended to try anyway, but merely to make you morally indignant at the idea that it is a widespread habit to drink intoxicating beverages here in Buenos Aires. Therefore I delete those.

The rest of that effort appears to be a compendium of absolutely useless and most uninteresting facts about this city, from which I extract the following characteristic items:

The right driving light of all motor vehicles is green, and occasionally the left is red, which makes one look (in vain) for the sleigh bells.

The Richmond Bar, most admirable rendezvous in our smartest street, has a sign the letters of which turn from side to side. Inebriates who try not to betray their condition by standing so that the letters do not seem to turn invariably land in the gutter. (Rather complex, but you can figure it out if you work at it).

The American (i.e. North American) Club here is very alcoholic and Prince George of England had to be carried out of it the last time he was here. He had never tried mint juleps before and he liked them.[15]

Very few of the inhabitants of Buenos Aires ever saw a gaucho, and most of them are not quite sure what a gaucho really is, aside from being a

15. A mint julep, a favorite in the Old South, is made with Kentucky bourbon, crushed fresh mint, a dash of bitters, ice, and sugar.

caberet entertainer who comes from Naples and speaks Spanish with a strong Italian accent.

Dulce de leche is made of milk and sugar and doesn't taste at all like either.

The two tanks of the Argentine army always bring up the rear so that they won't impede the infantry.

No oysters are produced in the Argentine. All oysters are said to come from Chile, but most of them actually come from Brazil.

It is legal to buy and sell roulette wheels in the Argentine, but it is not legal to use them for gambling. I suppose they must be sold to grind up and feed to horses afflicted with glanders.[16]

Most theaters have several complete and different shows one after the other, and you can usually pay for all or any given part.

There are 30,000 children in the Province of Buenos Aires (outside the city, which is not in the province) who are receiving absolutely no instruction and will have no opportunity to learn to read or write.

The last government purchased two million padlocks at approximately four times the market price, none of the padlocks being found when the new government took over.

To fix a policeman who detects one in the act of traffic law violation costs half to three fourths the usual fine for the offense. The policeman does not hesitate to ask for a bribe if you don't suggest, and is quite willing to bargain as to how much it should be.

Sale [dirty] does not mean sale. Remate, on the other hand, does, having nothing to do with second marriages. Desgracia [misfortune or clumsiness] does not mean disgrace. Use does mean use, but it is pronounced very differently. The Argentine pronounciation of the name of my Alma Mater [Yale] is almost like the English word jolly. Lider [lighter] and mitin [mitten] are English words, and so is futbol.

—You get the idea, it could go on forever.

My other letter to you, written only a week ago and still usable except that it has been carried around and had an accident—having had a long series of short and violent notes in Spanish written on the back, as I had no paper and had occasion to have a written quarrel with someone. One of my boy friends got tight on the party and I was trying to get him away without revealing his condition too fully to the rest.—Well, as I was saying before I betrayed myself into this too intimate glimpse of the famous Bonaërensian night life, my other letter to you was more narrative, more literary, more redolent of my innate charm and lucid style.

It talks, for instance, of a walk on a balmy day through a green park with red paths. Of an old leaf-raker who was worrying as to whether anyone would come to his funeral when he dies. Of a lovely dialogue between a taxi driver and a cyclist whom he nearly ran over, in which the cyclist, a master of invective and insult, reduced the taxiist to a quivering mass of rage with tears running down his cheeks and did it without using a single word that would be bad if removed from its context. Of a lunch at the ambassador's, thrown in to give the thing higher tone.

16. Glanders is a horse disease characterized by fever and swelling of the glands beneath the lower jaw.

Well, that's the sort of letter that was, so now I'll go ahead and write this one.

Last Sunday I arose at the unearthly hour of eight o'clock (and if you will allow me just this once more to indulge my passion for interminable parenthetical remarks, I may say that Buenos Aires has very sound ideas in the matter of time, breakfasting lightly or not at all about ten, lunching at twelve, teaing about six, dining any time from eight thirty to ten, going out for the evening at eleven, when most of the shows start, and to bed about two if there's nothing special to keep them up late; the only flaw in the program being the long time from lunch to dinner, which, however, they make up with tea, by which they are as likely to understand beer and sandwiches as the actual beverage tea) and was driven out to Luján. The driver was an Argentine, who as a nation are perhaps the worst drivers I ever heard of, but was quite good as long as he used only one hand to talk with. The flaw was that he would sometimes be tearing along at eighty kilometers or so and suddenly have to say something that couldn't be expressed without the use of both hands, such as Ah qué esperanza [not at all], or Qué sé yo [I don't know], and then the car would slither around and threaten to end four of the most promising careers in the Western Hemisphere.

Luján is a town some sixty or seventy kilometers from here over the flat and uninteresting pampa and is renowned for its Virgin (by the malicious said to be the only one in the Argentine), its historical museum, and its fossil deposits, which consist of a low muddy stream bank infested by old ladies selling holy medals and young ladies who should buy them. It happens that the miracle which led to Luján's becoming the chief place of pilgrimage in this part of the world took place just where the fossil deposits are. We rambled around and found a bone and watched them bathing horses in the holy and very foul stream. We then went back to town and had some refreshments, I very much wanting beer but asking for tea because I was rather ashamed to reveal my low tastes in such sober company, and the other three then proceding [sic] to order beer for themselves. We then drove home (and I immediately went off and had a very large glass of beer). And that was the high spot in my week.

I have just been enjoying, and slowly recovering from, a visit by Captain Buckham, an English gentleman, sir, and don't you forget it, and one Oliver Claxton, an American roughneck and proud of it.[17] The latter I knew in New York. Buckham is an air photographer and was hereon [sic] a trip around South America by plane, taking pictures for the magazine Fortune. Claxton was along as manager and dry nurse. Buckham breathes through a hole in his throat, talks by clicking his tongue, a language resembling Hottentot which only Claxton understands and he only occasionally, and blows his nose under the front of his collar, all of which makes him an interesting spectacle but a dull companion. They were complete babes in the wood and I devoted too much time to them, as they couldn't leave their hotel without getting lost and couldn't buy a postage

17. Oliver "Perry" Claxton (1900–1959), a journalist, editor, and author of several books. He and later his wife, Dorothy, were longtime friends of Simpson and Anne.

stamp or anything else except by signs which produced the exact opposite of what they wanted. It can be done without, but I do think anyone who is coming to South America for several months should at least learn a dozen useful words of the language. They had a terrible battle with Argentine red tape, which finally reduced them to such a condition that Buckham spent most of his time getting violently red in the face and whistling loudly through his throat and Claxton spent most of his time drinking. I enjoyed their visit very much.

They were my first contact with newcomers, and Claxton insists that I and everyone else who has been here long are no longer 100% North Americans, but half nigger (as the English and Americans playfully call the Argentines, who are as white as they are and many times more polite). He objects violently to being called a North American, doesn't like to have a hat tipped to him, and objects to the use of gestures in talking in spite of the fact that one cannot speak Argentine without gesticulating and that it becomes such second nature that one can't help doing it in English too. Now that they are gone I see only Argentines or people who have become very Argentinized, acriollados [nativized], and I suppose I will get that way too.

If you long for further comments on Argentine life, perhaps you would be interested in an almost literal translation of an obscure item in one of the most conservative papers this morning:

> "In Buenos Aires on the 19th of June, Señores Conde and Gálvez, representing Sr. E. B. Colman, and Señores Martínez de Hoz and Torino, representing Sr. A. H. Cabral, united in the Jockey Club (—the most exclusive in Buenos Aires—) and after exchanging their credentials the representatives of Sr. Colman made it known that the latter considered himself insulted by the expressions used by Sr. Cabral and demanded that they be retracted or reparation made by arms. The representatives of Sr. Cabral stated that he had neither explanations nor apologies to give and that in consequence they placed him at the disposal of Sr. Colman. The representatives of the latter then proposed an exchange of two pistol shots under the neutral direction of Sr. López, the shots to be fired at the count of 120 at a distance of thirty paces.
>
> "At a time and place agreed upon, in the presence of the above mentioned and Dr. Finochietto, the shots were exchanged in the conditions agreed upon. Neither combattant being seriously wounded, the duelists embraced each other, complimented each other on their courage and gentlemanliness ("caballerosidad"), partook of refreshments, and returned to Buenos Aires after signing the following document: 'The duellists are reconciled. Vale.'"

That is the sort of boys these are and the sort of country it is.

My triumph in coping with foreign languages to date occurred the other day when the Director of the museum asked me by phone if I would mind coming over to the administration building and showing some of my photographs and telling about the geology of Patagonia. All innocently I rambled over, to find that there was a meeting of a scientific society and that the bright idea was to project my pictures and have me lecture on them. This I did, impromptu, for over an hour, and apparently they understood it all. I am so impressed with myself that I now admit that I speak Spanish.

I see by the paper that our fossils have arrived safely in New York. Incidentally, whoever gave out the information there put me in a very bad situation here, for the item said that we had 35 boxes of fossils. That is not true, and it led the people here to think that I had held some out on them, an offence for which I could be sent to jail for six months.

Heaven alone knows when if ever I will arrive in New York. The day keeps receding. Now I am supposed to be there without fail by October, but quién [who . . .], in a manner of speaking, sabe? [. . . knows?] It can't come too soon for me. This smiling face hides a broken heart. I am completely fed up, bored, and annoyed with Buenos Aires and the Argentine and live only to leave here.

Hoping you are the same,

Much love to all,
Jorge

Buenos Aires
12 July 1931

My dear father*—

I'm going to write to you in Castilian because you read it very well and because it's good practice for you and for me, for me especially because I don't have anyone here to practice the informal "you" on. Perhaps it will be a little hard for you given that Argentinian, the only Castilian I know, is not genuine Castilian, and that I have learned to speak it, not write it. But you will understand.

Yesterday I received a letter from mother written on my birthday in the museum in New York. I am happy to know that my colleagues still remember my name. At times it seems to me that I have left the known world and that I have neither parents nor friends nor country nor religion, as our Argentinians say of deserters like me.

The work goes as always—little by little. Very often I believe that it will never be done, but the fact is that it is going sufficiently well and what I have to do here, in Buenos Aires, will be finished in about a month from now. Then I go to the La Plata Museum for a while, another month or six weeks. La Plata is not as interesting a city as Buenos Aires and I have no friends there except those of the museum, but it is near Buenos Aires, about an hour by train.

Also it will be more comfortable because my room here doesn't much suit me, there I'll have an almost luxurious room in the museum [illegible] not included! Here living is expensive. It costs me $220 Argentinian (that's $75 American) per month for a small bedroom and food, and in

*In Spanish; translated by editor.

Buenos Aires that is very cheap. Besides I've had to buy clothes, I drink a drink an apéritif from time to time, and more or less twice a week I eat out with my friends.

I celebrated the fourth of July with my compatriots (of which there are three or four thousand here) at a dinner and dance at the Plaza Hotel, the most chic in the capital. We were about 300 people—I went with the American ambassador's secretary, a nice widow but a little old (however, simple friends won't always be young) and with some of my usual companions, named Tobin, Chaple, Robertson, and so on, and I enjoyed myself a great deal. Then the 9th of July holiday with Argentine friends to a Creole barbeque. Argentina, which has too many holidays, has two days of independence, the 25 of May and the 9th of July. It was a deluxe barbeque.

Real Creole food, mutton roasted on a spit, veal roasted in the hide, meat pies, Mendoza wine—but very formal, with tables and chairs and everything arranged according to custom. There were also various speeches. I thought that Argentinians, being of a rather sad demeanor, could not make entertaining speeches, but it wasn't like that. I had a lot of fun there, too.

I received a letter from [Walter] Granger[18] (of our department in the museum at New York), saying that Professor Osborn also wants me to study the other part of the Ameghino collection[19] here. The part I'm studying now comprises more than 250 species! It's a terrible job. And now if I have to study the other part as well, I'll have to spend my life here. Not only do I want to leave here soon, but I also have to be in New York in the month of October of this year. If Osborn insists, I'll have to go there and then come back here again, but perhaps he won't insist because he doesn't want me to spend a long time here either. I am used to things here and I am much more content than I was before, but it still seems to me that the most beautiful view of Buenos aires would be from the deck of a boat heading back to the United States.

I have no more ink, nor more to say. I send all my love to you and Mother—I hope you are well.

Your affectionate son,
Jorge

18. Walter Granger (1872–1941), vertebrate paleontologist at the American Museum who collaborated with Simpson on several projects. Granger was a warm, avuncular person who took special interest in his younger colleagues. In his autobiography Simpson called him "virtually a loved second father." Among other museum responsibilities, Granger was chief paleontologist of the central Asiatic expeditions in the 1920s.

19. The Ameghino collection refers to the extensive series of specimens brought in from the field by Carlos Ameghino (1865–1936) and described by his brother, Florentino (1854?–1911).

[Buenos Aires]
July 25, 1931

Dear Mother, Dad, Martha—

Here are some of many letters I wrote & never mailed—Don't ask why, for I can't imagine. Just an old Argentine custom I suppose. I'm writing today by airmail also, so there is no news.

Best love,
George

[Buenos Aires]
July 25, 1931

Dearest Mother—

With any sort of decent connection this will arrive for your birthday—and carries my most heartfelt love and wishes for you.

A batch of ordinary letters is going of [sic] at the same time by ordinary mail—old ones I'd forgotten to mail.

Of news there isn't a bit—just daily living which is neither here nor there. I don't do much, & nothing of especial interest. I am used to Buenos Aires & rather like it at last, in fact do like it and except for anxiety to see you all again would like to stay here.

Nothing really happens, but all is rather pleasant. For instance last Sunday was a lovely day so I hired a horse & cab & had my very disreputable Jahee drive up and down the balneario for a couple of hours. The balneario being a long avenue along the river, outside the ship basins, something like Riverside drive [in New York City] but right next to the water, longer, not so high, without houses, & more Argentine—well, perhaps it isn't so very like! Juan Pueblo [John Doe] & his wife & daughters take the air there, in swarms. There are also merry-go-rounds, pigeons dyed every color of the rainbow, green grass, sunshine—anyway it was good clean fun. And then went to the City Hotel for tea & then walked home through the streets of the Centro. A very satisfying time but not really anything to write home about even though, in lieu of anything else to say, I am doing so at air-mail rates.

Thank's for L's [Lydia's] unknown whereabouts. Except for the children I wish they'd stay unknown. That's all over and done with—a year of peaceful contemplation of my folly has shown me that although seven years of strife never fully convinced me before—but for that very reason something has to be done about it, & I dread coming back to that.

I think Dad will be better off if I settle with him when I come back instead of getting him a Panama hat. I don't know his size, & anyway they are just as expensive here as in New York, being made rather more than half way from here to N.Y. Also one never knows how a hat will look without trying it on, & Latin styles in hats are often quite different from ours, for men—the women follow Paris, Latin men do too more or less, but American men follow London.

Oh, there is some startling news. I am learning to dance, or trying to, for my progress to date is not very marked. A very excitable little Italian dancing master has undertaken this herculean task. Everyone seems to enjoy dancing tremendously, & I couldn't think of any reason why I shouldn't enjoy it also. Perhaps Marthe will dance with me, & have her last illusions about her angel brother shattered as I am as clumsy as an elephant.

So here's a peso's worth of nothing in the way of news, but of a great deal in the way of love, gratitude, & birthday thoughts. I shall be moving to La Plata about then, & thinking of you very much—

Your son

[Buenos Aires]
Sept. 29, 1931

Dearest Mother:

I just sent off an air mail letter to you via Dad, but now your note of Aug. 25th came. You speak in it of a letter sent the previous day, and that did not come. The mails seem to be all upset. I can't imagine how this letter got here, for it has a B.A. postmark on a date when there was no U.S. mail and hadn't been for a long time.

Anyway, I also received two letters from Lydia this morning, both sent air mail and registered, but did not come air mail because like you she only put 5 cents on, and air mail to here is 55 cents. One is Sept. 15, which must somehow have come down the West Coast to get here so quickly by boat. That makes four letters I have had from her since I left. The tuberculosus [sic] scare is just rot, of course. Since I left she has had the children down with infantile paralysis, diphtheria, tuberculosus, and half a dozen other things, and yet in her last letter says that they are all very well as usual.

She has left New Haven with the three youngest and was in Kansas. She complains about having to travel with them, but does not say just why she had too [sic]. She mentions the fact that her father is very ill and not expected to live, but that is probably not the reason for her trip, for she goes on to say that she is going at once to the Boulderado Hotel, Boulder, Colo. She says the three youngest were in kindergarten in Madison [Conn.], and doesn't say whether they were going on to Boulder or not. She left Hélène at Shadow Lawn School, Cheshire, Conn., but says she is to go back to Mrs. Foote's in New Haven on Oct. 1st. She says nothing about her own plans or date of return to the East—I presume she is returning. She always does eventually. For some strange reason she doesn't mention money—a record. She closes with much love my affectionate wife, and wants to know my date of return. The date of my return to her is never. I have been taken in by loving phrases once too often already.

I very much appreciate your offer to care for Hélène in Washington. I also had the idea that I would try to get Hélène, hoping to put her in school in or near New York where I could see her and look after her and perhaps have her with me during vacations and perhaps have her live with me in a few years when she is older and I can afford a decent place and a servant at least part of the time. It seems only fair that I should have one child. I am lonely without them, and of course I prefer Hélène because I know her so much better, because she is older and needs her mother less, and because she probably gets less care than the others and doesn't seem to get along with them any too well. If I am to have her, it should be as soon as possible, before she is quite formed in character. I still can hardly believe that Lydia is not kind to the babies—it is almost her one virtue, or was.

What I am going to do as soon as I get back is to try to arrange a divorce, amicable if possible—but you know that is not much of a hope. If L. will divorce me but still grant me custody of Hélène that would be ideal and I would do anything she likes to facilitate that and make it easy for her. If all efforts for that fail, the other alternatives are divorce without custody, separation with custody, separation without custody—the latter extremely unsatisfactory, but better than the present status. Surely one of those can be accomplished. If only I had a rational person to deal with— but then, of course, if she were rational this wouldn't be necessary.

She will doubtless do her best to ruin me financially and professionally and she may succeed. Even if she does it will be worth it, and I can always get along. I can even come back here if necessary.

I want to get started on this the second I arrive in New York. I've waited and hoped too long as it is and this will probably take a long time, so the sooner it is started now that my mind is finally made up, the better. Does Dad happen to know a trustworthy and not too expensive lawyer in New York?

All that really comes before consideration of your extremely generous offer. It is out of the question to do anything unless I get legal custody of Hélène, because you know perfectly well that a mere promise from L. would not be any security at all. I will try very hard to get custody of Hé-lène as I can surely take care of her in some form at least as good for her as her present position. Then we can consider together what had best be done with her. The possibility of your caring for her seems to me now far too much to ask or expect, and in any event had best not enter into the matter or be mentioned until, or unless, she is legally mine and not her mother's. If L. gets the idea that Hélène would go to you, she would be just that much harder to deal with. The fact that Hélène is not actually living with her mother now should make the court more inclined to place her in my care. If after all efforts it proves impossible to be free from L. without giving up all the children, I will give them up. In that event it would even be better for them, and personally I don't see how I can go on living tied to Lydia.

Well, until I arrive in N.Y. that is that. I regret the whole thing terribly, and especially regret the worry and sorrow to you. I do hope no additional trouble comes to you during the coming battle. At least it will be an insur-ance for your future freedom from her as well as my own. My deep regret at having turned out such a terrible son is all the more incentive for putting an end to this now.

Much love—my next will be from the boat or from New York.

George

AMERICAN MUSEUM OF NATURAL HISTORY 1931-1932

THE FIVE letters that follow cover about one year, from the time Simpson returned from South America in October 1931 until a year later when Martha returned from France and they took an apartment together in Greenwich Village. During this year Simpson lived alone in New York City while Lydia was still moving from place to place with the children. Shortly after his return, Simpson filed for a legal separation from Lydia. This was easier to obtain in the era before the no-fault divorce than an outright dissolution of the marriage because of the limited grounds for divorce in New York State at that time. In early 1932 Simpson was granted a legal separation and custody of all four daughters. Helen remained in boarding school in Connecticut; Joan and Elizabeth went to live with Simpson's parents in Washington, D.C. Gay, the second eldest, had been diagnosed as having a serious congenital heart defect and went to live with Lydia's parents in Kansas.

In addition to the separation in 1932, Lydia's father Joseph had been seriously ill (he died the following year). These misfortunes aggravated her mental instability and she was committed to a hospital by year's end. She had been institutionalized several times before and would be yet again after this crisis.

Anne Roe was completing her aphasia research in Philadelphia, where she had moved following her separation from her husband. When the clinical phase of that work was completed, Anne moved back to New York City where she began analyzing the results of the study. By now she and Simpson were seeing each other regularly and somewhat more openly than before.

In the summer of 1932 Simpson left to do fieldwork in central Montana in the Paleocene-age Fort Union strata of the Crazy Mountains. These rocks contained fossils of the early Cenozoic evolutionary expansion of mammals. Earlier research had been conducted there by James W. Gidley of the U.S. National Museum, but after his death in 1931 the work was taken over by Simpson and the American Museum. Simpson also began a long series of publications on his South American fossil mammals as well

173

as continued writing the results of the work he had done in the southern U.S. several years before.

There are no letters for the period from October 1932, when Simpson and Martha shared an apartment in the Village, to October 1933, when Simpson departed once again for South America on the second Scarritt expedition. Presumably he kept up his weekly letter writing to his parents in Washington, but none of those letters survive.

174
AMERICAN
MUSEUM OF
NATURAL
HISTORY
1931–1932

[New York City]
Dec 12 [1931]

Dearest Mother & Dad—

I've been pretty well on the run, but things are perhaps clearing up a little.

L. arrived here Thursday afternoon—telegraphing to "meet without fail," which of course I didn't do. She brought the babies—told Mrs. Shreve (who had them in Boulder) that she had to, as she couldn't & I wouldn't pay their board. Mrs. Shreve pointed out that I *had* paid to date & said I would continue to do so, but L. just got mad & took them anyhow. She stopped in Kansas & her mother wouldn't let her stay there at all so on she came. Thursday she just phoned the Museum, Creighton, etc., but all denied any knowledge of my address or plans, so she made a date with the Museum director for Friday afternoon. I went off to my lawyer, & Friday (yesterday) morning made up a summons to serve on her.

Yesterday she came here about 1 P.M. I was trying to find Granger or Mrs. Lord to serve the summons on her, & ran into her in Mrs. Lord's room, but she didn't realize it was I until about half a second too late & I got away without a word being said.[1] That's all I've seen of her. I gave Granger the summons & when he saw her, he held it out & said that was all I had to say to her. She refused to take it & it fell on the floor, so *she* picked it up & threw it on a table. Granger left, & she picked up the summons & ran out with it to give back to him. He informed her that the summons was now legally served on her, and she proceded [sic] to say that according to the law it had to be served by a legal officer & she had to read it. No arguing with her, of course—she "does not consider that she has been summonsed." Unfortunately for her she can't make up laws to suit herself, and in New York State a summons is served if anyone who knows the defendant beyond any doubt (Granger & Mrs. Lord) see [sic] her with it in her possession (as it was twice, when she picked it up & when she left the room with it). That's all I wanted her in New York State for, & my purpose in trying to get her here is served.

They had quite a time with her. She started by sobbing about being deserted, penniless, starving, mistreated by everyone. I turn all her friends against her, lie to her parents until they turn against her, even her own children are turning against her as they get old enough to understand my

1. Mrs. Rolfe Lord was secretary to Walter Granger, curator of fossil mammals at the American Museum.

lies, she is ill, hungry, unclothed, etc. They told her they knew her stories were not true, so she switched to her other mood & accused them & all my friends of everything from lèse majesté to bigamy & from murder to stealing pennies from cripples. Whereupon even normally gentle Mrs. Lord got angry & told L. exactly what she thought of her. L. then decided to go get a policeman & have Granger arrested, but Mrs. Lord asked her on just what grounds she expected to have an arrest made & if she was sure the police wouldn't take her instead, so she thought it over & for a wonder realized that wasn't a good idea. So finally, about 4:30, she said, well, if I had anything to say to her I could come & say it & not think she'd go to my lawyer because of a silly unserved summons, & left.

175

AMERICAN
MUSEUM OF
NATURAL
HISTORY
1931–1932

Naturally I have nothing to say. She is now served. That gives N.Y. courts jurisdiction & gives her twenty days to enter a defense or countersuit. If she doesn't, she is in default. My own suit is being entered today or Monday. The twenty days dates from the moment of service of summons. I am suing for legal separation with custody of the children. If she returns & starts to battle here again, I'll get an injunction against her doing so.

Contrary to my lawyer's advice, I gave Granger $10 to give her if she hadn't any lodgings, but she refused to take it. She claims not to have a penny, but probably has plenty as she had $1500 a few weeks ago & seems not to be cramped. She spent Thursday night at a good hotel & had a woman caring for the children. When it was pointed out that she was known to have funds enough for several months, she said she had been robbed again. She probably has a lot put away somewhere.

I don't know where she is now or what her plans are. She talked some of going to New Haven & getting Hélène—I telegraphed at once that Hélène was not to be given to *anyone* except by court order. She said she would live in New Haven, then said she was going away somewhere with the children & earn her own living, another time said she was going abroad, another that she would stay on at the Wellington Hotel here until her suit is settled, and so on.

She said she went to Kansas because she was summoned as a witness in a lawsuit, then she went because her father was dying, then that she went for Gay's health, but she merely got angry when asked why she went to Colorado, & furious when asked why she then left the children & went on to California. She said she really left Kansas because Mrs. Lord & I had alienated her parents' affections!—And so on, the usual monstrous mess of lies & contradictions. She must be worse, because she didn't succeed even in telling a consistent story to the same person twice in succession, which she used to do.

While all this went on, I lurked on the floor below & received bulletins from time to time, leaving only to telegraph about Hélène when she announced that she was going to get her.

She doesn't know my address yet, & will have trouble learning it. Only a few people do know it, & they won't tell. Eventually she will probably trick it out of them or have me trailed, if she decides she really wants to know where I'm living, but for a time I'm safe there. Of course she can get me at the Museum whenever she really wants, but for the present probably will leave me alone, as I still think she believes the best thing is to keep out of my clutches, not aware that all I needed to do with her is already done, until judgement is given by the court.

I don't think she'll harm the children particularly. She knows that if she would leave them anywhere, I'd see they were cared for, & if she's really broke she has only to make application for temporary alimony pending outcome of the suit.

—I just called the Wellington—she left there last night or this morning, She is probably in New Haven.

Well, that's enough about her.

I enclose Dad's check for $3.00 to the Paleont[ological]. Soc[iety]. I find that I don't have to pay dues any more, because of being a fellow of the Geol. Soc. & of the Pale. Soc. Fellows don't pay dues in the latter, so that's that. You still haven't given me an accounting of what I owe you— I'd like to pay it now.

I've seen Lois & Creighton this week, had dinner with each. Aside from that I haven't done much of anything, not even Museum work, except lurk in hidy-holes, and dash back & forth between my lawyer & the Museum.

I'm quite well, although a little blue & can't sleep, but that will probably pass soon.

It looks very doubtful whether I can bring Hélène for Christmas. The poor kid will be heartbroken if I can't. Anyway, I'll come myself if possible, & there still is some chance I can get her.

Much love to you both, regards to Roes & Farringtons.[2]

George

(horizontal rule with a small black bar in the center)

[New York City]
Feb. 24 [1932]

Dearest Mother—

Lydia signed the [separation] agreement at 2 today, after a final three hour battle. I'll send you a copy later. The main provisions are:

1. That we are separated and neither can interfere with the freedom of action of the other in any way.

2. That no suit for restitution of conjugal relations can be brought.

3. That I pay Lydia $100 per month.

4. That my museum insurance stays in her name unless she remarries.

5. That all four children are forever in my custody, with no restrictions except—

    a. That L may visit them twice a week at most

        aa. That they shall never be in charge of her parents.

    b. That she may have them for one night at most once in two weeks.

    c. That she may have them one week at most in six months, all

2. Anne's brother Bob was a chemist working in Washington with the Food and Drug Administration. Marvin Farrington was a close friend and associate of Simpson's father.

176
AMERICAN
MUSEUM OF
NATURAL
HISTORY
1931–1932

expenses to be paid by her and she not take them over 15 miles from their place of residence, nor to interfere with their school.

d. Before having them or seeing them as in a and b she must give one day notice and ten days before having them as in c.

6. That I am not responsible for any further financial obligations whatsoever on account of L.

7. That if a divorce is obtained by either party, the terms of this agreement shall be incorporated in the divorce.

8. That if L remarries all financial provisions for her shall cease immediately.

9. That any further instruments necessary to give force to this will voluntarily be made.

So that is that. Every word represents a struggle. There isn't a single sentence we haven't had a long fight over. It isn't ideal, but it is infinitely better than what she wanted and pretty surely better than a court would have given me, and without the dirt & struggle and expense of bringing the action to a hearing.

She still talks of going to Nevada, but I don't urge her. I think she probably will. I'd have nothing to gain by it now, really, as I don't want to get married and nothing else would be changed. She has no dower rights.[3]

I am well & happy—more later—I just dash this off at my desk after returning from the combat—Love—

George

Helene: ☒ Gay: ☒ Joan: ☒ Betty: ☒

177

AMERICAN
MUSEUM OF
NATURAL
HISTORY
1931–1932

Harlowton, Mont.
Sunday, June 26 [1932]

Dearest Mother & Dad—

Here I am established in Montana & with 2 day's field work done. Today I'm writing letters & writing up notes & other information from my host & employee, Silberling.[4] Tomorrow we leave for a three day reconnaisance along the flank of the Crazy Mountains—good looking mountains, too, the highest in the state. We'll have to use horses & I'll be sore. The geology here is *very* confusing. I think I'll get it fairly straight, but it's lucky I came, as it would be quite impossible to do anything without studying it here in person.

This is not the usual type of fossil country—no badlands to speak of, but vast stretches of rolling hills & valleys, relief of about 3000 feet but everything so spread out that really steep slopes are few, and almost all grass covered & very green now, as this is a wet year. There are few trees, a few cot-

3. Dower rights is legalese for claims that a wife can make against her husband's property.

4. Albert C. Silberling was a native Montanan who had worked earlier for Gidley in excavating the Fort Union fossils. When Simpson took over the project, Silberling became his field assistant. When not collecting fossils, Silberling worked for the railroad in Harlowton.

tonwoods in the coulees and pines on some of the higher rocky ridges. Sheep & cattle, mostly sheep. Everyone broke. Very sparse population. These are great open spaces with a vengeance. Rock exposures are small & widely scattered, which makes study hard, & the rocks I'm concerned with are about 6000 feet thick. You have to travel thirty or forty miles to see much of a section —altogether the biggest sort of stratigraphic job, & it's no wonder it is so poorly understood.[5]

178
AMERICAN
MUSEUM OF
NATURAL
HISTORY
1931–1932

The Silberlings are nice people—Mr., Mrs., two boys & two girls. They are awfully worried, at least Mrs. S. is, that I am citified & will scorn their humble manners, but I think I'm rapidly teaching them better.

Silberling wastes no time. I arrived here Friday at 6:45 A.M. and at 8 we were miles out of town at a fossil locality.

After the Crazy Mtn. trip I'm going up north along & between the Belt & Snowy Mountains, then down to the Yellowstone River and over to Red Lodge, each a trip of several days. Then I'll come back here, make detailed sections through the best Lance & Ft. Union exposures, then prospect the best looking fossil prospects. That will round out my work all right.

As you know by now, I ran into Lydia on the station platform in Washington. She pursued me up & down & into & out of the train. I fled until you had time to get away, then I ditched her and concealed myself till the train pulled out. When I went to my seat, I found this note on the back of my Sat.[urday] Eve.[ning] Post, "Why don't you go to Reno yourself? I shall not write you *about this*. Send checks to me at address which I shall forward to you as soon as I have located in Washington."

I suppose everything is to be arranged by mental telepathy since she won't write about it. Now, perhaps I *could* go to Reno & spend 6 weeks there.[6] I shouldn't take the time, but I haven't had a vacation for several years and I probably could manage if it really would end things more efficiently. But only if I had more guarantee that she would file an appearance there. If she should or would make out & sign whatever is necessary to empower an attorney there to appear (but no contest) for her, I would probably do it. That is pretty well up to her. If she won't write me, I have no way of knowing what is what, & of course I won't go to Nevada on the chance that she would appear & not contest. If she does nothing further, of course I shan't either, but will simply return to New York, probably about Aug. 1st or a little later. Perhaps it would be well to get this information to her—possibly through Marvin?[7]—She obviously is anxious to have a divorce, but she'll have to let me divorce her, & she'll have to behave herself.

5. Because the sedimentary rock strata in this region are only very gently dipping, one has to go some distance laterally to move up vertically in the pile of layered strata containing the prized mammal fossils.

6. In the days before liberalized divorce laws, people commonly went to Nevada—where the laws were more lenient—to file for divorce after having "established residency" by living there for six weeks.

7. Perhaps Marvin Farrington.

Well, so it goes. I am well. Water upset me for a day or so, but now am fine, eating like a horse, & falling asleep as I hit the bed. Sunburned, of course, but not too painfully.

Much love to Mother, Dad, Helene, Gay, Joan, & Betty—

George

179

AMERICAN
MUSEUM OF
NATURAL
HISTORY
1931–1932

Harlowton, Mont.
June 26?, 1932

Dearest Martyr—

Thanks for your irregular birthday letter. This is one too, & very irregular, being written about four days after your birthday. Anyway, much love & many happy returns, of the *same* birthday or some previous one.

As you doubtless are aware, I have had quite a spell of alarums & excursions. I am a little embarrassed—ambiguities in a previous letter might well give you the impression that my last battle was well deserved. In fact, believe me, it was *not*. I don't know what happens next. You'd think everything possible had happened, but I've thought that so often that no[w] I only expect the unexpected worst.

Anyway, I fled to the great open spaces & here I am in them, once more enjoying sunburn, wind, & open plumbing openly arrived at. I'm getting all healthy again rapidly, climbing rapidly over sandstones & andesitic shales of Fort Union age all day & sleeping like a baby all night. I eat several bowls of oatmeal & stacks of pancakes for breakfast—that will give you some idea. This is an interesting & very difficult job I have here, but you wouldn't care about that, much.

You would, however, like this country. Vast stretches of bright green rolling hills (as high as mountains but with such gentle slopes & on such a broad scale that they look like hills) & in each direction great snow-clad peaks rising abruptly along the horizon. Tomorrow I'm going for a three or four day trip to the biggest of these mountain groups—the Crazy Mountains—on horseback. Will I be sore!

I don't know when if ever I'll go back east.

So, I hope you did have a happy birthday—

Your little brother

New York, Aug. 10, 1932

Dearest Martyr—

I have written you at least three times and mailed the letters, and various other times perhaps not mailed as I don't remember positively. Probably this will stray, too, as Rue Boulard is the address I have and you're at St. Germain sleeping in a different room every night and doing voodoo in a Mediaeval chapel.

For instance I did say, positively and at enormous length, that I wished that you could be here & we could have an apartment together this winter. I think my lease is up Sept. 15, but don't remember for sure. I'll find out. Anyway, we can manage whenever you come, as long as it is not more than a month or two later than that—the only thing being that I don't want to move twice or have a larger apartment alone for more than a month or two.

180

AMERICAN
MUSEUM OF
NATURAL
HISTORY
1931–1932

As I also pointed out two letters ago, there are certain disadvantages for you in such an arrangement. I am extremely untidy, & you are too, so our place would look like hell most of the time. I am moody, nervous, and (as many testamonials could be adduced to show) no picnic to live with. My habits are irregular and will doubtless continue so. I like to be free to come & go and behave as I please, but of course I would try not to interfere with you and expect you to be equally free with no prying or complaint on my part. Worst of all, my personal troubles have a way of intruding beyond my control, on everyone around me, & would be very trying for anyone living with me unless they are settled by then as they probably won't be. In short, it would be swell for me to have you, but I'm not at all sure you would like it. I do hope you decide to try it anyhow.

I'll do nothing definite till further word, but will make some enquiries. As I see it, we need two small bedrooms, a larger common room with good north light, some provision for boiling water and toasting bread, and a modern bathroom. I believe this could be had for $60–$70 per month now, with a little searching & luck. I don't want to be over 30 minutes from Museum, & gather that location is not very important to you, and that neither of us cares much about neighborhood so long as gangsters don't fight in our hall.

So I most urgently invite you to come and try to put up with me—

My humble apologies for insults to your intelligence—the most intelligent people I know take *no* interest in my bones. It's one of the signs of their intelligence. But if you really want to know, I now have five papers in press, *videlicet*: Some very nice horses, *Parahippus leonensis* and *Merychippus gunteri*, closely related and showing in a way[,] giving rise to some philosophic discussion of taxonomy, the transition between genera, with some sweet dogs of the genus *Cynodesmus*, etc.

Fossil Sirenia of Florida and the evolution of Sirenia: *Hesperosiren*, new genus, with little depressed toothless rostrum, and the whence, why, & whither of sea-cows any way.

Enamel in an Eocene Patagonian Edentate: The first known edentate with real enamel on its teeth, are you excited?

Dinosaurs and the Tertiary of Patagonia: There are *no* Tertiary dinosaurs in Patagonia, & so Ameghino and Roth & others are dirty liars, and there was no mountain building between Argiles fissilaires and *Notostylops* time, and there are Tertiary mammals in "Cretaceous sandstones" (everybody lied), and there is a queer mammal *Florentinoameghinia mystica*, new genus and species.

New & Little Known Mammals of the *Pyrotherium* & *Colpodon* Beds of Patagonia: *Micrabderites williamsi*, new genus & species; *Microbiotherium hernandezi*, new species; *Halmarhiphus riggsi*, new species; *Proschismotherium scarritti*, new species; and more complete specimens of Ameghino's species *Abderitas crispus* and *Perimys incavatus*.

And I am now finishing papers on some fossil Patagonian ungulates, on a new and big fossil snake, on three new skeletons mounted under my direction, on the nomenclature of the Monte Hermoso tucutucus, on *Cochilius*; on the Ft. Union of Montana, and others.[8] More fun! Wish you were here. The fossils sleep under blankets every night, but I don't.

Much love—

Tu hermano lejano y no muy intelligente[9]

181

AMERICAN
MUSEUM OF
NATURAL
HISTORY
1931–1932

8. Simpson, of course, is enumerating the various and sundry fossils that he was in the course of describing and interpreting. *Parahippus* and *Merychippus* are Miocene horses, the former an advanced browser, the latter an early grazer and the species named after Florida's state geologist, Herman Gunter. *Cynodesmus* is a Miocene carnivore on the main line of dog evolution. The Sirenia are an order of mammals—sea cows—to which the genus *Hesperosiren* belongs and which Simpson named in 1932. Edentates belong to one of several orders of anteaters among the mammals.

At the turn of the century some confusion was created by the Ameghino brothers and Santiago Roth who reported dinosaur remains in South American strata some fifteen million years *after* dinosaurs had become extinct elsewhere in the world. Florentino Ameghino (1854–1911) was an Argentine paleontologist who pioneered the study of South American fossils (collected by his brothers Carlos and Juan). He became director of the National Museum in Buenos Aires during the last decades of his life. Santiago Roth (1850–1924) was a Swiss geologist and vertebrate paleontologist who had settled in Argentina. He was director of the paleontological section of the museum in La Plata. Florentino Ameghino tended to claim that Argentine strata were older than they really were, thus also making the fossils they contained older and giving them greater venerability compared to similar fossils elsewhere in the world. Simpson therefore reported Tertiary mammals from what were once considered older, Cretaceous-age rocks, but in fact the strata were not that old. *Florentinoameghinia mystica* was a new genus and new species of an early Eocene South American marsupial of uncertain relationship; the mammal was named by Simpson after Florentino, who had previously named a fossil *Carloameghinia* after his brother.

Pyrotherium and *Colpodon* beds of Patagonia are of Oligocene age with new species, named by Simpson, of rodentlike and oppossumlike marsupials; of a ground sloth (named after his benefactor, Scarritt); and with better examples of rodentlike and chinchillalike marsupials. *Cochilius* is an Oligocene rabbitlike typothere, a South American hoofed herbivore.

9. "Your faraway and not very intelligent brother."

SOUTH AMERICA
1933-1934

S IMPSON APPARENTLY realized in 1933 that his life was
at a significant turning point, for in the summer of that
year he began his autobiography. Although he wrote only for a private record, his fifty-five handwritten pages did become the basis for the autobiography he published many years later.

In October 1933 Simpson left a second time for South America with his
museum associate, Coley Williams. Horace Scarritt was once again the
chief financial backer. The twelve previous months had been a time of
respite for Simpson from domestic problems. He was legally separated from
Lydia, the children were in his custody and being cared for by his mother-
in-law and parents, and he and Martha had set up housekeeping in a
Greenwich Village apartment.

Anne too was separated from her husband and would soon be divorced.
She completed her work in Philadelphia and took a position with Worcester State Hospital in Massachusetts as a psychologist in charge of intern-
ship training. Although paid less than $14 per week plus room and board,
she was grateful to find a job in these depression years and eager for the
opportunity to work with David Shakow, a distinguished psychologist.
Anne's health was not good during this period and she suffered another
severe attack of her still undiagnosed brucellosis during her stay in
Worcester.

Simpson and Williams made the trip to South America by passenger
boat, passing through the Panama Canal and then traveling south along
the west coast of South America. They then flew across the narrow waist
of the continent to Buenos Aires in the east. On the trip south Simpson
worked on the galley proofs of *Attending Marvels: A Patagonian Journal*,
which was published the following May by Macmillan. Keeping a travel
journal had by this time become a habit that he would continue for the
rest of his life. Besides enjoying the activity itself, Simpson used the journal to supplement his correspondence with his far-flung family.

When his work in Buenos Aires was finished, Simpson went to Paris in
the hope of traveling on to Mongolia in central Asia where so many interesting fossils of Mesozoic mammals had been discovered by the American

Museum's expeditions in the twenties. Before leaving for Moscow, where he planned to obtain permission to continue to Mongolia, Simpson rendezvoused briefly with Anne and Martha in Paris. He did go on to Moscow, but after six weeks of bureaucratic difficulties Simpson realized that he was not going to be allowed to enter Mongolia. He returned to New York City and the American Museum in the summer of 1934.

While Simpson was away, a stream of his publications appeared as manuscripts prepared the previous year made their way into print. Before he left for his second trip to South America, Simpson, now thirty-one years old, calculated in his autobiographical notes that he had published ninety-four articles and monographs, a total of two thousand printed pages—all in the space of seven years between the receipt of his Ph.D. from Yale and his departure for South America. During this second trip to Patagonia, the chief additions to this ever increasing body of scholarly writings were articles devoted to his previous findings in South America as well as those on the fossils from the southern United States.

[En route to South America on
second Scarritt expedition,
October, 1933]

Dearest Marthe:

I discovered your letter Saturday evening when I unpacked my things, and I was very pleased and, truth to tell, deeply touched. You seem almost apologetic about your housekeeping—if anything I should be apologetic. I am an extremely useless thing to have around the house, and I made many demands on your time, which is more valuable to you and to the world than mine could possibly be. You kept me elaborately comfortable and extremely well fed. What more could a housekeeper do, even if meant to be a housekeeper?[1]

But that is not why I feel so deeply grateful to you. There are more other reasons than I can say, and I guess you know them at least in part. I have never been so happy as during the last year. I doubt whether very many people are ever so happy in their whole lives. It was grand, really ideal, the way human beings are meant to live, and almost never do. I can think of no better wish for the future than that our household may be reestablished. Sister or no sister, you are a grand person, and also I am proud that I should have a really great artist for a sister.—This is running off into the vein of the admiration society, but it could hardly help it.

The slippers are swell, and my little Chinese steward has great fun hiding them every day.

Many thanks, and much love—

Geege

1. Martha had returned from France the previous year with the explicit intention of looking after her brother, who had been living alone for the year since his return from his first trip to South America. She cooked and kept house for him, but more important, she provided emotional support following what was perhaps the most depressing period of his young life.

[En route to S. America,
October, 1933]

Dearest Mother and Dad:

Your gifts are very deeply appreciated. Mother's, in case she doesn't know, is "A Traveler's Library," a collection of all sorts of novels, stories, essays, and poems in one volume. Practically all are new to me, and those I have read are extremely good. At the present rate this one book will last me the entire voyage. And my dazzling trousers are the envy of all. I wear them on all my public appearances, adding my new double-breasted coat for evening formality—no one dresses for dinner.

I miss you both very much. Passionately fond of travelling as I am, it is always a little sad, especially during these first days, because it means being so far away from all the people I love. If only I could bring everyone with me! And then too, it seems unfair that Dad, particularly, is not taking this trip with me, or even instead of me, because he has wanted to for so long.[2] I really will try to make him see it through my eyes, but at best that is a poor substitute for seeing it on his own.

I have the feeling that this is going to be the grandest experience of my life so far. It will be hard, possibly just a little dangerous in spots, but ceaselessly interesting, and I know that I am going to make it successful and that I am coming back safe and sound for a grand reunion next year. There is now practically no doubt that I am going on to Mongolia, or at least to Moscow to try to get through to Mongolia, for, as Dad knows, Scarritt has definitely pledged the money.[3] Surely no one ever had such a trip to look forward to as I have for the next year—to both of the two most remote spots on earth, Patagonia and Mongolia, the West Pole and the East Pole.

I know, Mother, that it is hard on you to have me go off this way, and that it is your sacrifice and your caring for the children that makes it possible.[4] I do not know how I can repay you, but I deeply appreciate it, I love you very dearly, and what I can accomplish for science and for the richness of human knowledge is due as much to you as to anything or anyone.

With much love to you both,

George Gaylord

2. Five years later Simpson's father did accompany his son and new daughter-in-law, Anne Roe, on the sea leg of their eight-month expedition to Venezuela.

3. Although the Mongolia trip did not work out, Scarritt had other opportunities to support Simpson's research, in particular his work in Montana on early Cenozoic mammals.

4. His parents were caring for Joan (just turned six) and Elizabeth (almost five) in Washington, D.C. Helen (almost ten) was at school in Connecticut and Gay (seven) was with Lydia's parents in Kansas.

Dearest Family—

Sometime soon we will be in Talara [Peru], & a north-bound Grace [Line] boat will soon pass & can take mail, so here's a note.

The Panama Canal resembles portraits & is a clean, instructive sight for young & old, which cannot be said of Colón or Panama City. They are picturesque, but like most picturesque places they smell. The population, to my surprise, is almost 100% negro, and aside from some elegant bars the shops, where they sell everything from diamonds to women, are tiny rooms without windows opening to the sidewalk by doors usually with dirty calicoe [sic] (or baze [sic] or dimity or something) screens. The upper floors all have balconies from which at any moment are thrown all sorts of things from sailors to—well, all sorts of things. This is not an exaggeration; a sailor was flung down into the street while we were there, & it was not considered unusual. We were disappointed at Panama City as we were there only a couple of hours & at night, so couldn't see the ruins or other sights. The famous night-life is not diverting, not even dashingly wicked.

Buenaventura, our Colombian port, was burned down not long ago & rebuilding has been haphazard & rather beside the point. It is a little dump, with one big, dead cavernous hotel, for passengers en route to Bogotá (reached by railway plus motorcar), & a lot of straw-thatched bamboo huts for the stevedores. Guayaquil, Ecuador, which we are just leaving, is a bigger & more interesting place. It used to be one of the pest-holes of the world, but now yellow-fever has been stamped out and they have only the usual diseases plus bubonic plague, leprosy, & a few other nasty things. From the river, where the boats anchor, it looks like a Hollywood set, dazzlingly tropical & clean & white. This is mostly sham, but it is an interesting town. We were there all day & most of the night, so did it very thoroughly, even getting them to open the museum for us, a funny little place, with a jumble of junk—a horse's tooth, a pen once belonging to [former president] Herbert Hoover, an obscene Peruvian pot, horrible chromos [or colored prints] of local patriots—all flung in together. There are no piers in Guayaquil & everything is lightered off, passengers going in boats with old automobile motors. The current is so fast that rowing is impossible.

At all these places everyone in town swarms over the boat, & the elite eat on the boat, which must be a pleasant change for them. At Guayaquil they also tried to sell everything conceivable, from shrunken human heads to an ocelot kitty. Someone bought the ocelot, & the poor thing is now whining its heart out in a cage on our upper deck.

These tropics are cold. Except in town along about noon with the sun out, the weather has been downright chilly. I needed a blanket the night we crossed the equator. I have not been sea-sick since the second day.

Stops cut into work, but I have completed Chapters IV & V, written VI, VII, & VIII, and am well started on IX.[5]

Much love to you all—
George

Talara is in sight—Looks like Patagonia!—Bare cliffs & tin shacks.

G

187
SOUTH AMERICA
1933–1934

Buenos Aires
Oct. 3, 1933

Dearest Mother and Dad:

Here I am, safe and sound and hard at it once more. We arrived here from Santiago [Chile] three days ago and since then have been extremely busy. I haven't time to write much, but must at least get a line off on tomorrow's air mail, the first since we got here. For the same reason, and also finances, will you please circulate this letter widely?

The trip down the West Coast was really splendid. We went ashore in almost every port, missing only one or two where we stopped only at night. In Lima I thought that we had run into a revolution, but nothing much came of it, just the usual street riots and only a few people killed. The city itself is fascinating, although I was disappointed that we nowhere saw any prehistoric ruins. Lima is about three-quarters churches, some of them very nice.

The boat was late so that we were in Santiago only a day and a half, but we improved that time. It is the best city of its size that I have seen in S.A. Clean and very active, and with the Andes laid out as a backdrop.

Last Saturday we flew over [to Buenos Aires], and I rate the experience as the grandest of my life so far. We left in a dense fog, then climbed rapidly above it, leaving the whole valley a sea of clouds below us. The plane cast a clear shadow on the tops of the clouds, and this was surrounded by a rainbow. The cordillera rose above the mist and was as clear as crystal. It was such a good day that we went over at 17,000 feet—sometimes they have to fly as high as 25,000. The altitude, the highest that I have ever experienced, made me a little dizzy but did not otherwise bother, and I did not even use oxygen, which is provided for those who need it. Three or four times we hit a pocket and dropped like a plummet several hundred feet, a nasty feeling, but it was smooth on the whole. The scene is indescribable, and so I shall not describe it. Just imagine the highest mountains you *can* imagine, rising from the planeover [sic] twice as high as does Pike's Peak [Colorado], a perfectly clear day, the mountains snow covered but with enough bare rock to bring out their forms. We flew very close to

5. Presumably Simpson is referring to galley proofs rather than the initial draft of *Attending Marvels*, for in the afterword of the latest printing of *Attending Marvels* Simpson says he wrote the book in 1932 and early 1933. Moreover, given that the book was published some six months later, it is more likely that the book was in galley-proof stage rather than first-draft stage.

Aconcagua, the highest mountain outside of the Himalayas, and right over the Christ of the Andes.[6]

Over the pampa the going was much rougher than over the mountains, for some reason, and I was terribly air sick, but even there I enjoyed the flight. Eight hours altogether.

Buenos Aires is like home and certainly looks good to me. Everyone is so cordial. It seems to me that I know everyone, and cannot stir out of the hotel without meeting friends, and I am constantly being embraced by my (male) Argentine amigos [friends] and socios [associates]. I have to finish this in a hurry because they are having a dinner for me this evening, and I have just come back from a long day at the Geological Survey—where, incidentally, I found that one of my best Argentine personal friends, Tomás Ezcurra, has been made director, sothat [sic] I own the place. Doello Jurado, director of the museo, is also extremely cordial.[7]

In spite of all this, there is delay about the customs and permits. The President himself cannot do anything rapidly here. At least I know the ropes now and its [sic] just a matter of paciencia—the most used of all Argentine words with the possible exception of mañana.

So—I am very well, happy but lonely, and I love you all very dearly. If you were only here I would be completely satisfied with life. I received mother's air mail in Valpo and one from the girls, Martha, and Anne here in B.A., and they cheered me very much. Thank you.

> Yours,
> George

Kisses for all (divide fairly)

xxxxxxxxxxxxxxxxxxxxxxxxxxxxxxxxxxxxxxxxxxxxxxxxxxxxxxxxxxxxxx

---

> [South America? 1933?]

Dearest Marty—

I've delayed a couple of weeks in getting the last of my journal off, but here it is.

No sense writing news here, as I'll get an air letter off to someone with all the news within a day or two, & that'll reach there a week before this does.

> Much love—
> Gg

6. Aconcagua, on the border of Chile and Argentina, is the highest Andean peak (more than 23,000 feet). Christ-of-the-Andes is a twenty-six-foot-high sculpture of Christ on a thirty-foot base by Mateo Alonso; it celebrates settlement of an age-old border dispute between Chile and Argentina.

7. Martín Doello Jurado (1884–1948), Argentinian invertebrate zoologist and paleontologist.

Comoro Rivadavia
Sunday, the day before
Christmas
[1933]

189
SOUTH AMERICA
1933–1934

Dearest Marthe—

I enclose an installment of journal which is, I think, especially juicy. Here I sit, in Bruzio's hotel, the old Colón, as if nothing had happened, just as I did three years ago today, & feeling rather desolate & somewhat sorry for myself. What a God-awful hole to be in for Christmas!

Well, late hours probably have something to do with my mood, as I was up till four, considered a very reasonable hour here, but not to me after going to bed every night with the sun for so long. Coley has gone off with Cameron & Tobey & they are probably tight again now in one of the local bars, so I am improving the occasion to dash off a few notes. Coley is— well, I'll control myself, besides even you would be surprised at some of the nasty words I know. He is a trial, let's say, & let it go at that.

I just received your air-mail letter of Nov. 18th, & with Anne's of Nov. 21st that is my most recent news. Yours reached B.A. on Nov. 27th, fair time, then went to Trelew, which it didn't reach till Dec. 10th, thence here on the 14th, & I just arrived to claim it. I am relieved to hear that my ms. ["Attending Marvels"] got there safely, & surprised, flattered, & delighted at your enthusiasm about it. I only hope it is so shared by a few thousand less biased people!

About the drawings, in the first place, as you see, I couldn't possibly get them to you soon enough. In the second place, homalodontotheres, pyro-theres, astrapotheres, & most of the local brutes have never been restored properly & I don't feel equal to making valid caricatures of them.[8]

It sounds like a full & interesting winter for you, with all your exhibitions & one thing [and] another, & I pant for news of them. You don't seem to have giving [given?] any dates, but I gather that the decorations, at least, are in full swing by now.

Well, much love, & que te vaya muy bien—

Barba roja [Red beard]

I reopen to include check for $50.00 U.S., saved by the lack of things to buy at the Tapena de Lopez. For the family if needed; if not, for you if needed now; if not, toward a reunion in Paris next spring.

G.

8. Apparently Martha had suggested doing illustrations of some of the fossil beasts discovered in South America. The ones named here are large endemic herbivores, tapir- and proboscideanlike of mid-Tertiary age. Reconstruction of extinct beasts, based on fossil bones alone, is always problematic. In this case there is the added difficulty that these fossils had not yet been studied in enough detail to make reasonable inferences about how they might have looked when alive.

Colonia Sarmiento
Feb. 7, 1934

Dearest Peg—

Surprise! I don't know why I am breaking my age-old habit of thinking that if I write to the rest of the family you are also written to. That isn't a good way to think, of course, & perhaps at last my conscience has risen up and kicked me.

One trouble with this once-every-ten-years sort of correspondence is that when you do write, selection is so hard. Should I try to write a snappy résumé of My Life & Times since my last letter? And if not, what news is important enough for such a rare & precious thing as this letter?

To begin with me, a bad beginning, & work outward. I am very well. Balder than ever. Bushy red beard. Flea-bitten & sunburned. I am settled at a table in the Family Parlour of the Hotel España, founded in 1920, Sr. Angulo proprietor. I have just had my breakfast consisting of black coffee and a slice of bread. All is quiet, the early birds (three Tehuelche Indians, a Basque, two Asturians, and a German) having departed & the late birds (a Uruguayan of Dutch descent, a Pehuenche Indian, a Galician, and a half Welsh and a half Portugese) have not yet appeared. Mrs. Angulo, a buxom dame with a mustache longer than mine, flutters about trying to give me more coffee, which I do not want. In the patio the very ugly, pock-marked Indian chambermaid is hanging out the wash, largely consisting of Mrs. Angulo's long woollen pants. In the street the sun shines intermittently and the winds blows steadily, the pebbles rattling against the front of the hotel like hail. All around, more or less peacefully, lies the town of Sarmiento, a remarkably heterogeneous collection of mud, tin, & brick buildings all, of course, one story high, scattered out over the flat floor of the Sarmiento Basin. To westward lies Lake Musters, named after an Englishman who never saw it, & to eastward Lake Colhué-Huapí named (in Indian) after a Red Island which apparently doesn't exist. Northward runs a wagon trail which winds up into the mountainous centro of Chubut & dies out in the wilds. From it runs a private road built by us which penetrates where never auto went before, to the site of our last camp, where we get a lot of swell fossils.

We are sitting around waiting, one of the most popular Patagonian sports. We are waiting, have been for three days, & may be here for three days more, for a repair to our poor Ford. The strain proved too much for it and it disintegrated. Until it gets pulled together again, we are tied, willy-nilly, to the back parlour, Room No. 3, and, perhaps especially, to the bar of the Hotel España. As we already have a collection, & a very special one, I am not as much upset at this delay as I was at similar contratiempos earlier in the trip, & can even enjoy the leisure somewhat.

The only person much worried is Justino [Hernández],[9] my halfbreed

9. Justino Hernández, just a few years younger than Simpson, was his field assistant on both South American expeditions. He and Simpson remained close friends throughout the years, and they met again, years later, as grandfathers when Simpson and Anne were on their way to Antarctica. (See postcard of February 1970.)

Lithuanian-Araucanian assistant who sits around cracking his knuckles furiously because I can't always think of things for him to do & it worries him not to be working. Coley Williams, on the contrary, does not have that fiery desire to labor, & is contentedly sleeping.

God & the nearest Ford Agency (a few hundred miles away) permitting, I hope to put up one more camp & get a few more fossils. We have only about three weeks more in Patagonia anyway. I have so much to do elsewhere that I cannot afford more time, & also summer is over already. We got caught in a snowstorm on our way into town the other day, although this corresponds to early August in the United States. And to think that almost everyone back home imagines that we are in the tropics! There was only a little over two months between the last snow of spring & the first of fall, & those two months were by no means all warm weather.

There isn't any very juicy local gossip, & it wouldn't mean anything to you if there were. Such items as the fact that Miss Oyanarte is that way about a German truck driver & her papa is sore about it, or that a drunken Chilean sliced up the older Gimilio boy rather badly with a machete.

Well, much love from your distant brother—

Gg.

American Museum of
Natural History
1934-1938

A FTER SIMPSON's return from South America via Paris
and Moscow in the early summer of 1934, he took an
apartment for himself and Martha, who was to return from France that
fall. Anne Roe later shared the apartment with Martha when Simpson
spent the following summer doing fieldwork in Montana. Anne was finishing her work at the Worcester hospital and traveling back and forth to
New York. Later she returned full-time to New York where she worked on
a research project dealing with newborn infants.

While Simpson was away, he wrote to both Anne and Martha, often
adding a short affectionate message for Anne in a code he had invented.
He substituted his own made-up alphabet for words first written phoneti-
cally in English, then transposed back to front. Thus, "I love you, Sweet"
became I LUV YU SWEET, then (transposed) I VUL UY TEWS, and then
(encoded) ⋎ ⋎∟7 ∟ʎ ∧\ϢΘ.

By fall 1936 Simpson had moved to Stamford, Connecticut, in order to
sue for divorce on the grounds of mental cruelty, grounds that were not
allowed in New York. Lydia's previous hospitalizations as well as her erratic
behavior—moving the children continually from place to place—gave
substance to Simpson's suit. Ironically, Lydia's own mother testified in
favor of her son-in-law at the divorce trial. Moving to Connecticut also
made it possible for Simpson to bring Helen to live at home rather than at
boarding school. Betty and Joan remained with Simpson's parents in
Washington, D.C., while Gay continued to be cared for by Lydia's mother
in Kansas.

Although Simpson filed for divorce in October 1936, the suit was not
settled until April 1938—a full eighteen months—and only after an
unpleasant court hearing necessitated by Lydia's countersuit. Compound-
ing his marital difficulties, the deepening depression resulted in stiff cut-
backs in staff and research support at the American Museum, including for
Simpson a two-and-a-half percent salary reduction.

Despite this turmoil, Simpson was as productive as ever. His travel narrative, *Attending Marvels*, appeared in May 1934. It was well received and
had many positive and prominent reviews, including front-page coverage

in the *New York Times Book Review*: "A fascinating book . . . . informal as it is, contains in essence the true scientific spirit." Against a solid background of articles and reports on South American and western U.S. fossils, Simpson began to produce more theoretical papers on evolutionary processes which would function as what he later called "door openers" to the full-length book, *Tempo and Mode in Evolution*. This shift in research emphasis was triggered by seminal work in the early 1930s on population genetics by such experimental and mathematical geneticists as T. Dobzhansky, R. A. Fisher, J. B. S. Haldane, and Sewell Wright.

194

AMERICAN
MUSEUM OF
NATURAL
HISTORY
1934–1938

Anne Roe's skills and insights as a clinician, working with relatively small statistical samples, had obvious relevance for Simpson, who was also attempting to make broad generalizations from his small collections of fossils. In 1937 he and Anne collaborated on a text entitled *Quantitative Zoology* (published in 1939) which explained simply the reasoning and techniques of statistics in making inferences about larger populations—whether fossil horses, living plants, or newborn infants—from relatively small samples of that population.

Simpson's chief fieldwork during the period 1935–36 involved continued collecting in the Paleocene-age Fort Union deposits in Montana, with additional reconnaissances to other Cenozoic formations in North and South Dakota, Colorado, and Wyoming. The first summer was unusually wet, and the collectors spent much time and effort keeping themselves and their fossils dry. The second summer Helen accompanied her father as he drove around the West looking for suitable collecting sites for the future.

[New York City, Summer, 1934]

Dear Babe [Martha]—

Thanks for letter. I can't write long, & this is "working" hours—I shudder at what would happen should a Boss catch me—probably he would make me send you his love.

Where are the first & last parts of my journal? Not here. Anne gave me the middle, but pp. 1–31 and all after 331 (last arrival at Comodoro) are lacking. I don't need them at once, as the other book [*Attending Marvels*] is not going well enough to warrant trying again so soon, but don't want to lose them.[1]

We have an apartment, & I enclose specifications. Hope you'll like it. He's supposed to put in an electric refrigerator, & if so that makes $52 per month, otherwise $50. Not bad. Relatively cheap because the building is very old, & the neighborhood is not so terribly good. They're repainting now & I can't get in. Bertha Ann's return three days ago ran me out of Creighton's—with their old-fashioned modesty the bride & groom didn't want me sleeping in the same room with them. So now I am sleeping in a camp cot in an otherwise completely empty apartment in the building next to the new place. I hope to get in in a day or two. I don't know whether I

---

1. Although well-reviewed (and still in print today), *Attending Marvels* apparently did not sell well enough to justify a sequel that Simpson had been planning.

can get the furniture, as I have nothing to show I am authorized, or whether it still exists, but I'll try.

A. [Anne Roe] was down a couple of times while I was at C's & may be down next weekend.[2] Seems well, except for heat which has nearly killed us all. Cooler today, however.

Your plans sounds swell, & I hope you have as good a time as you can & still want to return to our little love nest. How about bringing Jacques?[3] (I'm not matchmaking, which is none of my damn business anyhow, but it would be fun to have him visit us, I think).

Helen's at a day camp near D.C. Mother & Betty go off to Calif. this week, probably—I haven't heard exactly. Joan will stay with Agnes. No word from the Bête [Lydia]. She may not even know I'm here yet. Settled with Manton [my lawyer] for $50, & no other suits on call at this exact moment.

I'm sorry about Madeleine, especially because you probably are taking all the worry & trouble, but I trust she'll snap out of it all right. Why not marry the guy? There are fates worse than that. Anyhow, my love to her, to Jacques, to all the Tissiers, & to Maude if she's behaving, otherwise a kick in the pants.[4]

And of course my best to you—

Gg

430 W. 57th St., between 9th & 10th avenues—The studio has good north light. The two bedrooms have no light to speak of, but you're only supposed to go there to sleep. The dining room has light blocked by a wall, but wall is only as high as this story, so sun comes in during middle of day, & room is light enough. Cross-draft in hot weather by opening doors through all four rooms. Each room opens separately onto the hall & access to bath, or out of house, is without going through anybody else's room. There are more closets & things than even we need (but better hurry back before they're filled up). Electric refrigerator is promised. Neighborhood is one of those good-lousy patchworks. The lousy (like our dump) are being torn down slowly & new apartments put in. Traffic is moderate, & no streetcar. El is too far away (over half long block) to hear. ½ long + 2 short blocks to 6th or 9th Ave. Elevated station. 1½ long blocks to 8th Ave. Subway station. About 20 blocks to Museum. 2 long & about 10 short blocks to the theater district. About 1½ long blocks to river (if that appeals).

2. At this time Anne was still living in Worcester.

3. Jacques was a one-time Parisian beau of Martha's.

4. Madeleine was another Parisian friend who years later had a restaurant in the Latin Quarter. The Tissiers were Royalist-inclined friends who, when they met Anne and discovered she was a psychologist, exclaimed, "Ah! You must be reading our minds," which Anne thought very funny as her French was so obviously at the schoolgirl level. Maude was an art student friend.

5-7th Street - Wide and light.

↑
NORTH

Brownstone steps

Stoop
outside
Doors

Window   Window

Tall Mirror

M's Studio

Big double doors, kept closed. Come in here

House Hall

Fireplace (no go)

Stairs to upper apartments

Wardrobe

Apartment Hall (not as wide as it looks here, but thin people can pass each other in it.)

Apt Door

M's Room

Closet

Wardrobe

G's Room.

Three closets

Bathroom (all essentials, but no cat-swinging).

DINING - LIVING ROOM

Little window to poke food thru.

Wash tub

KITCHEN

Dishes

Stove

Sink

Fireplace (no fire)

Windows

WALL 1 STORY HIGH

Dearest Marthe—

Your letter, received 5 minutes ago, plunged me into the depths of
despair—which is about as good a paradoxical opening as the famous
" 'Hell,' said the Duchess.' ["] What is a poor man to do? It being pretty
decisively settled that the place was to be such & so, you not being at
hand & the case urgent, I literally camped on the job, in a vacant apartment two homes away, & haunted the painters & repairers. Every morning
I would tell them what to do & (as soon as he found out what went on)
every afternoon the landlord would come & raise hell because they did it.
Result: they made the place liveable, although not perfect, & the landlord
has been tearing his hair ever since because it cost him, he says, $100 more
than he anticipated. Probably means $50 but even that is a month's rent.

It is gray, by God, & a time I had getting it that color. Pale, very light,
not a battleship gray. I like it. But gray it is, & cold until relieved by pictures & what not. Obviously it is no use asking the landlord to do it over,
& we can't afford to have it done. And it won't do for you.

All is not quite lost. Naturally, I won't take the place until you come
(we weren't going to buy any furniture or drapes till then anyhow), & then
we will either find another place & move again at once. We have no lease.
The landlord would be pretty sore, with justice, but he isn't the one who
has to live there. Or else we will get paint & I will repaint the thing to suit
in the course of the winter.

I'm terribly sorry, & had no idea you wouldn't like it or that you felt so
strongly about it. I had to have some place to live at once, was pretty tired
of hotels & unable to afford one as it would cost at least as much as the
ap't, & I much want to start getting settled & a roof to call my own.

The thing is pretty messy now. I am tired evenings & have little time.
A. [Anne] was down last week & we both worked hard all weekend, but only
got the heavy arranging done. I must finish putting things away & cleaning, but naturally won't do anything else until you arrive, except old curtains temporarily for privacy & because it is a little cheerless with nothing.

The place is fundamentally unusually satisfactory, but I can see now that
it's no go, & I'll look about for another place, but of course will do nothing
definite till you arrive. Three times would be too much. We must get a
place where we can settle down happily & with at least the idea of staying.

Miss Francis[5] sent a card to Creighton asking what other work you have
& saying that in September she can use a lot more. I'm writing her saying
that there are some things here, which I can fix up for exhibition if
needed, but that I must hear from you about them first. I guess you know
what there is, & can tell me what ones to let her have.

No other news. I am already pretty sick of living alone, but well. Much
love. Be a good girl. Don't be upset about the ap't, because we can easily
fix it or get another.

Gg

197

AMERICAN
MUSEUM OF
NATURAL
HISTORY
1934–1938

5. Martha's art dealer in New York City.

198

AMERICAN
MUSEUM OF
NATURAL
HISTORY
1934–1938

[New York City]
Aug. 4, 1934

Dearest Mother—

Here your birthday has sneaked up again, & I have just learned that you are not in Washington as I thought but in California already—and now I find also that I do not have your address there, but I'll write this & mail it as soon as I can get the address.[6] Anyway I send you the deepest love & more than I can say of gratitude for all you have done for me & meant to me for these many years. I have never known or heard of anyone who was more thoroughly what anyone would want in the way of a mother, or of a person for that matter, & the more I learn & see & experience the more I appreciate, value, & love you. You deserve a much better son, but whether you realize it or not, this one loves you very deeply.

I am in my new apartment, but far from settled yet. Anne helped me, but of course didn't have more than a few hours. The kitchen, living room, & my bedroom are livable now, but even they require much more work & Marty's suite isn't even started. I'll have it as nice for her as I can before she comes, but won't pick things like new curtains which depend too much on personal taste.

Coley is definitely & finally fired & it is a relief, but he goes around glowering at me & plainly has decided I am a viper. The idea that he's just no good, at this job anyway, will never occur to him. I didn't fire him, or even suggest it, but there has to be a vilain [sic] & he picks me as the one who is blocking the advance of virture & sterling worth. My sympathy with him is worn slightly thin by the fact that his last act here was to ask Andrews (director now) for *my* job![7] Andrews told me as a joke, & told Coley he couldn't even rattle around in my shoes. Coley's coolness toward me frosted at that moment. Well, I hate to make enemies, but it can't be helped, & in fact he'd have been out long ago if I hadn't been too soft-hearted.

I am toying with the Chicago idea but only because I need more money.[8] They may raise me here, although it's hard to see how they can, & in that case I'll certainly stay here, & possibly will anyway.

Speaking of money, where is the bank pass book? I must get a separate account in order before things break loose.[9] They haven't yet—no news at all.

Much love to everyone, but especially to you—

George

XXXXXxxxxx. . . . . .

6. Simpson's oldest sister, Peg, was married and living in Los Angeles.

7. Coley is Coleman Williams, Simpson's South American field assistant and museum associate. Roy Chapman Andrews (1884–1960) was an American paleontologist, zoologist, and explorer who spent more than a half-century at the American Museum. He is perhaps best known as the leader of the museum's spectacularly successful expeditions to Mongolia in the 1920s which yielded so many primitive mammals, dinosaur eggs, and other interesting fossils.

8. Apparently Simpson was entertaining a job offer from either the University of Chicago or the Field Museum of Natural History.

9. Presumably Simpson is referring to a bank account held jointly with Lydia, which he suspected she would empty once she learned of his filing for divorce.

[Montana field camp]
June 17, 1935

Dear Anne—

It's hailing hard & this cook-tent is thin & lets a little spray through, so if this is moist, it is not with tears.

The enclosed check was an unexpected windfall from Natural History [magazine] for writing a review, & comes very apropos. Spring it on Marty on the 22nd (I hope it makes it) for dinner, a show, & remainder for something frivolous.[10]

Horse stripping[11] is finished in the quarry, for the time being, & we are doing hard work, between thunder showers, but have to wait for a little better weather to work the actual bone level. All are well & everything is fine. EXCEPT THAT I MISS YOU SO. I LOVE YOU—[12]

Gg

199
AMERICAN
MUSEUM OF
NATURAL
HISTORY
1934–1938

[Montana field camp]
June 20 [1935]

Dearest Marty—

Surprise! You'll be expecting daily letters from me. We are somewhat marooned by cloudbursts, but the rural carrier is going to try to make town tomorrow. The immediate occasion is to ask you to send Mother $30.00 for Helen, if you have that much of mine. If not, there must be considerably more still in the savings account. I was going to send a check, but my check book got left in town in what I laughingly call my good clothes, & I can't get it very soon, & mother needs the cash.

Journal has gone on to Mother & Dad every few days, so I waste no time on mere news, save that I am well but, like everything else in Montana, wet.

Much love—
Gg

10. Simpson wrote many popular articles for which he was paid modest but handy sums. In this case, the money came in handy for Martha's thirty-seventh birthday.

11. Refers to using a team of horses to scrape off the overburden of soil and barren rock strata to expose the underlying fossil-bearing rocks. Today bulldozers perform this necessary function.

12. Words here (and in later letters) in capitals are those written in code to Anne.

200

AMERICAN
MUSEUM OF
NATURAL
HISTORY
1934–1938

[Montana field camp]
[June, 1935]

Dearest Marty—

Happy Birthday to You! And much love.

The Reverend Lekban[13] (I called him "legband" to his face by accident) paid us a surprise call & is champing at the bit to go back to town, so off this goes—

Gg

[Montana field camp]
Glorious Fourth [July, 1935]

Dear Ones [Mother & Dad]—

We have a rainy afternoon, which I should spend on numerous other chores, but am going to spend in part writing letters, as the rural carrier can take them tomorrow, if he makes it. Journal enclosed, with all the news and much that isn't news, and as usual there is nothing really to add, save that I love you all & am lonely for you. Here is a good brainstorm: if Mother goes to California (& I much hope she does) why not go via the Milwaukee—it's a good railroad, all air-conditioned cars, & goes through interesting country new to her. Then she could stop off in Harlow for a day (or of course more if she would). We'd meet her train (6:29 A.M.), then run her out to camp (an hour's easy drive), she could spend the day here, then go back that evening, spend the night in Harlow (a good hotel) and depart at 6:36 the next morning. It's the best chance she's likely to have to see her favorite son at his favorite occupation. Wouldn't that be swell? Do consider it seriously & try to do it.

So no more for now—

Much love,
Gg

[Montana field camp]
Fourth of July [1935]

Babes[14]—

The rain is falling gently but firmly all around. Johnny is doing an Achilles in his tent (classical allusion), Al is sewing iron rings for ropes into the cook-tent flaps, & I am writing this letter (in person). There probably isn't a soul within a hundred miles of us, except of course Jimmy Wyn, who is doggedly setting off firecrackers in the rain, his family, about 101 other ranch families, & the inhabitants of the towns of Harlowton,

13. The wife of Simpson's chief field assistant was a Seventh-Day Adventist; her pastor called at the camp from time to time.
14. Martha and Anne, who are sharing the W. 57th St. apartment.

Melville, Bigtimber, Livingston, Two-Dot, & a few others. The grass is surging up gladly from the warm soil, & the grasshoppers are surging it right back where it came from. The little woodticks are all out drinking in the beneficent dew from heaven & nourishing each his little colony of spotted-fever germs. The Black-Widow spiders are quietly multiplying in all dark crannies, & the rattlesnakes are sulkily wagging their tails in their holes. The alfalfa is blooming, & the gophers are rapidly eating it. The loco-weed, camas lilies, & larkspur are green & luxuriant, and the sheep are dying in convulsions from eating them. A normal summer day in Central Montana, & so I sit down to send you love & kisses, (no fooling).

201
AMERICAN
MUSEUM OF
NATURAL
HISTORY
1934–1938

I have just sent a large and practically illegible batch of journal to Mother & Dad, from which you will learn little that is to your advantage. The Hunters[15] left us three days ago, & I miss their company, but find going slightly easier in some respects. We have had a lean streak in collecting, getting only 5 or 6 jaws a day, but hope to do better when, or if, the rain stops.

I am in fine-fettle—whatever that is, I'm in it. My job is ¼ complete, but I'll be happier when it is 4/4 complete, if you can follow that.

Be good girls. Have a good time & come home early. Brush your teeth twice a day & see your dentist twice a year.

Love,
Gg

I MAY ALSO REMARK THAT I DO NOT BELIEVE THAT I HAVE EVER BEEN SO CONTINUOUSLY PASSIONATE IN MY LIFE, AND IT IS SHRIVELLING MY NUTS. I HAVE BEEN THAT WAY PRACTICALLY ALL THE TIME FOR TWO OR THREE WEEKS, AND I HOPE IT STOPS SOON. YOU HAD BETTER BE GETTING LOTS OF PEAS AND BUILDING YOURSELF UP FOR WHEN I DO GET BACK, FOR I SUSPECT THAT IT WILL TAKE A STRONG CONSTITUTION TO STAND WHAT I HAVE IN MIND.

I LOVE YOU, DARLING, AND I AM ALREADY IN QUITE ENOUGH OF A FRENZY WITHOUT GOING INTO ANY MORE DETAIL ABOUT YOU—

I am always yours.

[Montana field camp]
7/7/35

[Dear Anne],
I think we'll dash in to town for supper tomorrow, & I'll mail this, also that. Yours of July 2, & also a bundle of New Yorkers (& *many* thanks) arrived yesterday via Mrs. Silberling, her selected son-in-law (who seems

15. Mr. and Mrs. Fenley Hunter, Flushing, N.Y., who helped finance Montana fieldwork, along with Horace Scarritt. The couple worked with the field party during June 1935.

blissfully unaware as yet of his selection, but hasn't a chance in the world), & all the little S's (of B's). They went right back to town, hallelujah.

I'm all right except for the perpetual continuance of the difficulty I have mentioned. We have 177 jaws now, & 250 isolated teeth, which is pretty good, but not perfect. They certainly don't collect themselves. We're lucky to get 7 or 8, three of us working for ten or eleven hours.

202

AMERICAN
MUSEUM OF
NATURAL
HISTORY
1934–1938

I just wrote to Jimmy [Holland, my lawyer[16]] about his fee. AND ALSO TOLD HIM THAT I WANT TO MARRY YOU, AND THAT I *MUST* HAVE A DIVORCE, AT ALMOST ANY COST.

Most of your news seems good, & I am glad.

I don't get one thing in your letter—about Farley[17] taking anti-venim (incidentally, not anti-venom) [sic]. Did I give the impression he was bitten? He wasn't. They've left us now & all is calm & serious male labor for a change, although I still like them both very much & miss them personally, but not expeditionarily.

I MISS YOU SO, DARLING. I LOVE YOU LIKE THE DEVIL. I NEVER WANT TO LEAVE YOU AGAIN FOR MORE THAN FIVE MINUTES, AT MOST. I LOVE YOU, SWEET.

Give my best love (OR SECOND BEST) to Marty—

AND MY VERY BEST TO YOU,

[Montana field camp]
7/12/35

Babes—

A hurried line. Please send enclosed checks—no message necessary—to 210 St. Ronan St., Care Mrs. Percy Walden, as dated.[18] I have sent for June 30, July 7, & July 14, not covered by checks left there.[19] L. asked for a conference on legal status, setting a date long after I left N.Y., & since then only stony silence—probably cooking some mess.

Did the little adding machine ever arrive?

Are you well? I am, but not very happy.

Journal to Maw 'n' Paw, much love to youse.

Gg

I LOVE YOU. I AM SO BLUE. BUT I LOVE YOU.

16. James Holland, Simpson's lawyer, knew Anne at Denver University when they were both students there.

17. Farley was a local entrepreneur who was interested in the work going on at the field camp.

18. Mrs. Percy Walden was a Yale faculty wife who had befriended Lydia and who would testify in her favor at the divorce trial.

19. Simpson undoubtedly refers to child-support payments.

[Montana field camp]
[August, 1935]

Dear Babes—

Had a swell visit from Mother last Sunday, her birthday, & saw her off early Monday morning. I wish you all could casually drop in! Incidentally, Mother insists on journal going to her, in Calif., first, & so it is. There isn't much news anyway except continued quarrying (461 jaws to date) and arrival of Dr. Granger who is at the moment very stiff & sore (physically, not emotionally). Bill Thomson[20] is coming next week—we'll have quite a party yet.

Not to be playing any favorites in my haremlik, enclosed is 7 (7) berries to blow on Anne's birthday, & many happy returns of the day, Anne.

So must close, with much love—   OVER

Gg

TIME WILL OVERCOME THIS AGAIN, AND RACING OR CRAWLING, DOWN THE HOME STRETCH—TOO LONG A STRETCH. I WILL BE THERE, MY DARLING. I LOVE YOU, SWEET. I AM CRAZY ABOUT YOU, AND HALF SICK WAITING AND WISHING [FOR] YOU. I LOVE YOU.

[over]

Also enclose two checks for the bitch [Lydia]—Better put 50 bucks in my checking acc't., also send me a new check book filler—these are my last blank checks.

Gg

203

AMERICAN
MUSEUM OF
NATURAL
HISTORY
1934–1938

Gidley Camp [Montana]
Rainy Day [late August]

Dearest Babes—

Herewith 11 (eleven) berries, earned by writing about *Thomashuxleya* on museum time.[21] Pretty soft for us scientists, huh? I was counting on this for A's [Anne's] birthday (Natural History [magazine] having kindly paid for M's [Marty's]) but it came too late. Now 7 [dollars] can replace that sent, & the other 4 [dollars] you divide between you, without quarreling, as unbirthday presents to be spent on fun, frivolity, gin, gingerale, lipstick, licentiousness, or new brake linings.

20. Albert "Bill" Thomson was a highly skilled excavator of fossils in the summer; in the winter he carefully and delicately removed the surrounding rock matrix for subsequent detailed study by the paleontologists. His skills both as a collector and preparator were highly valued by his American Museum colleagues, so much so that he accompanied several of the expeditions to central Asia.

21. *Thomashuxleya*, a large South American fossil mammal named by Florentino Ameghino after the nineteenth-century British biologist Thomas Huxley.

(The sun just came out, hurrah, but too late to save the day for dear old fossils).

You probably get news of me tens days late, now Mother is in Calif. & I promised her the journal first, but just pretend I am in Patagonia. The news is of no interest anyway. Just quarrying & an occasional visitor— quite a flock of the latter yesterday, with several Princeton boys from Jepsen's camp,[22] the whole Martin family, and the Eneboes [sp?]. This week we expect Shea (proprietor of the Northern Hotel in Billings, & world-famous as the only man who always knows where Barnum Brown[23] is) & also Bill Thomson, who will stay probably as long as I do.

Mail carrier is due to arrive in a moment, if he can navigate in all this mud—

Gg

DARLING,

I WROTE YOU ONLY A COUPLE OF DAYS AGO. THE MESSAGE IS STILL THE SAME—I LOVE YOU LIKE HELL. I TRY, VERY UNSUCCESSFULLY, NOT TO THINK WHAT I AM GOING TO DO TO YOU WHEN I GET YOU ALONE. YOU REALLY ARE SO DAMN SWEET, DARLING.

TIME IS PASSING. AND I AM COMING.

ALL MY LOVE—

Miles City [Montana]
Aug. 16 [1936?]

BELOVED, DUCKLING, SWEET—

I LOVE YOU LIKE THE DEVIL. I AM SICK WITH LONGING FOR YOU.

This is for your birthday, & sends such wishes as you know, & hopes. Not knowing your plans in any detail or present circumstances (for mail seems to have all gone astray) and expecting to see you in two weeks or *less*, I plan such birthday celebrations as we may be able to stir up for you on my return, & trust that you do not mind the delay.

We have not encountered any fossil field in this god-forsaken region, so will leave it sooner than expected & soon turn back East—sending Al on to collect in the Scarritt Quarry while I stop a day or two in North Dakota, then Kansas, then home, probably (but not surely) a day or two before Sept. 1—will let you know more definitely when I can.

Address mail to c/o Mary Pedroja [Lydia's mother], Buffalo, Kan., for last chance—probably be there by about 23 or 24.

22. Glenn L. Jepsen (1903–1974), Princeton professor of vertebrate paleontology, especially known for his studies of early Tertiary mammals.

23. Barnum Brown (1873–1963), vertebrate paleontologist and curator of fossil reptiles at the American Museum.

AND REMEMBER HOW VERY MUCH I LOVE YOU. SURELY THIS MUST BE
THE LAST BAD TIME WE HAVE APART—AND NO TIME IS BAD
TOGETHER—

I LOVE YOU.

205

AMERICAN
MUSEUM OF
NATURAL
HISTORY
1934–1938

[Montana field camp]
Sept. 8, 1935

Dear Babes,

Many thanks for letters written Labor Day. I'll send this off by our dirty-eared rural carrier, who passes tomorrow. We go to town, to bid Dr. G[ranger] farewell, day after tomorrow.

Your question regarding my return I answered as best I could in my last, I think it was. I have to go via Kansas, & spend at least a day there I suppose, & take the kids to D.C., so although I haven't trains worked out yet, I think it will take all of 6 days from Harlow [Montana] to N.Y., maybe a week. As for when I can leave here, I am obligated to run the expedition a full 4 months, i.e. to return to Museum not before Oct. 1, without personal time out, & also to Al to keep him at work through Oct. 2. But both obligations could be discharged & I can still leave a week or so *before* Oct. 2, by staying only until the actual digging is complete, leaving the raising of camp, transport to town, packing & shipping of camp & fossils—probably about a week's work, to Al, who fortunately is competent to do this routine work, which, incidentally, I also dislike. That is how I plan to work it, & I *think* I can plan on being in N.Y. on Oct. 1—you can be sure it won't be for lack of trying! I haven't sprung it on Dr. G., or elsewhere yet, but when I do get there I hope to find out how much vacation I have coming, which must be 2 or 3 weeks, possibly more, & take it there before I go back to work. I *must* do some writing, & this would give me a chance, & also it wouldn't be bad to have spare moments to lavish on you-all.

About the car, if already purchased, read no further. My opinion is definitely *not*. We would get practically no use from it all winter, & it would cost probably about $15.00 per month just to keep it standing still, in running order. Actually to run a car of that description would cost certainly 10¢ per mile, probably more, which is very expensive transportation—we could go farther, stay longer, & come back drunken by almost any other means. If Marg. has had light repair bills, that means she has made no replacements & that we'd be almost certain to get them at the most embarrassing moments. I've run around some in old cars, & it's the devil. It really is cheaper to get a new car, or none. Furthermore they are decidely not safe, especially in the sort of traffic on all roads around N.Y.C. And they aren't a pleasure. Passable, perhaps, for ferrying to & fro a given spot, as M. used it, but not for pleasure jaunts in the country. If we can afford to run a car next year when roads are open, it would pay us to get a better one.

As for Mme. Lemaître,[24] no me kicking her out till Oct. 1, as it is practically impossible, I fear, that I make it before then. After I return, be she as charming as [the Queen of] Sheba & as self-effacing as [Vice-President] Garner, out she goes, by God, or I won't pay my rent! Our ménage à trois is swell, but one à quatre would positively require more or different space & a more considered, & preferably not female, fourth!

206
AMERICAN
MUSEUM OF
NATURAL
HISTORY
1934–1938

Of course nothing is new here. I am well fossilized myself, settling into a sort of hopelessness at the endlessness of it all. We have 589 jaws. We work quite a lot. Haven't been to town for over a week, so have heard no scandal, & scandal about people you don't know is not very exciting anyhow. Of course I know everyone there by now, & with a few exceptions we view each other with suspicion. While quarrying I make up lewd limericks about the townspeople, a childish occupation (I always go childish after a few months in camp) e.g., a mild one—

> The harlot of Harlow is Hattie.
> She's homely & often quite batty.
> Though her morals are nil,
> You can say for her still,
> She's the one girl in town who's not catty.

You can see how hard up I am for painless subjects to think about. And each night, "Thank the lord, another day done," which is a shame, because it's a grand place (camp, not Harlow) really, & a terrible waste to be disliking it, & all for the reason of being lonely with four other people in camp, three of them excellent company.

Well so it [illegible scribble] goes (note aphasic error: my left frontal or is it temporal? lobe got sunburned today)—Much love to one & all, or each & every one—

> Yours,
> Gg

AND LOVE AND LONGING TO YOU, ESPECIALLY, MY SWEET—

More or less the 8th [25]

Dearest Marty & Anne—

This is a pretty good place—coral strand, coconut palms, blue sea, & one thing another. I am enjoying it. The first day or two everyone was well plastered, but with that out of their systems have sobered up somewhat & all goes merrily. Three days we've gone out to sea in a little motor-boat which bobs around like nobody's business, but surprisingly enough I was only squeamish once & the rest of the time have enjoyed it thoroughly.

24. A Parisian friend of Martha's staying at the New York apartment.
25. According to Anne Roe Simpson this letter was written by Simpson when on a fishing trip to Florida with Horace Scarritt; probably written about this time.

Fortunately the fishing isn't taken really seriously, & we quit before I get tired of it. I've caught a couple, just enough to see what it's like.

But I'll tell you all the dirt when I return. Needless to say, I'm very well & getting all rested & full of vim & vigor.

I only wish you all were here. You'd like it even better than I do.

Much love—
Gg

207

AMERICAN
MUSEUM OF
NATURAL
HISTORY
1934–1938

Small note for Anne—over—

BELOVED—

I MISS YOU TERRIBLY AND THINK OF YOU ALL THE TIME. WAIT TILL I GET HOLD OF YOU, MY HEARTY WENCH. AT LEAST, I DO HOPE THAT YOU ARE HEARTY AGAIN BY NOW. I LOVE YOU SO, SWEET. IN THE WHOLE WORLD THERE IS NO ONE ELSE SO DARLING. I WILL BE WITH YOU SOON AGAIN.

—The only clothes I haven't needed are the warm ones! It's almost hot, but not unpleasantly so.

Stamford [Conn.]
Feb. 7, 1937

Dearest Marty:

It is shocking even to me how poor I am at writing to anyone. I love them, miss them, think of them, but I don't write to them, and I guess the only way out is to call it a disease and give it a Greek name and get sympathy instead of curses.

In the now very remote first place I very much appreciated Christmas remembrances from you-all. The book money has now turned into Owen Lattimore's "Desert Road to Turkestan" which is a very swell book, and I recommend your reading it. It is an unusually good travel book, even regardless of prior interest in the region it covers.

Apparently you were duly informed of my having flu, and unduly alarmed at it. I am now perfectly well again, and never was really seriously ill, being sat on and prevented from becoming so by the combined efforts of numerous females, even the doctor being of that stern sex. Of course this sitting upon was a very good idea, and I appreciate it even though I felt slightly restive at the time.

Aside from this break in schedule, nothing of any interest has happened, the commuter's life continuing its boring and fatiguing way.[26] After Christmas and just before flu I went to some scientific meetings, which were pleasant in different ways, and there will be some more in April and

---

26. Simpson was commuting daily from Connecticut to New York City, a round trip of some 60 miles.

May—I look forward to these little breaks in routine more now than when the routine was less rigorous.[27]

Work at the Museum proceeds, but I get a little less done than formerly, partly because of my decreasing ambition and increasing weariness, partly because of the more rigid hours which no longer permit the most practical system, for me, of working more when I am good and less when I am not, and partly because the internal stresses and strains of our department are now so great that there is no esprit de corps and no cooperation. I love the place, it worries me to see our department, once incomparably the best in the world, going steadily down hill, and it will be a dreadful wrench to leave, but if things continue as they are it will also be a relief to get away from there. Granger is a broken man, Brown is a jealous egomaniac, Frick[28] rides roughshod over us all, and no one even tries to guide the rudderless ship and simply snipes at anyone who does try to. We have also lost about a quarter of our staff in the last couple of years, and not a single person has been replaced. My own work, however, is still pleasant and satisfying to me.

Helen is still doing very well in school, and mostly gets A's, although she slipped down to a B in English, which happens to be taught by the one teacher she doesn't like. I enjoy very much having Helen here, and while things are not ideal for her they are probably the best now possible and do seem to do her more good than harm.

My [divorce] business up here is quiescent at the moment. The various lawyers involved still think it possible that an arrangement for an uncontested suit, which would settle things in a couple of months, but I haven't a grain of hope for that. However, we have a good chance of winning the contest, and it can't drag on later than next Fall. Everyone assures me there is nothing to do at the moment, so I am just sitting around wishing something would happen. Eeven [sic] an application for temporary alimony has been let drag on in abeyance.

I still have lunch with Anne almost every day and an occasional dinner. She has been mildly ill, but is better now. Perry Claxton, who becomes a solider citizen by the week, was threatened with gall stones, but got out luckily with only having to reduce weight, which he is doing surprisingly. Bertha Ann and Creighton are the same as ever. I very seldom see the Peets or Claxtons now and almost never anyone else in New York—or anywhere for that matter as hours and other things have prevented my making any friends in Stamford.

I have decided not to attempt to get out in the field this summer, or indeed to go much of anywhere or do much of anything until my affairs are settled here. Then, if satisfactory, I am going to paint not merely the town but the whole U.S.A., at least, a peculiarly brilliant shade of red.

27. The post-Christmas meeting was that of the American Association for the Advancement of Science in Atlantic City, where Simpson read one of his first "door-opener" papers that foreshadowed many of the ideas more fully developed in *Tempo and Mode in Evolution.*

28. Childs Frick (1883–1965), scion of the wealthy Frick family, who had a research position at the American Museum; his specialty was fossil mammals and he used much of his inherited wealth to acquire an extremely valuable collection for the museum. The Frick Wing of the American Museum is named after him.

208

AMERICAN
MUSEUM OF
NATURAL
HISTORY
1934–1938

The best news that I have had since I can remember is that you are so well, putting on weight, and generally blooming. The shape you were in when you left here has been haunting me ever since, and desperately as I have missed you, I am delighted that you did go since it has proved to be so good for you.

My best love, and also suitably salute Benson, Peg, Mac, and Jack, and tell them I am flourishing, or show them this, as I fear I will not immediately write to anyone else out there.[29]

As always,
Gg

I think I remember Peg's address, so I'll send this there and hope for the best.

209

AMERICAN
MUSEUM OF
NATURAL
HISTORY
1934–1938

[New York City?]
June 18 [1937]

Dearest Mother—

Helen & I, with Jimmie Holland & Mr. Walker, (my lawyers), spent an unpleasant but finally rather satisfactory afternoon in court today. Lydia was there with her two attorneys, Mrs. Cook & [Mr.] Cressy, also with the obnoxious & ubiquitous Mrs. Walden & with two pretty queer dames who, it turned out, were planning to take Betty & Joan for the summer at $140 a month (I'll say immediately that this did not work). The hearing was, of course, on L's demand to visit the children, but Cressy started off demanding that the children either be turned over to her for the summer or else brought to Connecticut & taken away from both her & me (or my agents in the matter)—just what had been asked in the motion for custody that he withdrew. The judge was one who had nothing to do with the case hitherto & knew nothing of it, & he started out very indignant at the idea that I had these four children & was keeping them from their mother & was on the verge of just ordering that I turn them all over to her forthwith, but Walker tactfully & promptly scotched this & got the thing turned back into the discussion of how & when she might have a visit with them. The idea that Gay be brought East was soon discarded by the judge, & she was dropped from the discussion. Helen's case was more difficult but after an hour or so of wrangling, it was finally decided that she go to camp as planned, leaving L. speechless (for just a moment) with fury. The real battle was over whether Joan & Betty could go with the Farringtons to Maine & this was the only point on which testimony (by L. & by me) was taken & that was argued out in open court.[30] L testified that her mother (is this true?), you, Dad, & the Farringtons themselves had all been to the place in Maine & had all reported to her that it was unfit for human occupation & that they would have to stop going there for fear of disease & disaster.

29. Sister Peg was married to Duncan ("Mac") McLaurin, who adopted her son Jack from a previous marriage.

30. Marvin Farrington was a close associate of Simpson's father. He and his wife Edith lived in Washington, D.C. and had a summer home in Maine, which Simpson and Anne had visited several times when on vacation.

Further that the Farrington home life was a continuous pitched battle, that Mrs. Farrington was not morally fit to be in contact with small children, and that Helen & Edith, while she didn't want to go into things that would ruin their reputations, were just at the age when they are man-crazy. Of course I testified the opposite, how long we had known the Farringtons, that they were really part of the family, how nice they are, & that I had been to the summer place, which I described, & found it most pleasant & desirable. Then one of these dames was placed on the stand by Cressy, a Miss Little, & testified that she had had the children for some time in 1932—if true, she must have been connected with the place on Long Island where L had the children hidden out before we got them—& that she now had a cabin in Maine where she took 6 or 7 children for $70 a month apiece [sic] summers. By this time the judge was somewhat fed up & he didn't even let her finish her testimony, said that Farringtons sounded all right to him, & Joan & Betty could go there if they visited their mother first.

So the final order of the court was that Gay & Helen do just as planned, & that Joan & Betty visit Lydia for two weeks in New Haven before going to Maine. This is, on the whole, remarkably favorable, especially in view of the way things started out. I am sorry that J. & B. have to be with L. even for a short time, but I know they will come to no harm & think they will even enjoy it. The order will include stipulations that I am to know exactly where they are, that they will be in New Haven only, that their personal effects are to be turned back to me exactly as received, that no medical treatment is to be given without my consent, & other safeguards. Also if L makes one false move, her chances of eventual custody are completely gone, & Mrs. Walden & Mrs. Cook know this, will be seeing the children daily, & will see that no harm comes—I hate Mrs. W. & dislike Mrs. C., but in this respect I know they are dependable.

Another good aspect is that the judge, who may be the one to hear the trial & certainly will be interested in it, was by the end very obviously impatient with L's unreasonable & nasty attitude & most favorably inclined to us & to our willingness to do everything possible in the situation. His only ruling against us (& on advice we agreed to this) was the 2-week visit & this under Conn. law & precedents was the least that could be granted L unless we had proven her unfit to associate with children at all—this we could not do in the time & with the witnesses available or without practically proving her insane & having the divorce action thrown out of court.

So on the whole I think we did well, & I hope you will think so too & will not worry about it.

L's two weeks are June 28–July 12. I will come to Washington with Helen June 25, bring Joan & Gay to Stamford on the 27th (Sunday) & to New Haven on the 28th, or perhaps through to N.H. on the 28th. I will get them on the 12th & see that they get to Maine—I suppose I have to take them up, & will if there is no other satisfactory possibility.

I think it might be well to have the doctor see them before they go & write a statement of their health then, to forestall any claims. Also I would

210
AMERICAN
MUSEUM OF
NATURAL
HISTORY
1934–1938

like to get some photographs of the Farrington place, as detailed as possible, for I foresee that the question as to whether it is fit for children may very well rise again in the trial next Fall if not before.

Helen & I are well, & I enclose a letter from her. I much appreciated your birthday greetings & check—I do not think I need the latter & do not feel that I can take it. I owe you so much already. Of course I really believe that it is a pleasure to you to give me so much, but I feel uncomfortable about it & fear that it means that you are deprived of things, much as I treasure the love that makes you want to do so much for me, & strongly as I return it.

Love also to Dap[31] & to Joan & to Betsy, & thanks & kisses to them for their greetings.

211

Gg

[Postscript]

The "story" Helen "told to get away one day" was to tell her teachers that her mother was not waiting outside the school for her, when Helen knew her mother was there—the purpose being to avoid her mother, which she did by slipping out a side door. I told her that I disapprove of lying, of course, but that we sometimes were almost forced into it until we developed the tact & force of character to evade gracefully when absolutely necessary to do so, & that even so we were sometimes attacked by others in a way that made small prevarications necessary in order for the larger truths to prevail—to be sure that she was acting for the truth & for right, & then not to worry if it was absolutely necessary to mislead on rare occasions. I don't know whether that is proper doctrine for a [13-year-old] girl, but I believe it, & I don't want her all tied up with guilt feelings if she sometimes has to fight Lydia's fire with fire. I don't believe Helen ever told me a lie in her life, even to avoid punishment, & she says it's because I am fair to her & she doesn't need to lie.

[Stamford, Conn.]
Jan. 19, 1938

Dearest Peg and Marty—

My apologies for not having written often before, but with many things on my neck I haven't really felt equal to the task, & after hearing most of my letters for the past 15 years read in court I resolved *never* to write one again, but here goes for just a brief word. Mostly belated, most sincere thanks for your swell Christmas presents—both books I wanted & enjoy very much & didn't have before. As usual your judgment was perfect, & I've had & am still having hours of pleasure from them. I suspect the slips in "Overland through Asia" were meant to check on whether I really do read all these books: I do, & the slips prolong the personal touch. Knox was a quaint guy & is unusually diverting in many ways.[32]

31. "Dap" and "Dappie" were affectionate family names for Simpson's father.
32. Thomas Knox (1835–1896), American traveler and author, best known for his young people's books in a popular series, "The Boy Travelers in the Far East."

The [divorce] trial is pretty bad, of course, but at last it is moving along. We daren't think of the outcome yet. Probably 2 or 3 weeks more, I estimate. Mother was on today & she was *wonderful*—Before her cross-examination was over the opposing lawyer certainly wished he'd never started it, as she turned every trick question to our advantage in the calmest, most sincere & truthful way.

I love you both very much & think of you often in these trials (pun)—

Gg

212

AMERICAN
MUSEUM OF
NATURAL
HISTORY
1934–1938

[Stamford, Conn.]
Sunday, April 3 [1938]

Dearest Marty:

Still no decision on the case and no indication when there will be—I wouldn't believe it if there were as we have been promised it so often before. It was promised last week in the most definite and circumstancial [sic] terms, and nothing whatever happened. All the other news is equally annoying, although none of it desperate, so I shall skip it for this time. We are none of us either well or really ill.

Anne, Helen, and I all read your murder mystery with the greatest pleasure and I meant to write about it long since, but have been most decisively out of the mood for writing, or for anything else for that matter. I know Anne wrote, and I agree in general with her opinion. The story is fine and certainly has real merit, even if you are my own sister. Of course it needs to be either a lot shorter or a lot longer, preferably the latter, and I understand that you are making or have made it so. The one general suggestion I had that Anne may not have thought of or have repeated was that the first murder, which is the more interesting of the two in itself, seems to fade out without having its qualities fully exploited and to have relatively little bearing on the bulk of the book. My idea—which is only very tentative and may well be all wet so discard it if it doesn't appeal— would be to lengthen the investigation of this a great deal, clues in the class, questioning the students in detail, and so on, perhaps for about half the book, and then to bring in the second murder when this has run up entirely against blind ends and interest begins to flag. The bulk of the present ms. would then be the second half of the full length book. This would take care of length and would also get full interest and excitement from there being two murders, which seems almost unnecessary at present. Alternative (supposing that there is anything in my feeling about it) would be to make two separate stories out of it, an art school case and a wine cellar case.

I am so glad to hear that Jack seems to be improving and to be happier, and I hope that this relieves you and Peg of all serious worry about him. I also hear that you have sold another picture which is swell and I do hope that you are beginning to flourish and will continue to do so—you, yourself, I mean and not only your external success although of course you know how much I hope for that too. It would be so nice to see you and Peg again. There is, however, little prospect for this year. I have no idea when the wedding [to Anne] will be—if ever, I almost added, but that is just pessimism at the delay, for of course it will be soon. If I didn't think so I am afraid I would just give up. And when it is we will have rather less than no money at all and two weeks vacation or less, so I guess California would be too much to attempt.

Anyway, much love to you all from us all.

Gg

213

AMERICAN
MUSEUM OF
NATURAL
HISTORY
1934–1938

[Stamford, Conn.]
Sunday, Apr. 3 [1938]

Dearest Mother and Dad:

I have been thinking almost daily that there would be news of some sort to give you, but there hasn't been and isn't. We were assured that there would be a memorandum from the judge of his decision early last week and that he would enter it legally probably by last Friday. Neither happened, and there is no sign that the judge is doing anything or really contemplated giving a decision at all, although of course he must sometime. All of which is very trying, of course, and my nerves are not so good.

I have heard from probably reliable, but unofficial sources, that the job at Yale has been filled. This may not be true, but it almost surely is. The Museum fired many of the staff last month. My yearly contract has not come through, but I have not been fired and am assured that it is improbable that I will be. All of us that are left have had our salaries reduced, only 2½% but that hurts in the present condition. The Museum will not accept any scientific manuscripts for publication this year or indefinitely into the future, which is ominous in itself and also because it robs my particular job, 90% writing scientific papers, of any real reason for being. Anne is badly upset, of course, and also is continually ill with her brucellosis and with the radical treatment for it—it is yet to be seen whether this will do any good.[33] Helen continues to have frequent very severe headaches that the doctor thinks are largely nervous in origin, but there is a chance that it is her eyes and I am having that carefully checked this week.

Will you please either sell those stocks or send them to Anne to sell? They keep on getting lower and lower and I have to have the money quickly anyhow as I am still running behind and have had additional legitimate but unexpected bills. I don't want to handle the stocks myself for known reasons.

—Well, that's all the bad news in one large dose, and I was hoping not to have to send it until it could be balanced by good news. The only good news now is that I do still feel that all will clear up before so very long, deferred as hope is.

We saw the Peets not long since and they asked about you and send

33. Anne was still suffering from brucellosis, but at least by now she had the satisfaction of a correct diagnosis. The ailment is caused by bacilli, usually transmitted in raw milk, from cattle, pigs, goats, and sheep. The bacilli tend to lodge in one or another organ; common symptoms are nonspecific fatigue, backaches, nausea, and fever. Until the discovery of antibiotics the disease was difficult to treat. Anne's treatment here was a series of painful serum injections.

regards. Bertha Anne is again pregnant and it looks as if she would be successful this time—I think you know that she lost a baby last year. To her
surprise she still has her job, but it is so insecure and unpleasant that she
will be glad to leave. From where I stand their finances certainly look
shaky, but they must know what they are doing. The Claxtons are doing
very well. Dr. Granger has been having a touch of lumbago and in general
his health seems to me to be going downhill, but he is not really ill and is
as nice as ever and a great comfort.

214
AMERICAN
MUSEUM OF
NATURAL
HISTORY
1934–1938

Anne had a nice note from Marty. It seems as if things were brightening
up somewhat out there, and about time. There really seems to be every
prospect that Jack will be cured and in the meantime he is really happy
and the worst of the burden and worry is off Peg and Marty.[34]

Helen is about due home from church, so I will quit and stir up some
waffles for lunch.[35] My deepest love to you both, and I will let you know
immediately when anything definitely happens. Oh, Anne perhaps didn't
tell you that I got the cigarette holder, and I haven't thanked you for it,
which I do now. I like it very much and have not smoked a cigarette without it since it came.

Gg

[Stamford, Conn.]
[April, 1938]

Dearest Mother & Dad—

I still have not seen the decree & do not know all the details, but here
are the essentials:

The divorce is granted unconditionally & finally to me as the injured
party. It dismisses L's charges against me, cruelty, desertion, unfaithfulness,
& all as baseless. It does not give her any alimony.

I am given Helen's custody unconditionally & Gay's on the understanding that she will, for the present at least, remain with Mrs. P. [Lydia's
mother in Kansas].

L. retains Joan & Betty for the present, subject to review *by this same
judge* (a very important point I think) if there is any evidence of neglect or
improper care. Some provision is made for my seeing them & checking on
this. I have to pay L. $130 a month for their care—that's one point that's
hard to take, but it's in lieu of & better than alimony.[36] For one thing, it
ceases if she does not care for them & anyway when they grow up, so I am
not saddled with her support for life.

Appeal is possible until April 22, but Walker now says the chance of
appeal is slight. Cressy will not appeal unless he is paid handsomely, of
which there is little prospect. The judge has been so extremely careful that

34. Peg's son Jack was suffering from a neurological illness at the time.
35. Simpson regularly made "Mongolian" waffles for Sunday breakfast. "Mongolian" because the recipe he used called for cream of tartar; the recipe itself had been
painted on the kitchen wall by Martha in one of the New York apartments.
36. This amount was about one-third of Simpson's gross salary.

an appeal by L would be practically sure to lose, but of course its nuisance value will attract her.

This is about as sweeping a victory as a man ever won in a contested suit, vastly better than really seemed possible. Perhaps there *is* some justice! Anyway I'm not mad at anyone now. The remaining problems are all capable of solution & seem entirely unterrifying now that I am free of the unbearable ones.

215

AMERICAN
MUSEUM OF
NATURAL
HISTORY
1934–1938

We cannot make definite plans till the appeal period is passed, but we tentatively think of being married at Jessie & Chet's on April 30 (Walker says no need to wait after the 22nd for any reason). That is Saturday when most people can get away easily. Would this be all right with you? We would like to be married in Washington, & will if for any reason you cannot come to N. Jersey, but that would cut out most of the friends we would like to have there.[37]

I don't think you have any doubts or misgivings, but I do want to say again that this is one marriage that is really fool-proof.[38] I have loved Anne deeply for years & only become more fond of her as each year passes. Through all our troubles we have never had a serious misunderstanding & have never exchanged cross words. We know each other as well as any two people ever did, including the faults & weaknesses, so that there can be no disillusionment or unhappy discoveries. I *know* that it is impossible for anything to break up this marriage or for any further sorrow for you or me to result from it as long as we live. I am certainly not rushing into this, & I have certainly learned to recognize signs of difficulties & troubles.

The other thing, really the first thing, that I wanted to say & that I can't quite find a way to express well or fully is my everlasting gratitude & appreciation for my parents & all they have done for me. Your constant forbearance, love, & help in every way have been the most sustaining & blessed thing that life could hold. You have not only helped me materially, but also have enabled me to remain sane & have given me continued faith in human decency, love, & kindness. There aren't words with which to thank you or to tell you how much I love & appreciate you.[39]

> Your unmarried (but happily
> engaged)
> Son

Dear Dappie and Gramma;
    Ain't love grand? I'm going to get a new dress out of it and be bridesmaid.

> Love & Kisses
> Helen

37. Jessie Read was a physician who knew Anne at Denver University. Her husband Chet was an industrial chemist. The wedding did take place the following month in their New Jersey home, and Simpson's parents came up from Washington, D.C., to attend.

38. The year that Simpson died, he and Anne celebrated their forty-sixth wedding anniversary.

39. On the back side of this letter Simpson's mother wrote that she "wept buckets" over it.

Dear Keeds[40]—

The battle is over, & all troubles henceforth will be seen like child's play I imagine. In short, as Dr. Granger would say, I have no more wife than a jackrabbit. L. did not appeal or obtain an extension of appeal period, & now never can appeal & the business is absolutely finished. There is no interlocutory period in Conn. & divorces are final forthwith, so mine is as of April 8, but curiously enough they can be upset nevertheless on an appeal, which must be filed within 2 weeks—this can be extended indefinitely. Our two weeks ended Fri. afternoon. We didn't hear & didn't hear & were both—Anne & I, for I was able to go to her place for the first time this year—we both were jittering & finally Jimmy turned up & he didn't know. Then about 11 I managed to get Walker on the phone & he thought it was OK but wasn't sure. He called me at last yest. morning, Saturday, that all was entirely well & a certified & final copy of the decree [was] in the mail.

Anne & I were all for getting married the 30th, but were talked out of it by Jimmy, who was shocked & does deserve a lot of consideration after all he's done, & by the fact that Mrs. Roe will be here late in May & not before. So now we think May 27, Friday evening. Anne is going to Phila. with her friend Cathy McBride[41] & then perhaps to Washington for a while, in the meantime. We'll be married at Jessy [sic] and Chet Reed's [sic].

I must dash—this was only to give you this news. Probably Anne has anyhow, but I want to also.

Much love,
Gg

40. Addressed to both Peg and Martha who had by now moved to southern California, where she was to remain for a number of years.

41. Katharine E. McBride (1904–1976), Bryn Mawr Ph.D. in psychology who recruited Anne for the Philadelphia aphasia project. They later collaborated on a book dealing with adult intelligence (1936). McBride subsequently became dean of Radcliffe College and then president of Bryn Mawr, 1942–1970.

216
AMERICAN
MUSEUM OF
NATURAL
HISTORY
1934–1938

VENEZUELA AND NEW YORK CITY 1938-1941

IN 1938, one month after his divorce was final, Simpson and Anne Roe married. Following a short car trip through the southeast for their honeymoon, they returned to Stamford, Connecticut, where Simpson had been living with Helen, his oldest daughter. Gay stayed in Kansas with Lydia's mother and Joan and Elizabeth remained with their mother, although the following year Simpson was to get full custody of them as well.

In the fall of 1938 Anne and Simpson took an extended trip to Venezuela where Simpson was a guest of the government to study some recently discovered fossil mammals. Anne, unable to pursue her own psychological research, learned to trap and skin various mammals for the American Museum collections while her husband searched for fossil bones in a nearby quarry. Misinformed about the weather, they landed in Venezuela just as a prolonged rainy season began, which greatly handicapped Simpson's work. Later, they went down to the Gran Sabana, in the southeastern part of Venezuela, a rugged, hilly plateau region where Angel Falls, the world's highest waterfall, lies. Anne and Simpson were among the first to fly over the falls. While there, Anne continued her collecting, but Simpson became interested in a remote tribe of Indians, the Kamarakoto, whose way of life and language he studied. He subsequently published a monograph on these Indians, and Anne published a paper on the living mammals she had collected.

The Simpsons returned to New York City in May 1939 and settled, together with Helen, in an apartment across town from the museum. Before long they also had custody of Joan and Elizabeth, who joined them. Except for Gay, who continued to live with her grandmother in Kansas, this was virtually the first time in years that the whole family had resided under one roof.

Although Simpson's domestic life had taken a turn for the better, various administrative and financial matters at the museum began to plague him. Not only were the various departments to be reorganized in a manner that diminished the distinctiveness of paleontology but Simpson was to remain associate curator, a rank he had held for thirteen years and that little reflected his reputation in the field. He had recently been elected to the two most prestigious scholarly societies in the United States, the American Philosophical Society and the National Academy of Sciences. Thus, by 1941 Simpson was restless and beginning to consider leaving the museum.

Simpson's research was focusing more and more on the theoretical aspects of the history and evolution of life. By late 1942, just before he joined the army, he had completed the manuscripts for the book *Tempo and Mode in Evolution* and for the monograph *The Principles of Classification and a Classification of Mammals*. (Both were published during the war, while Simpson was overseas.) Simpson was also developing a theory of historical biogeography to explain the dispersal of terrestrial vertebrates during their evolutionary history. Many of his ideas came from his study of South American mammals, and he concluded that mobile animals moving across stable continents had produced the geographic distribution of mammals in the past. He thus became a strong opponent of the theory of continental drift as proposed earlier by Alfred Wegener, who argued for immobile organisms subsequently separated by mobile continents.*

Anne's preoccupation with establishing a home for her new family precluded her full-time involvement in research. She did, however, find time to do some editing, some research on alcohol education, and consultation on a foster-child study.

[Stamford, Conn.]
Monday
May 7, 1938

Dearest Mother—

I saw Joan [age 10½] & Betty [9½] yesterday & they are well. I turned up early, a-purpose because I thought L. couldn't stop me if they were home & that I would surprise her before she took them away, if she planned to. I don't know what her plans may have been, but they were home. She sputtered when she saw me, told me to come back at the proper time, & slammed the door, so I got a chair & sat down on her porch in front of her windows (she pulled the shades) & read a chapter of Gibbon's Decline & Fall of the Roman Empire—a chapter much about Faustina, one of the less delicate empresses, as it happened & much to the point! She stood this about 5 minutes & then bounced out to order me off the porch & tell me I couldn't see the children for another hour. I pointed out that I was under no obligation to set a time, that now was a reasonable time, so did she or didn't she intend to obey the court order, then (if she'd only try!) & [sic] disappeared, then popped out and said "You lousy pest!" in a *very* sincere tone, & again & finally disappeared. It had begun to be a

*Of course, Simpson was wrong and Wegener right. For more on this, see Léo Laporte, 1985, *Wrong for the Right Reasons: G. G. Simpson and Continental Drift*, Geological Society of America, Centennial Special Vol. 1, pp. 273–285.

lot of fun to be pestering her a little for a change, & yet with entire courtesy & the awful dignity of the law to support me in action perfectly correct & yet almost unbearably annoying to her! Almost at once the children came out, so I took them for a walk & we sat in a nearby parked place for an hour or so. Then I took them back, as we were talked out, I had nothing to occupy their time pleasantly for longer, & I didn't want L. to get so mad she'd take it out on them. But the house was all locked up, so I just left them sitting on the porch—I'm sure they were all right, & I suppose L. will say I should have kept them longer so she could have a vacation!

They both seemed very friendly & loving toward me, only the least bit subdued as was inevitable, & told me all about their school & friends, but practically nothing about their home life. Mother was well. They were happy at home. No details, & I didn't push them too hard as I didn't want to trouble them overmuch or make spies of them, & of course they had been daily threatened if they said a word. I told them all the provisions of the decree that concerned them—it was all news to them & very welcome as they were under the impression that they were absolutely taken away from any contact with me except at their mother's sweet pleasure. Betty did break down & say that Mother seemed angry at something!

I gave them your letter & they both read it with delight & sent you much love & talked of happy memories of Washington—no fear that propaganda has changed their attitude toward you & Dap. They also asked me to thank you & everyone for their Christmas presents, because it was the first chance they'd had. I gave them some stamps & they said they'd try to write to you, but weren't sure they could. They wanted very much to see you, & I told them I'd fix it so they could some time later this year.

Joan had a cold a week or so ago, but was over it. They both looked well & hadn't had to see a doctor recently, although not gaining much weight. Betty has had no acidosis or ear attacks since the spell last year.

They were keen about their school & both passing all their subjects. Joan got a D+ in I think it was cooking, some such thing, a B in arithmetic, & all the rest A's. Betty got a D (that's below average but passing) in spelling, & mostly B's for the rest—she wasn't so very explicit! But in any case she is passing all right. They have at least one good friend at the school about whom they talked a great deal & with whom they had been to the circus. They behaved very well & were sweet & nice.

This relieves my mind, & I hope will yours, in showing that the children are not in any serious need & are generally happy. It doesn't give much to tie into in the way of getting them back, but this will come in time, probably quite soon, & of course we don't want it to arise from any actual suffering on their part.[1] What pleased me most was the obvious fact that their minds are not poisoned against you or Dap or me—It was apparent from their whole manner, much more from their manner than from

1. Lydia received custody of Joan and Betty on the condition that she care for them properly. Nonetheless, Lydia kept herself and the children on the move, at times even leaving the girls in care of others and keeping her ex-husband in the dark about their whereabouts. Before long the court awarded permanent custody to Simpson.

what they actually said although they said so too, that they still love us, think of us, & count on us in the background even though not desperately unhappy as things are.

I'll soon write to them & see whether they get it, & then of course will go see them again too, although perhaps not again before the wedding as I am so busy & active worry about them is relieved for the time being.

Much love to you & Dad—It will be so nice to see you again soon—

Gg

[Stamford, Conn.]
June, 1938

Dearest Mother—

Thank you so much for the briefcase, which is perfect & exactly the sort I wanted, except that it is better than any I had in mind. Also thanks for the check, likewise perfect! About the car, though, I wonder whether you really want to spend much on it or put another summer's wear & tear on it before selling. I can use it very little & no one has had money's worth out of it for a couple of years. Of course I shall do just as you wish about this, but won't get the car out till you have thought it over & decided definitely.

There is still a chance that I will be going to Venezuela this year, per-haps in the Fall. This is vague now, & we'll discuss it further when things shape up.

Helen wants to be with us in Stamford most of the Summer, but thinks she might like a couple of weeks in camp, in August, & if this can be man-aged it would probably be good for her.

Anne shipped Marty's pictures to you today, by your friend as recom-mended. He sent them collect because he did not know how much they would be. I would be glad to help pay this if & when, however.

I feel awfully well & tremendously happy.

Much love to you both—

Gg

July 4—noisy, too, because prohibition of fireworks seems to be just as effective as prohibition of liquor [1938, Stamford]

Dearest Mother:

We had a too brief, passing visit from Mrs. Roe and by now you have had news of us from her. We were glad, in turn, to hear of you from her and to get more details about your accident. She told us your wrist was very painful, as of course it would be, and we do hope and pray that it is less so now and knitting satisfactorily. It is some consolation, although not very much, that it could have been so much worse.

We went to see Joan and Betty again yesterday and had a very pleasant visit with them. They are still perfectly well. They could hardly talk about anything but you and Dap and plans to see you, to which they look forward most eagerly. The present idea, which I can still modify if it does not work out well for you, is to have them here from the evening of Friday July 22 to that of Sunday July 31. I thought that would get it in before you are leaving for California and at the same time give you a chance to recover more fully from your accident. Please let me know if this is not satisfactory. We have located some beautiful drives and picnic places, beaches, and so on and look forward to a grand visit then. I hope you can spend the whole time of their visit here, or more, but of course suit your own convenience and plans. Of course we also are eager to have Dap for as long as he can manage. Perhaps Helen can go down with you, him, or both when you go, as her camp will be starting soon after.

There is no more definite news of Venezuela yet, and I suppose this may be another of those last minute propositions, but it looks probable and we are tentatively going on that assumption. Granger is coming for a short visit if we go, and would love to have Dad travel with him. Of course we want him one way with us if that can be managed.[2] Sorry we cannot yet be more definite.

Mrs. Roe probably told you that we have all three been a little under the weather. Helen and I are perfectly well again and Anne, who got it last, is nearly so, and there is nothing to worry or even think about. We are really flourishing and very happy.

Much love to both of you,

Gg

Much love & I'll see you when you come up

Helen Frances S.

Fourth of July
[1938]

Dearest Marty:

Many new leaves as I have turned over of late, none seems to have resulted in prompt replies to letters or recognition of important anniversaries. Anyway, I am celebrating today by writing this letter, which will be a hummer if this feeling of lassitude doesn't get to me before I'm through. It is a) a reply to half a dozen from you, b) thanks for the extremely slick present you sent me, as well as greetings, c) the same and many of them for your birthday, now well past but not unremarked even at the time, and d) just a plain letter, and about time, too.

The song book is perfectly swell and is just about my greatest treasure. I don't see where you found all the hours to lavish such labor and loving care on it. It is beautiful, amusing, and altogether grand. I am slowly decipher-

---

2. Simpson's father, then sixty-eight, did join Simpson and Anne on their boat trip to Venezuela, where he remained with them for a brief vacation.

ing the tunes—deciphering not because they are unclear, which is not the case, but because you know how I am with new pieces, especially the few in flats, and not being able to use my little finger.[3] There are some swell tunes I have never even heard of before, and where on earth did you get them all? We are all taking a whack at them and having many a happy moment and many more to come.

Belatedly we did get off a book for your birthday. We were not absolutely sure that you did not have it, and if you did please return and we will send some other. If you did not have it already, we hoped it would give you some pleasure, and in any case it is loaded with love and good wishes of everything. In short, happy birthday, happy 75th anniversary of the battle of Gettysburg, happy Yom Kippur, happy Bastille Day, and just plain and fancy happy everything. You must be quite a big girl now and it seems only yesterday that you were only so high and needed to wipe your nose.

Perhaps you heard that I got married recently. My new wife is a lanky blonde hoodlum name of Anne, and I find being married to her curiously exciting and soothing at the same time. It's sort of fun. We went on a grand [honeymoon] tour of the central southeast, including Blantyre, Charleston, and more other places than you could conveniently shake a stick at. Some fun. Since then we have been desperately settling down for a final two months in Stamford, caning chairs (we really have) and raising cain. In spare moments we have explored all the little and big back roads in this part of the world, finding them amazingly beautiful. Roses, roses all the way this time of year, also greenery of every variety, nice little streams and lakes, lovely old homes, old churches, and all.

We have been to see Joan and Betty a couple of times and had nice visits with them. With mixed feelings, we find them very well and apparently quite happy with only slight reservations. I don't know how long this will last, but at any rate they are currently suffering no harm, and they are still very friendly and loving to all of us, including Anne, which probably shows considerable sales resistance.

By now you have heard of Mother's accident, which is certainly tough luck. Aside from word from Dad, we have recently had firsthand news through Mrs. Roe, who assures us that Mother is really doing and looking well aside from severe bruises and much pain from the wrist, which is knitting and should not be painful much longer. Toward the end of the month I am having Joan and Betty here for a visit and Mother is coming up then to see them, probably Dad too for a week-end.

Mrs. Roe was to stay over today, but her office has been having its usual shake-up and as usual she is the fall-guy and had to dash off again to Washington to work over the holidays. Ed has a job in Denver which may be permanent. Ethel Mae married her very nice boy friend and they stopped with us on their way to Europe, distraught but happy. In general the Roe family seems to be in pretty good shape.[4]

3. Simpson was a self-taught mandolin player and often accompanied Anne on the piano.

4. Anne's mother, Edna Roe, was national field secretary for the Parents-Teachers Association. Edward and Ethel Mae ("Pat") were her younger brother and sister.

Bertha Ann is apparently getting along well this time and the little Peet will arrive early in the Fall. She has been up in Maine for a while, escaping heat and stairs, but returns this week, and then she and Creighton are going to stay in a friend's house in New Jersey for the rest of the summer. The Claxton's continue to flourish. Perry has just dashed of[f] a book for tots, of all things, and also a play so dismal that no one will produce it lest all the audience commit suicide.

You have probably been told that McGraw-Hill, who or which is the big time publisher in our lines, snapped up Anne's and my book [*Quantitative Zoology*] on very fair terms and with pleasant comments. We will make no fortune, nor even pay labor wages for the hours put in on it, but every little [bit?] helps, and we expected a battle to get anyone to take a chance on publishing it at all. This is going to press right soon now, and we still have a bout of [page] proof and indexing to look forward to. I suppose it may get out about the end of the year.[5]

The Museum gave us a tea-reception, to which some eighty-odd, odd people flocked to vet Anne, all deciding that she was sound and I could keep her. To our surprise we rather enjoyed it.

Helen had a severe bout of poison ivy, but has recovered and is now being immunized, which can be done with pills, curiously enough. Her school is over and she did fairly well and passed everything, although she got a rather low mark in English to my and her surprise. In a way the year was hard on her what with all the publicity and strain which she naturally felt also.[6] Then she had a little attack of flu which I caught from her and Anne caught from me, but we are all about well now and flourishing. Helen's going off to Camp in Virginia for a couple of weeks in August.

The chances of our going to Venezuela this Fall look fairly good, but still nothing is settled. Dad will probably go with us if we go, which will be a lot of fun. Dad has always longed so to go to South America and it will be swell if he now gets the chance.

I guess that about brings you up to date. We all love you very much. You are a swell girl. In fact we are all unusually nice people.

Gg

"Les Robles"
[Venezuela]
Nov. 23, 1938

Dearest Marty—

I suppose one trouble of our journal system is that it keeps people informed in such exhausting detail of our activities that we feel free of any necessity to write personal letters, & the catch to that is that few bother to write us. Mother has told us news of you occasionally, but not enough, &

5. *Quantitative Zoology* appeared in 1939; a second edition, coauthored with Richard C. Lewontin, was published in 1960 by Harcourt, Brace.

6. Simpson's divorce trial had been sensationally described in the New York City tabloids.

we haven't heard direct from you, as we long to do. Anne wrote you some time ago, but we haven't your address now & we never know whether anything will arrive or not.

Just now the sun is shining brightly & you never saw a more placid Venezuelan landscape, but lately we have been having rain, rain, rain. We were completely misled into thinking this the dry season here. It happens to be the rainy season, & the tropical rainy season is something terrific. However, it is short in this relatively dry area & there is hope that it is nearly over. My quarry is a sea of mud & I finally had to stop work a few days ago. If tomorrow is also dry, I'll be able to resume then. Meanwhile I am writing letters & generally catching up on things while Anne, who is the most earnest little mammalogist you ever saw, is engaged in skinning a very handsome fox next to me. She has a beautiful howling monkey, which has bright red whiskers & looks not unlike me, an anteater, lots of opossums, spiny rats, & all sorts of curious tropical critters which, incidentally, smell something fierce. You can say lots against fossils, but at least they don't stink. Incidentally our donkey boy has provided us both with slingshots, but so far we haven't even scared anything, let alone killed it. Two of my dark working gentlemen are earnestly discussing anatomy, squatting on the ground next to me. The dusky damsel who cooks & washes our cans (of which Anne has a hoarded assortment) is up in the cook-tent making ayacas (an extremely tasty local dish). Several filthy but nice small black children are prowling around. The donkey boy & donkey are up in the next mountain getting a load of wood. All the local buzzards are down in the canyon below us eating the body of Anne's fox. In short it's a placid afternoon in camp, & we hope you are the same.

We have become really attached to this spot & are unhappy when we have to be in town. Despite minor ups & downs we are both well & find this a pleasant mode of life in general. I have eaten thousands of words about women in camp & heartily approve of them (or her) in this one.

We have been looking for spot news of the art world, but have achieved few items. The pulpería "Los Milagros de la Divina Pastora," between El Rodeo & Villa de la Capilla de Santa Rosa "Tin-Tin," has two new paintings on its façade, the blue one generally believed to be a lion and the red one either a man or a monkey.[7] The Hotel Washington in Barquisimeto has a new poster on the wall of the dining room, exhibiting a blonde in a pink slip who owes her beauties to a popular laxative on sale cheap in all farmacias. In the higher art world of Caracas, the Museo de Bellas Artes was only erected some ten years ago & so of course is not yet ready to open, but it is evident elsewhere that Venezuelan artists (who all study in Paris if at all) either paint battles with a few Royalist corpses in the foreground & Bolivar on a white horse in the background or else magnificient classical scenes like the Death of Nero. It apparently has not occurred to anyone to paint Venezuela, which is the most colorful &, to my skilled eye, paintable country I ever saw. In towns we particularly like the colors of the houses, every conceivable hue & shade & yet giving a nice gay &

7. Simpson is describing mural paintings on a countryside inn (*pulpería*) with the name "The Miracles of the Divine Shepherdess."

not terribly clashing effect. Then, city or country, there are flowers every-where, greater in variety & quantity even than California at its best, from tiny sky-blue flowers hidden away on the ground to enormous trees simply blazing with big scarlet trumpets. With the blue sky, the usually red earth, & the infinite varieties of green, there is so much color it almost hurts.

(Slight pause while we debate "Resolved that a lady mammalogist who gets a beautiful fox skin owes it to herself to make it into a neck-piece instead of a study skin," with the undersigned upholding a firm negative).

It does seem like years since we were together on 57th St., & we talk & think about you so much & do hope things work out for us to be together again. By the way, you will have heard before this that the Peets have a large son, which delights them & us, although we can hardly imagine the household with a baby in it & wonder whether it or its Pa will get the upper hand as regards strange noises & the like. We have no news of the Claxtons.

We hope to get to Caracas for Christmas & then to buzz around a little seeing Venezuela. We are not sending any Christmas presents except a cou-ple of bits for the kids for several reasons, among them that the country is almost without arts & crafts & everything worth giving is imported, with a 200 or 300% raise in price, from the U.S.A., that packages are lost with great regularity, & that the customs regulations are very difficult. Some-place [sic] we hope to find something Venezuelan to take with us as belated presents when we leave, but I don't know just what or where.
[Anne adds the following:]

This seems a fairly adequate summing up of the present situation, except that I have 3 rodents waiting to be done & I think I ought to train G[8] into this, things sometimes pile up. Fossils have another advantage besides not stinking (of course G wouldn't mention that our usual stinks to date have been when he wanted skeletons of some beast or other to compare with his old fossils—nice mammalogists don't have anything to do with nasty skele-tons below the neck).

I enclose a Xmas present (ed. note "sic") of a few Venezuelan recipes or approximations of them which I hope you'll try thinking of us. We think Peg is in Honolulu is that right? Or is it you?

Much love & kisses
Anne
Ditto
Gg

[Venezuela]
Dec. 13, 1938

Dearest Mother—

We are still marooned on our mountain, with impassable torrents all around us, as we have been for three weeks. Our work & our belongings are in hopeless shape—my quarry a deep morass of mud with a few broken

8. Anne's nickname for her husband.

bones floating around in the muck, everything covered with mold, etc.—but we are safe & well, ourselves, & still eating regularly.

We have had no news from the outer world except one letter and a package of magazines that got through in some curious way. Tomorrow my [illegible] is trying to get out with a donkey, & he will start this letter on its way. We are somehow going out ourselves next week, but so far have not seriously needed to. The whole country is disrupted by floods & I couldn't work anywhere, & here we are as happy as elsewhere.

I enclose a check for Helen's expenses.

We really are all right, only a little tired of rain & anxious for news of you. Perhaps Trino will have some for us when he gets back, which will be day after tomorrow with luck.

Gg

[Anne adds:]

There really isn't any news. We spend time going down & weeping over the quarry, writing, and looking to see which direction the next storms are coming from. But we love you all—

Anne

Caracas
Feb. 7, 1939

Dearest Family—

Here we are safe & sound in Caracas once more & reveling in another dose of luxury. We alternate between the very primitive & the very luxurious, & it'll be hard to settle to a balance between them again. We are at the Phelps' once more, & as usual here are treated like king & queen.

Last week in Barquisimeto was pretty hectic—so much so that we are behind on the journal & can't get it off on this boat. I'll have to return there again to polish up, but meanwhile got most of the bones to Barq. & about 500 of them packed—in the end they began to pile up enormously & I almost felt, for the first time in my life, that there were too many bones.

Now we are awaiting the high-sign to join the boys on the Orinoco [River] & parts south. This seems a little disorganized, as usual, but will probably work out. Aguerrevere [a Venezuelan geologist] flew to the Gran Sabana yesterday in order, among other things, to get the Indian chief's safe conduct for Anne! He'll be back next week & let us know definitely & we'll probably flit down soon after. Zuloaga is on the Orinoco & I hope to talk about that with him by radio tonight.

Meanwhile we are catching up on both rest & work here & glad of a few days without pressure. Anne had a touch of the old brucellosis, which is discouraging as we thought it was gone, but not bad & she is generally in very good condition.[9] I lost 10 or 15 lbs. in my bout with anemia but am

---

9. The serum injection treatment Anne had received did not effect a cure of her brucellosis. Only during the war did she finally shake this illness.

practically well & feeling my oats again—I was very lucky to pull out as well because it sometimes takes years to conquer. They had a bad outbreak of blackwater fever in Barquisimeto but we both skipped that & are well out of the infected zone now. In short, all is well & the goose hangs high! [Anne adds:]

He really is much better, I'm finding it difficult to keep him under control again, so you see— Give my love to my parents, too, and our daughter, there. I'm not going to write any letters until I finish this (ms) —I hope today—

[GGS adds:] This cryptic sign means "manuscript". We are both so delighted at our little sister's [Martha] getting hung (in the Metropolitan [Museum of Art, in N.Y.]). We know it's only a question of time before she'll be as famous & appreciated as she deserves, and are glad that the time is not to be much longer. We don't remember "Suppertime"—is it one we have seen?[10]

Our dearest love to you all—

Anne & Gg

This is forged, too, just to keep up the family skill.

I hope your forging was better!

Anne

[GGS P.S.] By the way, it's MINERIA, not MINERLA as everyone is beginning to write on Mother's authority.[11]

[New York City]

July 15, 1939

Dearest Mother & Dad—

I enclose a letter from Betty which pleased us because it sounds happy. The camp really is a nice place and while they do need love and a home, it is infinitely better than where they were and a good second-best to having them here.

Anne's & my book, "Quantitative Zoology," is out at last, and a great relief. We'll send you a copy soon, when we get an extra one. You won't want to read any of it, but perhaps will enjoy having one around & will get a kick out of "by George Gaylord Simpson and Anne Roe," as we do!

Last night we had an elegant buffet supper with Dr. & Mrs. Granger, Dr. & Mrs. Colbert, & Father Teilhard (the French Jesuit paleontologist, a swell guy, now visiting here) to celebrate our book, to warm our now completed new home, and just for general fun.[12] It was very pleasant.

10. There is no current record at the Metropolitan Museum of Art of any of Martha's works in their collections.

11. Simpson was working with the Servicio Técnico de Minería y Geología, the equivalent of the U.S. Geological Survey, Department of the Interior.

12. Edwin H. Colbert (b. 1905), curator of fossil reptiles and amphibians, was a longtime associate of Simpson's at the American Museum and succeeded Simpson when he resigned the chairmanship of the department of geology and paleontology in 1958. Pierre Teilhard de Chardin (1881–1955), French Jesuit and vertebrate paleon-

Among other things, it was nice to see how natural & easy Helen [age 15½] was with guests. She seems almost to be over her adolescent awkwardness in some respects & seemed very pleasant & mature. She helps a lot around the house and while obviously (& naturally) not fully grown-up

is really a great pleasure and aid in the home. She and Anne get on so well, like unusually fond older & younger sisters, which is naturally a delight to me as well as to them.

My Venezuelan bones are all here at last and I will have a man at work on them this week. There are nearly *six tons* of them! I had to stow them in the courtyard because we have no store-room big enough, so no one complains that I didn't get a collection—quite the contrary. In the meantime I am working mostly on Indians, just now writing a brief dictionary & grammar of the Kamarakoto language, which is fun because I never did anything quite like it before & for that same reason is a lot of work. I had to knock off this week to write a paper on Antarctica for the Pan-Pacific Congress—another subject I knew nothing about & so had quite a struggle with![13]

Anne has had a painful stiff neck since our drive last week-end but it is better now. She took her driving test a couple of days ago & was scared to death but passed it easily & now has her license. We were going camping this week-end but decided not to until Anne has quite recovered from her neck & her trichinosis, both now nearly well. Helen is bouncing, as usual, & I am also entirely well again, no dizziness at all now, suffering from the heat but generally feeling fine.

Greetings to Jack and my deepest love to both of you—

Gg

[New York City]
July 15, 1939

Dearest Marty—

I think it's the limit that I haven't written you for so long and I haven't even a decent excuse to offer. We think about you & talk about you all the time, but that's just another reason why I should have written.

In the first place I have been enjoying the St. Nicholas, Pompeian Frescoes, & other tidbits to an extent that is positively immoral considering that I have never thanked you for them. You knew I would love them and I do, and thanks a million for them.

You may have noticed that you received no birthday present—there is a reason, although inadequate, for this. I intended & do intend to send you a new photograph of me because I now look so unlike all the old ones, but

tologist, best known for his work on fossils associated with Peking Man and for a view of a goal-oriented, purposive evolution. Teilhard de Chardin built up a loyal following of believers who found much that was appealing in his nonmaterialistic evolutionary views. Although Simpson disagreed strongly on this latter point, he and Teilhard de Chardin were quite friendly during those times that Teilhard made extended visits to the museum.

13. This conference convened in San Francisco and it was there that Simpson presented his first formulation of his theory of historical biogeography.

with rush of work & a bout of illness this didn't get taken until recently & even now I haven't the prints. You'll get one someday, if you want it.

We're well settled in our new place now & thanks to Anne's efforts it look very swell.[14] We had a party last night for the Grangers, Colberts, & Father Teilhard which was a great success. Helen has matured a great deal & it was a pleasure to see how much more at ease & more adult she is with people.

Father Teilhard, charming as ever, is here for a brief visit & we've managed to grab him for a couple of drinks & for the dinner last night. The competition is heavy, as usual, & we have to snag him from all the women in town who follow him around in flocks!

We were so sorry that Mora Phelps hadn't time to look you up there. It seems that she found an old friend of hers in sad difficulties there & spent all her brief time trying to help her. Now they, the Phelps, are in Rochester, Minn., where Mr. Phelps has been getting some repairs made by the Mayos, but I think they'll be back here soon.[15]

Joan & Betty are in camp in Conn. now & we saw them last Sunday. Their tribulations have had visible effects, sadly but inevitably, but their inherent sweetness is still there. The camp is a fine place & next to being with us it seems best for them. We still think we'll have them next Fall.

The Peet's baby is a darling and incidentally is almost as crazy about the picture you sent him as his parents are. Creighton's "Dude Ranch" book is out at last—Oh, by the way Anne's and my statistical "Quantitative Zoology" is also out, & what a relief! Despite swearing "never again," Creighton is working on another photographic book, on building a house. The Claxton's continue to prosper & Perry to be the picture of clean-cut American manhood. I still don't think you can reform a man by marrying him, but everything has to work sometime & that did in that case. He's working on "Cue" [magazine] now, which has developed into a very popular "Paris Semaine" for New York, rewriting a play, & dashing off occasional fiction for the slick magazines, and altogether is becoming one of our more successful journalists & authors.

Walter Granger is still his jovial self & has been well, although he is aging slowly. He's going to South Dakota for a while next month & may get on to California, I'm not sure.

We had a pleasant visit from Jack, whom we like and who seems to us to have a good chance to pull himself together & become a useful citizen.[16] He's driving Farringtons to Maine next week—but of course you are well up on Washington news.

I'm having fun writing about Indians, doing a brief grammar & dictionary of the Kamarakoto language at the moment, which fascinates me but annoys my colleagues who want to know when is a paleontologist not a paleontologist.[17]

14. This was the Simpsons' new—to them—apartment on East 81st St.
15. Mona and Walter Phelps whom the Simpsons had visited earlier in Caracas, Venezuela. The Mayo Clinic is in Rochester, Minnesota.
16. Another reference to his nephew, Jack McLaurin, who had suffered some psychological problems.
17. Perhaps Simpson's talents, made even more evident by his Kamarakoto monograph, stimulated a little envy among his less talented colleagues.

Anne had a severe bout of trichinosis (parasites from bad pork with horrible symptoms), which is very widespread in N.Y. now, but is about over it. I am entirely well again & Helen is bouncing. They both send you their best love, & so do I and many of them. We do long to see you.

Gg

New York City
[June, 1940?]

Dearest Marty—

I hate having people give me checks instead of presents for birthdays, unless they are $100,000 checks, of course—But that's the way it is this time, I fear. Anyway lots of love, a million birthday wishes, & I hope this reaches you in time.

With one thing and another we have become a trifle confused. That's one reason why when Anne went to Boston a few days ago we figured she'd be back in lots of time to gift you & greet you at leisure, & only now I suddenly realize that such is not the case but contrariwise. Hence this hurried note & check instead of the diamond stomacher we had planned.

Anne went to see one man about the evil effects of alcohol and another about some spiny rats—Believe it or not it's true, so help me.[18] And speaking of believe it or not we were on Ripley's radio program, b.i.o.n., which was rather fun although involving about four lines which we had to rehearse for three hours & which were then entirely rewritten two minutes before we were on the air. He's an amusing, incredible guy, complete with the most amazingly junky apartment you ever dreamt of, housing, among a million other things, his ravishingly beautiful Chinese girl friend.[19]

Museum situation remains wholly in the air & my nerves are twanging.

What I do like to get for birthdays is books like the fascinating one that you sent me for mine. I am crazy about it & thanks a million.

I am currently (when I can bear to work) clearing up all the mysteries of evolution, you'll be glad to hear.[20] This aft. I got bored with it though (who wouldn't) & took Joanie to see the beloved Brooklyn Dodgers wallop the nasty St. Louis Cardinals. More fun! And hoping you are having the same—

Much, much love & birthday
kisses & hugs—
Gg.

18. Robert L. Ripley (1893–1949), American cartoonist who created the newspaper feature "Believe It or Not!" and later appeared on the radio and television.

19. The Simpsons were on the radio program to recount some of their Venezuelan adventures. Afterward they went to Ripley's apartment, crammed with exotica from around the world, where Ripley's Chinese mistress confided to them that "there wasn't a damn thing to read in the whole place."

20. Simpson is here referring to the manuscript of *Tempo and Mode in Evolution*.

[New York City]
[January, 1941]

Dearest Marty—

As usual, we are crazy about your presents, & I especially about the
swell Gauguin which is a feast both as to words & pictures. Incidentally
the translator, Van Wyck Brooks, talked recently to our pet society—the
Amer. Phil. Soc. in Philadelphia—& he is a charming man. We've just
come back from my fossil-bone meetings, where we started a new Society
of Vertebrate Paleontology—Anne & I are developing a string of meetings
that we go to which we enjoy very much, most of my colleagues in various
fields being a lot of fun & far from fossilized.[21]

We do wish that you lived here & miss you a lot, as do all our friends.
Leaving New York would in many ways be an awful wrench, but I may
have to as it begins to look as if there were no future at the Museum & I do
think I should advance a little beyond my present position, good as it was
for a young man when I was one. All is very, very tentative, but it is barely
possible that I will go to the Univ. of Michigan to teach. Also, but still
less, possible that I might return to New Haven. I won't stay here, if I can
get work elsewhere, unless they remodel my department radically & make
me boss, which is extremely unlikely—All very tentative & not exciting
for you, but I thought you should know what's in the air.

Now I have a nice, clean sheet of paper & nothing more to say except
more thanks and lots of love.

Gg

[New York City]
Mar. 2, 1941

Dearest Mother—

Before I forget, could your owl-woman be a harpy? They were part
woman & part bird, anyhow. I'm not very good on mythical birds. The
Simurgh in my article is one.[22] So are the anka, amru, & sinamru, which
may be varieties of simurgh, & the roc, another, is sometimes confused
with them. Vishnu's bird was the garuda, as mythical a bird as there is.
The most famous is of course the phoenix. I can't think of any more by
name, although birds & part birds keep occurring in myths.

You will be pleased to hear that Betty finally cut her hair (i.e. asked &

21. The Society of Vertebrate Paleontology was organized at the Cambridge meeting of the Paleontological Society in December 1940. Simpson became its first elected
president the following year.

22. "How dost thou portray the simurgh?" appeared in *Natural History* magazine in
February 1941 and was written by Simpson in connection with the exhibition "The
Animal Kingdom" at the Pierpont Morgan Library in New York. The article discussed
the portrayal of animals in art through the ages. The mythical simurgh is shown as a
cross between a parrot and a peacock in a thirteenth-century Persian illuminated
manuscript in that collection.

permitted Helen to do it) & looks much better & more grown up. She's been on a rampage for some days, sometimes hilarious & sometimes not.

Thanks for forwarding Mrs. Mastich's note—you doubtless remember that I do know a good deal about Florida fossils, having dug some & written about them.

Anne is overdoing getting her monograph ready for publication.[23]

We had dinner last night with the Claxtons & Peets. Young Peet (not there) has a cold but generally is blooming & a bright little toughy.

[Anne adds incomplete note about coming Easter holidays for children. Apparently page missing here; w/GGS's signature?]

23. "Intelligence in Mental Disorder," 1942, *Annals of the New York Academy of Science* 42:361–490; written with David Shakow.

WAR YEARS
1942-1944

SIMPSON'S SITUATION at the American Museum looked bleak enough for him to consider moving to Yale. Perhaps because of his dissatisfaction, he was promoted to curator (approximately equivalent to a full professorship in an American university) and received a twenty-two percent raise in salary. Then the United States entered World War II. Although Simpson was forty years old and had five dependents, which would have protected him for a time from the draft, he nevertheless volunteered for the army. In December 1942 he was given a commission as captain and joined military intelligence. He skipped basic training and instead went to a series of schools before shipping overseas. His military records indicate that he completed a six-week course of instruction in one week! Given his fluency in French and the November invasion of French North Africa by the Allies, Simpson soon found himself in Algiers. Among other duties, his intelligence group had the task of recovering enemy documents as soon as opposing forces' headquarters and command posts had been taken. Although Simpson was always reluctant to discuss his war experiences, they must have been dangerous, for Simpson was awarded two Bronze Stars.

From Algiers Simpson was transferred to Tunisia, then Sicily and Italy in the wake of the advancing Allied armies. He was promoted to the rank of major in 1943. The following year he suffered a chronic attack of hepatitis, and his condition eventually worsened so that in late 1944 he was sent back to the United States for treatment. He was placed on inactive status and although not fully recovered, he returned to the museum. He was appointed chairman of the department of geology and paleontology, the prewar reorganization plan having been scrapped in the meantime.

Helen was about to enter the University of Michigan; Joan and Betty were living with Anne and Simpson in New York City; Gay remained in Kansas. Anne had begun a study of alcohol education in the United States and published a monograph on the subject in 1943. Simpson's enlistment in the military had put a financial strain on the family, so Anne accepted an assistant professorship at Yale's laboratory of applied physiology. This meant a move from New York City, but at the time it also seemed likely

237

that Simpson himself would end up at Yale after the war rather than return to the museum. Nonetheless, Anne and the two daughters moved back to New York when Simpson left the army.

Martha had moved to Los Angeles before the war and she remained there for the duration. Simpson's father retired about this time, so his parents also moved to California from Washington, D.C. Sister Peg was in Los Angeles as well, so much of the family was united there. Martha ran a successful business in ceramic crafts and was doing far better financially than she had ever managed to do as a painter. In the fall of 1943 Martha married a young man nineteen years her junior—William Eastlake—who after the war had considerable success as a novelist and screenwriter.

Simpson had completed two major works before entering the army: *Tempo and Mode in Evolution* and *Classification of Mammals*. Both appeared in print toward war's end. Simpson had also recently completed a long essay on the beginning of vertebrate paleontology in North America, which was published in the *Proceedings of the American Philosophical Society.* While overseas Simpson was awarded the society's Lewis Prize for this essay. He was also honored twice by the National Academy of Sciences: in 1943 he received the Thompson Medal for his work on Mesozoic mammals and his contribution to evolutionary theory and in 1944 the Elliott Medal for *Tempo and Mode in Evolution*. Though out of sight, Simpson was hardly out of mind.

New York City
[January, 1942]

Dearest Marty—

I am adding a line while pinch-hitting for Anne (temporarily ill) at Air Raid Zone Headquarters,[1] where we keep someone on duty 24 hrs. a day— a lot of damn foolishness because no one has the slightest idea what to *do*, except to sit here & look important. Like all such volunteer services, at the start at least, this is run by a lot of stuffed shirts & self-important nincompoops who haven't sense enough to know a bombing if it hit them. Anyway, it is something to do & relieves the feeling of civilian uselessness in the emergency.

I really meant to write mainly to reinforce Anne's thanks for the lovely, wonderful books you sent—we have all devoured them & gloated over them, from me down to Bets.[2]

The Museum situation is very unsettling, but despite Anne's pessimism I have no real fear about having a job—only I am not entirely pleased with it. I am, however, managing to continue more or less & to accomplish something, as well as to enjoy life.

Our Cambridge meetings were very pleasant, including one of the Soci-

1. In the early months of the war, when the United States felt vulnerable to air attack after the disastrous blow at Pearl Harbor, the country took extreme precautions against further surprises on either coast.
2. Nickname for Elizabeth, the youngest daughter.

ety of Vertebrate Paleontology, which I founded and which is a great success (& a lot of work), if I do say it. We have 228 members—who would
have thought there were so many bone-diggers? Now that the organization
is running well, I am stepping out (or into the presidency, which involves
none of the work) & passing on the torch.

Except for the grippe now running through the family—skipping me so
far—we have been pretty well & had a grand holiday. We do miss you
greatly—you seem almost fabulous, as even the most dearly beloved do
when they are long away—and we long to see you.

Gg

[New York City]
Saturday
Jan. 31, 1942

Dearest Mother and Dad:

This is your errant son weighing in with a load of news for a change—
pleasant?

To get business over first, I received Mother's list of securities, etc., and
it is stowed away safely inside my desk at the Museum.

Of course we are all deeply interested and concerned with developments
regarding Dad's possible retirement, and anxious to hear as soon as anything is decided, as of course we will. I know how extremely trying the
long uncertainty of waiting for such decisions can be.[3]

Some of our uncertainty has been relieved, although perhaps the most
crucial point of all remains and may drag on for another year. What is
decided is partly good and partly bad. The staff is being very sharply
reduced, and our department has suffered out of its fair proportion. We are
forced to discharge the two last people hired, who happen to be an artist,
Alastair Brown, and a preparator, R. T. Bird. The artist deserved to be
fired anyway, but we badly need someone to replace him, which of course
we do not get. The preparator is good and we very badly need him. Then
they have put in forced retirement at 68, as in many universities, effective
on June 30 for those already past that age. In our department this shoves
out Barnum Brown and Bill Thomson. In general the department is left
less than half as large as it was only a few years ago. The scientific staff,
recently with six active members, is reduced to only two: Colbert and me.

I have been kept on and my salary has been raised $1000 (making it
$5500 per year) which was certainly unexpected and is as welcome as you
might imagine. Now we come nearer to seeing our way clear to paying
increasing taxes and girl-expenses and still getting out of debt some day.
On the other side of the picture, I have not been promoted and some
opposition has developed to what would appear (to me) the obvious move
of making me head of the department. This will not be decided, in all
probability, until Brown has actually retired, and perhaps not then. This

3. Simpson's father had turned seventy-two the previous November.

promotion was promised me, it is due me, and if it does not materialize I will raise hell and quite likely get bounced out—but it will probably go through all right and there is no use getting too excited ahead of time.

In the meantime I am plugging along at odd jobs and getting something done to justify my delightfully larger wages (for fun I worked them out on an hourly basis—about $3.40 per hour—which I often do not feel that I am worth when I consider what I did in some particular hour!).

I am enclosing Joan's and Betty's report cards, so you can see how they did, which is not brilliant but generally satisfactory. Betty proposed to write you only the good marks and finally surrendered her card for this purpose only reluctantly and on my promise to emphasize she has a B-plus average and is on the honor roll of her class. Bets has had a little cough for a few days and some sinus trouble but this seems to be under control again—she has not really been ill with it anyhow. She is taking serum to see if it won't help these recurrent difficulties.

Joan is a very busy little bee these days. With some leisure from school because of examinations—unlike my days, when they have examinations they have almost no work to do for a couple of weeks—she has been helping a good deal around the house and got taken to Porgy and Bess matinee for a reward. She also keeps up her choir singing at the church[4] and her girl scouts and now she has started to do guiding at my museum on Saturday afternoons, or at least has arranged and is learning to do it—the actual guiding has not started yet.

Anne has been having a seemingly endless round of sewing for one and all—I can't give technical details and will let her tell you when she writes. As you can imagine she is more than busy and has wanted to get a little farther along with the wardrobes before even hunting a job. Her big monograph is on the verge of appearing, as she has finally received and corrected the final proofs—also quite a task.

Almost everyone seems to be having illnesses, although our family avoids anything actually serious in that line. For instance Perry Claxton suddenly came down with bronchial pneumonia and is in the hospital, but has been getting sulfa-this and that and is doing all right. Bertha Ann and little Creighton both had flu or something of the sort, but are both up and around once more. Our social lives seem to be very limited these days—we have such enjoyable society in the home, and so much to do, that I suppose we do not make as much effort as we might. Anyway we are planning to go out to see the Reads tomorrow unless something prevents—at the moment there is a raging wind and rain storm outside which is almost approaching the proportions of a baby hurricane.

Feb. 12 I am lecturing at the University of Pennsylvania and we will stay over in Phila. to attend the Am. Phil. Soc. until Feb. 14, when I am giving one of the papers there in a symposium on early science in American history. Mine is on paleontology, of course, which I found had a long and fascinating history before 1842, all unknown to me previously or to any of my colleagues, apparently. Then I have to dash right back to New

4. Although not believers themselves, Anne and Simpson encouraged their daughters to attend the Unitarian church.

York because Jewett, President of the National Academy[,] has asked me to a dinner here that night.[5] The Academy itself next meets in April, and then I will have a chance to get down to Washington.

That's all the news—and quite a lot, too, now I look it over. I love you both very much, and so, I need hardly add, does the whole family.

Gg

New York City
July 19 [1942?]

Dearest Marty—

Many happy returns, happy birthday, & many thanks for your ditto. The box arrived before your letter, so I had two whole days of thinking all those riches were mine alone before I had to shell out to my troupe of (more or less) trained females.

It was sweet of you to send everything & the candied fruit was not at all spoiled by des[s]erts.

Anne, as she doubtless has told you, has taken a job here in N.Y. for 6 mo's or more, & we will be in town all summer save two weeks vacation. I am doing just lots of bone business & nothing amusing to you artists. We are all pretty well except when, as usual, the [war] news makes us sick & disheartened at trying to be civilized in a world that doesn't prize civilization.

It is so long since we saw you & we long to, but hope that you may be happy there. If only we could all be together!

Anyway, the fondest embraces and love to a sweet sister from her unworthy brother—

Gg

New York
Oct. 11, 1942

Dearest Family:

This is a circular letter copies of which will go to all of you, and I will add separate notes.

I will probably enter the army within the next month. I have received word that my application for the Army Specialist Corps has been approved for the rank of Captain and a tentative assignment made to the Censorship Branch of the Military Intelligence Division of the General Staff (G2). There is much more red tape before final approval and induction so that this is not to be considered certain, but it is so highly probable that you will all want to know about it now—When it does become final things may be expected to move very fast.

While Anne and I were vacationing in Vermont I received a wire that

5. Frank B. Jewett (1879–1949), electrical engineer and president of Bell Research Laboratories, and later chairman of the board of Bell Telephone.

G2 wanted to interview me and so I dashed down to New York, was elaborately interviewed, and returned to Vermont rather convinced that nothing would come of it. Then to my surprise I received notice on Oct. 9, almost immediately after our return, that they do want me and that I am to start through the mill. The first steps were completed yesterday: letter from my Selective Service Board, Nativity Affidavits, and Physical Examination.

The physical is something; two solid hours being examined, head to toe, inside and out, by 13 different specialists, complete with chest x-rays, urinalysis, blood analysis, and everything else. You will be glad to hear that I am in perfect physical condition. They found no defect of any sort and I am in better shape than most of the kids half my age that they are running through. Even my eyesight without glasses is 20:20 for both eyes, which means perfect. The nystagmus, the one physical defect I do have, was found to be so completely compensated and under control that they did not even enter it on my record.[6]

If and when I am inducted I will be sent to camp for "indoctrination" for about six weeks—this simply to learn the basic military procedures, get shaken down into uniform, etc. Being an officer in non-combattant service I will have a minimum of drilling and no combat training. After their indoctrination most of the officers in this particular branch are given an assignment in the U.S.A. to break them in for three or four months, then are given a foreign assignment, sent to special school for advanced linguistic and other study pertinent to the given foreign field, and then sent off. Of course they send you where and when they need you and there is no guarantee in any given case that this procedure will be followed. It is not certain that I will be sent abroad and if I am, I haven't the slightest notion of where it will be except (note, Mother) that it is practically impossible that it will be Mongolia.[7]

The work called censorship in G2 is not quite what we usually think of under that term. The purpose is to acquire information, not to delete it. In general they study intercepted enemy mail and other foreign mail and try to read between the lines for intelligence useful to our army, and they check domestic and allied mail to try to catch code communications or other evidence of spying or sabotage. As a captain I will be well up in the organization—the chief of the branch is only one step higher, a major—so my duties will doubtless be more in the way of organization and direction than the duller routine of the job.

Of course I did not seek the job on this account (in fact I did not specifically apply for this job at all; they picked me), but it happens that this is just about the safest job in the army, short of being on the staff in Washington. It not only is non-combattant, but also is better done away from any objective that is likely to invite attack. Yet it is very definitely military work and just as important and useful as driving a tank or flying a bomber.

6. Nystagmus is a rapid, irregular, and involuntary twitching of the eyeball which Simpson had displayed since childhood.

7. Perhaps Simpson's mother hoped her son could finally complete the abortive trip to central Asia.

I am applying for leave without pay from the Museum for the duration of the war and will expect to step back into my present job after the war, so my career is not really jeopardized and only interrupted for a while. Our income will be drastically cut, of course, but Anne and I figure that we can manage all right with stringent economy and her working part time, at least, as she is now and seems likely to be able to continue doing. We were looking forward to getting out of debt this year, and now we will have to delay this until the end of the war. This is annoying of course, but not too serious as our creditors are very decent and patient and no interest will accrue.

It is still possible that all this will not eventuate, so do not consider it final until I am actually inducted, but you may take it as highly probable. Although leaving home and family and breaking off my work to learn something entirely different will be very hard indeed, it is less than millions of others throughout the world are having to do, and I will be proud and happy to serve my country and civilization.

Much love,
George

Dearest Mother & Dad—

I send this copy to Dad & ask that he forward it to California [for mother]. I am sending another copy to Kansas [for Gay] and one to Helen. I know that you will be pleased at this development. I could not bear just to sit in the Museum & study bones with everything worth living for being attacked, and I could hardly have a better break than to get into such interesting & important army work with a good rank. There is nothing in this to worry you & I know you won't worry.

I love you both very deeply—Gg.

[New York City]
Nov. 9, 1942

Dearest Marty—

We were so happy to have a letter from you (not to mention ceramics). I am afraid that we, like you, think that all information is forwarded, & of course it is, but we realize from our pleasure at hearing more directly that direct information is more satisfactory.

Well, for point (a), the ceramics arrived in perfect condition (who does the packing?) & to put it mildly we were overcome. They are *wonderful*, toots! There was a riot and—I hate to say so—by the time Peets' choice came around there was no choice to speak of. Joan grabbed the dancers & (she being of a suspicious nature) we have never seen them again. Anne, though torn, had to have at least the Madonna plus, of course, the two lovely toilet water bottles. And then I could not give up "In Church" or the pair of flower vases (now full of little pom-poms & looking perfectly wonderful) & so, the poor Peets are left with little choice. But blood is thicker than water, I always say.

In case you haven't grasped it, we like them. And thanks a million. We have *never* seen nicer things.

We were at the Peets' last night. Poor Peetie (Creighton to you) had an earache, but was sweet anyway. He seems so grown-up & is a fascinating little boy. They all and always ask about you & miss you as we do.

I do not know whether I am coming or going—or staying or going. Military Intelligence tells me one minute to stand by for induction in a moment, & the next minute, next two weeks, tells me nothing whatever. So I have no idea whether I am a vertebrate paleontologist or what, & am a little high-strung about it. But that's all right, too.

Anne is trying a new treatment for her brucellosis, also working terrifically all the time. Results are unclear as yet & we can only hope. The rest of us are very well & bouncing. Of course you do have the dope on the family distribution: A. [Anne], J. [Joan], & G. [George] here, Gay & Bets in Kansas [with Grandmother Pedroja], & Helen in Ann Arbor [at the Univ. of Michigan], forlorn without her boy friend but otherwise thriving.

Anne wrote a tremendous letter to Mother c/o you this morning, so this note is not for news but for thanks & love, of each of which there is more than I can tidily put on paper.

> Your very attached brother,
> Gg

> [New York City]
> Nov. 28, 1942

Dearest Mother—

I have just been informed that I am now a captain in the Army of the United States & have been ordered to report for duty at Fort Washington, Maryland, on Dec. 3—next Thursday. This had dragged on so that I had about decided it was all off, but this is official & final. I do not know details yet, which will arrive later by letter—my present notice is only a telegram from the Adjutant General.

Of course this is what I wanted, & I am very happy. I hate the war & everything involved in it & I am desperately unhappy to be separated from my family. But that is exactly why I want to & must do what I can, even a little, to help end the war our way. I know you feel so, too.

I do not know my proper army address yet, so continue to write to 151 E 81 [St.] & Anne will forward, of course—

Much love to you, my sisters, & all—

> Gg

Dad says to add he may change his plans & stay here till I leave—He'll write you Monday, when I have more details—

> Gg

[Ft. Washington, Md.]
Dec. 5, 1942

Dearest Mother—Here I am, an old army man with two days of longevity, as they say. Naturally I am still somewhat confused at the abrupt & radical change in my life, but the adjustment is not proving as difficult as you might think or as I feared. The worst, of course, is missing my family, but I just keep busy & try not to dwell on that more than I can help.

I am the only new officer (I mean new to the army—some have just become officers but have been in the army for some time) in my school or, as far as I have discovered, in any of the schools—there are several schools here, quartered & eating together but taking different courses. They skipped all the normal routine of giving me basic military training & I am assumed to have it. This is a difficulty, as you can imagine, but I am learning fast from reading, observation, & advice from the old-timers. Everyone is most cordial & helpful. As an average, I have never encountered such a keen, fine body of men as the officers at these schools. Of course all are selected men.

My current address is:

Capt. G. G. Simpson, A.U.S.
Class #2, Military Censorship School,
Fort Washington, Maryland

This class takes only three weeks: we graduate on Dec. 26. In the meantime we have to go through 29 text-books, ranging from small pamphlets up to good sized books. After that I haven't the slightest idea where I will be. I might be run through another school here or sent almost anywhere. Until you hear different you'd better send mail that would arrive after the 26ᵗʰ in care of Anne to forward when she knows where I am.

We had a fine visit with Dad, although in all the excitement I fear he didn't get the attention he might have otherwise. I was so happy that he was there when I finally got my orders & that we could have such a visit & he could see me off. I hope fate may decree that I may see the rest of my family before I am sent anywhere very remote, but that seems unlikely.

When we really buckle down to studying I may have no time to write to the family separately & I will trust that whoever does hear from me will pass any news around. At the moment, as you see, there is no news except that I am here.

Much love to Marty & Peg, to Dad when he arrives, & my best to you—

Your loving son,
Gg

Dearest Mother & Dad—

I am on the move, having been graduated from Fort Washington a week or so ahead of time. Just now I have been at Fort Meyer for a couple of days, & have stayed at Farrington's—they have not required that I stay at the Fort nights. I had Anne come down because we may be moved to another camp without [getting] leave to visit her in N.Y. We have sandwiched in a pleasant visit between spells of getting equipment, being inoculated, packing, etc.

This afternoon Anne came out to Fort Meyers & helped me with my packing. We had quite a bee, because several other officers in my detail also had their wives visit here & all of them had the idea of spending Sunday afternoon packing. We then came in & had dinner with a couple—Lt. & Mrs. Greene. The New York wives, of whom there are several, have exchanged addresses & plan to keep tabs on the group as a sort of rooting section for us.

Just where we will be next is quite obscure & I cannot say at present, or when we move on. In the meantime until we get a "permanent" address again they have given us an A.P.O. [Army Post Office] number that will automatically trail us about & keep our mail following us. So until further notice address me A.P.O. #4015, care Postmaster, New York, N.Y.

I am sorry to say that this very unexpected move has prevented my receiving your package at Fort Washington. It will doubtless catch up to me soon, & in any case I greatly appreciate your sending it. Since I am likely for some time to move hither & yon at an instant's notice, do not try to send anything perishable or bulky for the present.

Much love to the best of all parents—

Gg

[Note added by Anne]

And love from me too and Christmas greetings—predictably belated—and New Year's ones, too—to all the California contingent.

Anne

[Algiers]
Mar. 13, 1943

Dearest Mother & Dad—

I have been writing quite regularly to Anne, in the hope that she would get them more promptly, being in N.Y., & would send the news of me on coastward [to California]. The mail service seems to be extremely bad, but I am sure that by now she has received some of my letters & has passed the word on. As for me, I received nothing whatever until day before yesterday when I got a letter from Yale that Anne had forwarded! That is all I have yet heard from the U.S.—It is rather annoying that the letter she forwarded got here before any of Anne's letters. Incidentally the letter from Yale asked me whether I would like to go there after the war as professor & curator of paleontology—I thought that possibility was *fini*, & am quite surprised. I don't know yet what I will do. "After the war" looks pretty indefinite from where I stand, & I do have obligations to the Amer. Museum. Nevertheless that would be a very fine job & I may take it, if they really hold it open until I can leave the army. Don't tell anyone about it for the present, but I knew you would be interested.

Things here are the same as when I wrote you, aside from the reports to Anne. Happenings are either censorable or dull, as a rule, although we do manage to have some mild diversion. I am near enough to headquarters to get my mail through it, as you know from my official address, so of course I do not sleep in a foxhole or take pot shots at Germans, although you would think we did to hear some of the swivel-chair soldiers around here talk. We're in the war, of course, & within easy bombing range (although it honestly amounts to nothing), but we live in relative luxury.

My C.O.,[8] Capt. Hereford, and our 3rd in rank, 1st Lt. Bollag (respectively Charly & Larry) & I have, after a week or two of discomfort, wrangled a delightful apartment, furnished, with a private room for each of us. No bath or hot water, but you can't have everything & we have a gas plate to heat water & a tin washtub to bathe in. Our landlady & family live right across the hall & they are truly charming to us—Keep the place spotless, look after all our little needs before we know we have them, & have us in to Sunday dinner (which is excellent) as members of the family. No one else I know here has had quite so good a break. As you can imagine, the town is crowded beyond capacity, & not all local people are so nice, to begin with, or so well disposed toward the strangers in their midst, to go on with. For relaxation, aside from gabbing & swilling the good local wine, we study Arabic, which is enough to keep us out of mischief all the rest of the evenings of our lives. Aren't we good boys?

Much, much love to you both, to Marty, & to Peg. You are all dear to me & I hope to see you again soon, & to hear from you sooner—

Gg

[Algiers]
April 1, 1943

My dearest Mother & Dad:

Mother's letter of Mar. 7 reached me on Mar. 30 & was the first word I received from the California wing of the Simpson family. I have written you several letters from N. Africa & I hope some of them, at least, are getting through all right. I write almost daily to Anne & have had 7 letters from her, by which I learn that my earlier ones are beginning to arrive & that she does pass on news of me to all of you from time to time.

---

8. Army talk for "commanding officer."

The news we are allowed to write is not very extensive. You will see by the papers that things are beginning to go well here, even though I can't tell you about my own very small but necessary part in this.[9] Incidentally I am *not* connected with the Postal Service as a New York newspaper said. I can't say what I am doing, but it isn't that.[10]

I would really enjoy life here if it were not so lonely & I did not miss all my loved ones so desperately. It is a beautiful & interesting place & I do have good friends—but not like my own family. Also I couldn't quite bring myself to acquire a temporary wife here as—you will not be surprised to hear—some do.

My Arabic lessons are coming on well although I find it hard to find time for studying.

I have to scram right now. I am very well & delighted to hear that you all are, as of Mar. 7. I think of you so often & pray that you will continue well & that I will soon see all of you again. I love you very much—

Gg

[Algiers]
April 4, 1943

Dearest Mother & Dad:

Two days ago I received Mother's V-mail of Mar. 13, yesterday Dad's cable of Mar. 22, & today Mother's V-mail of Feb. 24 & 27 & Dad's of Feb. 28. Service is erratic, to say the least, & apparently something happened to hold up the late Feb. & early March mail as it seems to be later than mail from the middle of March. Cables do not come here direct but are mailed from a place part way here & usually take ten days or two weeks after being mailed.

It was grand to hear from you & swell of you to write & cable so promptly &, although I know I haven't all of them yet, so frequently.

Does the Caldecot son have the same APO number that I do, & what are his rank & unit?

Herb Irwin is not particularly a friend of mine at this point. Tieje is OK, however, & I would be interested to know how he is. I think he's at U.L.A. or U.S.C. & not U.C.L.A.—All those schools in L.A. confuse me.[11]

It sounds as if you both were finding many occupations & interests & I am pleased. I know the adjustment [to retirement] is a bit difficult, espe-

9. The British and American armies were pushing the German and Italian forces into the Tunisian peninsula after having repulsed strong counterattacks by General Rommel's panzer units.

10. The *New York Sun* printed an interview with several soldiers in Simpson's unit and gave the impression that they were postal censors reading G.I. mail.

11. Herb Irwin had been a classmate of Simpson's at Boulder; he also worked on *Dodo*. Arthur J. Tieje (1891–1944) had been Simpson's geology instructor at Boulder; he later went to the University of Southern California where he became head of the geology department.

cially for Dad, & I do so hope that you will have there much of the happiness you so richly deserve.

As for me, I am not 100% happy, so far from all my loved ones, but otherwise am really just fine & you need not have a minute's worry about me. I'm not crazy about this climate, but I do love the town & am very comfortably situated. (Have just returned from a 3-hour meal with our landlord & landlady—swell food & much gaiety.) *Much, much* love to the best of parents.

—Gg

[Algiers]   V-Mail Letter
April 4, 1943

Hi, Keed! [Martha]

I haven't heard from you yet, but I bet you wrote me, so I'll answer it anyway. Under the new censorship rules we are not even allowed to tell anyone the correct time. (We never know it anyhow; one of the generals is said to know what day of the week this is, but I think that's just a vile canard or a latrine rumor as we say in our ordoriferous military patois). I can tell you all about my love life, however, which I will now do in one word: zero. How's yours? Excuse me while I get a verre [glass] of vin rouge & cry into it a bit—(that proved easier said than done because our landlady had borrowed all our glasses for a party—glasses are unobtainable here & worth their weight in gold—it was OK though because she invited us to the party, which was a lot of fun—Love found a way & I did scrounge a glass & am now degusting the wine with gusto).

In spite of the wear & tear of war you would love this town which is, just now, a great deal more like Paris than Paris is. In fact no one would notice if my street were set down on the Rive Gauche [Left Bank]. It's pure French in this quartier, although some parts of town would remind me more of Bagdad.

I hope you are well & a good girl, but not too good. Aside from being too good, I am fine. If I just had my family within reach I think I'd almost enjoy this war.

—Your ever-loving brother
Gg

[Algiers]   V-Mail Letter
April 4, 1943

Cutie [sister Peg]—

This is in answer to your letter that I haven't received. Mother told me you wrote one but the Army Post Office hasn't delivered it yet. By the way the N.Y. "Sun" said *I* am in the A.P. service. *NO, NO,* a Thousand Times *NO!* What I'm doing is none of your business, but it has nothing to do with the postal service.

In my spare time (which consists of 21 minutes every other Wednesday), when I am not guzzling wine (which I do whether the time is spare or not) I study Arabic language & customs. Some customs! I read in the Coran where as a man grown & bearded I am entitled to four wives, legal, & as many concubines in addition as I can handle (= 0). I'm looking the field over for my other three. Only trouble is, you can't lift the veil until you've already married & then you pay cash for the privilege (no foolin!). However I *did* see a real sweetie-pie unveiled. I was visiting her husband, Abulqasem, "Just call me Abu," Tejini when someone began dropping bombs in the neighborhood & she galloped out all atwitter without her veil. Was she embarrassed! (*No*). Was he sore! (*Yes*). Anyway I'm asking Santa Claus for her. Since then old Abu, who is my Arabic teacher, doesn't have his heart in his work because the evil-minded old Moslem thinks I want to learn the language in order to make improper suggestions to his youngest lalla. Maybe he's got something there.

I'm swell, everything under control although I'll be ready to go home when this is over, not to mention long before. Hoping you are the same & love with kisses—

George, the Keed Brudder

[Algiers]
April 19, 1943

Dearest Mother & Dad—

My mail is slowly catching up to me &, aside from letters previously acknowledged, I now have Mother's letters of Jan. 29 & Feb. 13, sent before you had my present APO number, & a more recent letter, without date & postmark illegible, sent direct to APO 512.

Please date letters & please put "INC Section" in my address. Without the latter, it has to go through the directory service which takes some time. The Force Hq. is so big that addressing there without the section is like sending a letter without the street address.

It is wonderful for morale to hear that you are all getting along so well. I hope you would say so if you weren't, although I'm not sure I can trust you in that respect. Do give me bad as well as any good news—if there should be any bad, if only so that I will know the good news is *bona fide*!

I am not really surprised that Mother is actively gardening; she's such a spry young thing![12] I hope she doesn't overdo & must trust Dad & the girls to see that she does rest occasionally. Marty seems to be a grand success & probably she is the one most likely to overdo. The news of Jack [Simpson's nephew] is swell, too—Please give him my best & tell him I'm proud of him. I haven't heard direct from anyone but Mother there in Calif., although I expect some of the others have written & their letters will reach me eventually, maybe. I write almost daily to Anne, all the news that's allowed, & I know she passes it on. I don't write to you on schedule but

12. Simpson's mother was seventy-two years old.

average about once a week so you'll get my love directly once in a while. The love itself is not weekly but continuous, as is my gratitude for having the finest parents anyone could possibly have.

—Gg

[Algiers]
Apr. 29, 1943

My dearest Mother:

Mother's day is not celebrated in these parts, so I don't know just when it occurs, but I'm pretty sure it is in May sometime. I hope that is right & that this letter will reach you somewhere around that day. If not, no harm is done because I need no special day to think of you & no special occasion to assure you of my continued love & gratitude. No one was ever more fortunate in his parents than I have been, & I sincerely thank God for you, & as sincerely thank you for all you have done for me. I have tried your love & sympathy beyond the ordinary, but have never reached the end of them. Aside from a really happy childhood & training that has stood me in good stead, in all my life you have helped & stood behind me with steadfast devotion that helped so much to pull me through & that did not even consider whether I was right or wrong. I haven't always been very demonstrative about my appreciation, but I have never failed to realize this, to count on it, treasure it, & return it, in my own measure, with a very deep affection. This lovely & remote life here in Africa, not without moments of terror, draws heavily on one's inner resources. Among my resources, standing up in every severe test, is the love I have for you & my knowledge of your faith in me. Believe me, it is one of the mainsprings in my life. I love you very much & I hope that you always know this, even when my expressions of it may be occasional or may seem perfunctory.—

Your son, George Gaylord

[Algiers]
May 11, 1943

Dearest Mother & Dad:—

Mail coming in bunches, as it usually does, I have just received Mother's letter of Apr. 3 & two V-mails with no dates (please date them—they do not even have postmarks & time in the mail varies from 4 months to 2 weeks so when they are undated I have no idea when they were written), & Dad's V-mails of Apr. 4 & 18.[13] It is such a delight to hear that you are

13. To expedite the huge volume of mail the Army postal service created "V" (for victory)-Mail. The standardized pages, on one side of which messages could be written and folded in such a way that the page formed its own envelope, resembled air letters. The messages were photographed and reduced to smaller negatives that were then mailed overseas, where a somewhat larger positive was made. The reduction and uneven quality of photographic reproduction often made such V-Mail letters difficult to read.

WAR YEARS
1942–1944

well, Dad hard at it with Marty, her [ceramic] business booming, Mother's garden blooming, & all.

This is my half day off (we are reduced to 4 hours off per week, including Sunday). I was bathing in the kitchen when the Fat'ma (Arab cleaning woman) arrived & she pounded continually on the door until I had finished my leisurely bath. Since then she has been trying to "do" the place while I tried to finish my toilet & change clothes, which has involved our harrassing each other all over the place. Since she is almost as un-beautiful as Arabs come (which is pretty un-beautiful), this is all good clean fun, & amuses me although it annoys her no end & she occasionally calls me names in Arabic, which I understand but don't let on to! I forgive her because she has christened me "the sheikh" (not being able to pronounce such an outlandish name as "Simpson"), which likewise tickles me.

3½ days ago Tunis & Bizerte fell & now I have just heard that the last enemy resistance in all of Africa has ceased. You can imagine the rejoicing here, although we are all keenly aware that the hardest struggle is still ahead. But it is a tremendous victory & really marks the turning of the tide.

I am very well, very busy, & very homesick for my loved ones. If only they were here, I would be contented, however, as I like this place very much & would not at all mind living here.

No air raids for a long time.

My best love to you all—

Gg

[North Africa]
May 31, 1943

Dearest Mother—

I just received two letters from you which are without dates (a bad habit of yours!) but which, from other evidence, I suppose to have been written around the end of April & beginning of May. I probably haven't kept my resolution to average a letter a week to you, but lately there has been quite a period when I had no way of mailing letters—I'm not sure I can get this off, but think so. I have been with the British 1st Army for a while & in places where mail is only beginning to catch up. It's been a lot of fun & extremely fascinating, as you can imagine. I'll rejoin my own outfit eventually.

I am well, as usual, & seldom hear a gun nowadays—everything very peaceful at this stage of the war.

I haven't yet learned where 899 T.D., A.P.O. 302, is with respect to me so haven't located Caldecot but may run on him. Strange encounters do happen & I am now getting about a little more than before.

I expect that Anne has told you that I won another medal—scientific, I mean,—which is pretty good. I never won any in my life before & now as soon as my back is turned I get two! Perhaps I should stay here & they will appreciate me more than back there!

I can't say that the army over-rates me, exactly, but they do give me work to do & I do it, & that is all I want from the army anyhow. That & an honorable full discharge the day the war ends.

This is written on my knee, so excuse even worse writing than usual. Much, much love to you all—

Gg

[North Africa]

May 31, 1943

Dearest Marty—

Please let me know *immediately* if you do not receive this letter. It is written with booby-trap ink & I suspect that it either fades out entirely or goes up in flames after a day or two, so I want to know whether to keep on using it. "Booby-trap" ink is, of course, ink "won" from our foes the Jerries [Germans] & the Eyeties [Italians]. This is a mixture, 50% Jerry & 50% Eyety, & it's bloody awful as we say (I'm attached to the British 1st Army at the moment). I am acquiring quite a reputation for smoking booby-trap cigarettes, some of which aren't bad, the ones that contain tobacco, that is. The only line I draw is when someone stows a box of Jerry hand grenades under my bed, without even telling me, as someone did. They're in the garden now where they'll only blow down the staircase, so don't worry about them. Oh, yes, I do draw another line—I won't eat the Eyetie meat ration, which my British opposite number quite graphically refers to as "tinned cat." We trade it to the local populace instead of eating it.

By the way, I have met the 2nd Tournier son. He confirms the information I previously sent you, obtained 2nd hand. He still carries on the book business although stocks are low as Jerry even went for books, but he expects to be called into the army. The 3rd son came through the campaign alive & unwounded. They have a sister who is now married & has 2 children. She is in Tunis.

Your [ceramic] business sounds marvelous. Only trouble, I couldn't make out what your net is, whether $77.80, $778.00, or $77,800 per month. The latter, I trust.

Much love from your errant
brother—
Gg

[Algiers]

June 10, 1943

Dearest Mother—

They have passed a law that we can say where we have been, as long as we do not say where we are or any army unit is. So, not in this order, I have been in Algiers, Maison Blanche, Telergma, Souk-el-Arba, Le Kef, Le Krib, Madjez-el-Bob (what's left of it), Mateur, Tunis, Nabeul, Hamamet, Oran, & have seen but not been in Casablanca. You probably can't find all those on a map, but some of them. I've been in a few other places, too, although on the whole I am very sedentary here & have spent about 4/5 of my time in one place.

I have recently been off on a special mission. I wrote you while on it, but there was no mail service & I'm not sure the letter ever reached you. I was gone almost a month, so there's another gap in my correspondence & I'm terribly sorry, but naturally I go where I'm wanted & don't complain (much!) if I find the mail not operating.

Anyway I love you all very much & on my return I've received your letters of May 10, 13, & 20 (all dated, too!) which made me *very* happy. Of course I am, we all are, very happy too about the military success here in Africa, but you're tough enough not to need false visions to buoy you up. The end is not even in sight yet. The tough part begins now. We'll win, never fear, as long as we all fight & realize that we have a long war ahead still. All my love to all of you—

Your son, & proud of it.

[Algiers]
June 14, 1943

Dearest Mother—

Your birthday letter reached me on time, as you see, with a couple of days to spare. I also have V-mails from Dad (May 8 & May 23), Peg (May 11), & Marty (May 22), all *intensely* enjoyed.

I was off on my special mission nearer a month than two weeks, but now I'm back. Temporarily I have a new job, in the same old town, and I hope it becomes permanent because I like it much more than my regular assignment. Only a slim chance, however, & I probably will be shot back to my old job & then transferred, in it, to another city which I do *not* like. Time will tell. Meanwhile, use the old address & ignore any change you may have been given—I wrote one or two people a new address that is now invalidated by my shifting about, but won't delay mail much.

Incidentally, I have *not* been in Turkey or Egypt. I am now allowed to say where I have been, as long as I don't say where I am or what unit I was with or what doing in various places. OK, among others, I've been in Oran, Casablanca, Gibralter, Tunis, Mateur, Algiers, Nobeul, Medjez-el-Bob, Telergma, Teboursouk, Souk-el-Arba, Le Krib, & parts east, west, north, & south. That'll keep Dad busy in the atlas for a while! I've purposely mixed up the sequence, not to annoy you but to abide by the spirit as well as the letter of the rules.

Peg's new place sounds wonderful & I only hope your own housing situation works out well.

I'm glad you got your teeth fixed up. How lucky you have been not to need a full set even earlier. I know so many people still younger than you who have lost all their teeth.

I am beginning to suffer from the African summer, which is all it's cracked up to be for heat, but otherwise I am fine & I still love you, & the rest of your family, very much—

Gg

[Algiers]   V-Mail Letter

June 14, 1943

Dearest Keed—

Your address doesn't look right, but it's all I can make out so I hope it goes well. As a matter of fact your writing is very photogenic & I can read almost every word even when reduced to the size of a sheet of extra small or Liberty toilet paper—Your V's are photographed now, although mine aren't.

Your "Happy Birthday" on the margin hit it very well, as you see. This here is a birthday letter to you, although alas it won't make it quite in time. Couldn't be helped, as I'm run ragged. In the last week I have been assigned on 5 minutes' notice, to three totally different jobs, one after the other, each more complex than the last, so I'm going in circles. Anyhow much love & the happiest of happy birthdays to you.

Your business fascinates me & I did enjoy your account of it. I wish I could see the line (how commercial your jargon is, grandmother!) but I will after the war. I intend to visit *everyone* then before I settle down, if it breaks me.

It is OK for me to say I was in Tunis on a visit once upon a time & I talked to Etienne's cousin's son, or whatever it is. I was to go back & have lunch with him, but never made it—I'm sure he understands that army officers only make engagements subject to change without notice.

So lots of love & stuff, Toots, from your bearded, doctor, captain, sheikh, kid brother aged quite a lot as of day after tomorrow—

Gg

[Algiers]

June 18, 1943

Dear Dad—

Your V-mails of May 8 & May 23 came almost together. That was a few days ago, but I wrote to Mother first & so you have had to wait for a reply. As I work from about 8:30 A.M. to 6:30 P.M. or later, 7 days a week with occasionally a half day off (only once in three weeks lately), you can see that my time for eating, chores, & letter-writing is not too abundant. However, I am very pleased with my work at the moment. I have had quite a variety & just today have been transferred with a view to some extremely interesting work coming up. This is especially appreciated as I was on the point of being sent to a city I detest to do very boring & unimportant work, & now that is canceled.

I *think* father's day comes around about now. I have probably missed it—
I tried to find out but nobody knew just when it was. Anyhow you know
you're my favorite father all the time & that I love and admire you
tremendously.

The weather here has turned warm, occasionally very hot, & I must
confess I dread the summer because as you know I don't stand heat so well,
& this promises to be the busiest summer I ever had, but I'll manage all
right.

The very best to all of you—

Gg

[Algiers]

June 23, 1943

Dearest Peg & All—

This is a reply to yours of May 11. I did receive it a couple of weeks ago,
but my letter writing is rationed so I drew lots to see who'd get answered
among my California family.

You certainly sound as busy as a little bee, or even a medium-sized bee,
& I sometimes think you all work harder at the war than I do. Then other
times (such as when I've been at it 10 hours & realize hopelessly I'll have
to come back after dinner to finish) I don't think so. As a matter of fact I
have two new jobs which are very interesting & although each is a full-
time job I do not have to work too terrifically, in fact rather less than on
my old job, which was beginning to be a killer. I was getting a bit disgusted
& am delighted at the transfer & the envy of all my former colleagues,
every one of whom had applied for transfer. All were refused except my pal
Bollag & me. (Bollag = 1st Lt. who has been with me or near me ever
since I joined the army & now shares my apartment—most unusual for 2
officers to stay together so long, & among dozens I started out with he is
the only one I'm still with).

My letter writing is a bit hampered by our Fat'ma (Arab maid) who
dashes in with little tidbits in Arabic every two minutes. I can't even take
the occasion to practice making improper advances, as you recommend,
because she might accept & that would be horrible. If she was ever anyone's favorite wife that was in the days of Harem al Rashid.

So I guess I'll go eat. The same to you & many of them—Love—Gg

---

[Algiers]

July 1, 1943

Dearest Mother & Dad—

I'm writing to you together because I have Mother's regular letter of
probably June 3 (she says June 5, but the postmark is June 3!) & Dad's V-
mail of June 6, & I fear won't have time to answer separately. I certainly

look forward to the photos you are so diligently taking & can hardly wait for them—but I guess I will wait!

I'm sorry Peg is moving—It sounds grand but seems rather to strand you. I had a note from Bob [Roe] who said he had seen you & was going to L.A. It will be nice if he can be with you, near you, I mean.

I don't have details, but I know it's settled that Joan & Anne will both be in New Haven next winter. Not ideal, but what is? And I quite agree that the responsibility would be too much under all the circumstances & that this is the best arrangement for now, even though I know you would also have enjoyed having her. The following winter, God willing, I'll be able to take over. It's tough on Anne & all of you to have to look after so much without my assistance & backing.

In answer to Dad's query, I never mention radio because I never hear one. The Belpaumes (landlord's family) have one, but it is no good on long distance & besides we don't like to bore them by listening to programs they can't understand. I get the news fairly promptly, however, although not in much detail. Of course there are aspects of it that I know considerably better than the radio or newspapers, anyhow.

I'm enjoying a bout of dysentery, which is jolly because I'm head over heels in work & *have* to keep going, but it will soon pass, I know.

Much love—Gg

[North Africa]
[Postcard postmarked 19 July 1943]

Dearest Mother & All—

I have Mother's letter of June 11 & Dad's & Marty's of June 16 for which most heartfelt thanks. [Simpson's 41st birthday was on 16 June.] I certainly have a family worth fighting for, & I'll send better answers as soon as I can manage. In the meantime this card [with painted chrysanthemums]—which has a strange history—carries a world of love to all of you. Life is developing some tough moments, but I'm doing fine & I love all of you—Gg

---

[Tunisia]
20 July 1943

Dear Folks:

Look, typing! This is one result of being in command of my own unit (and having an efficient supply officer). Unfortunately I do not yet have an efficient secretary, so excuse the mistakes that are already appearing in great numbers.

It has been so long since I had a moment to write that I am terribly behind in my correspondence and owe everyone several letters. I hope you all keep on writing anyhow, as I have time to read them even when I haven't to write them. I expect you all know by now that I have moved again, and this time for good, that is, as far as can be foreseen I am not returning to the old place where I was for so long. Thence a totally new address and this one, although always subject to further change, is not a false alarm like the last one I was erroneously instructed to give. (Note: the row of X's is not part of the address but just the first of the mistakes.)

Even before I left I received and I think briefly acknowledged Dad's and Marty's letters of June 16 and Mother's of June 11. Just today I received Mother's of June 24—she remarks that [it] was the third letter that week but the others haven't come yet.* I certainly look forward to them.

You all seem to be well under control there, with your new houses, your visiting relatives, your feasts, your booming ceramic business, and all. I was sorry to hear of the trouble with Marty's eye but apparently that too is better. I do long to see you all and can hardly wait until I do—I guess I will wait, however, under the circumstances!

My business is booming, too, even [if] I can't go into detail. I assure you that I have all the work I can handle and sometimes almost more. The last days I [have] been putting in 12 to 14 hours although I am beginning to get organized now and hope to get back into a leisurely schedule of 10 hours. I am in command of a new unit and a new type of activity and naturally I am anxious to make it go. Circumstances are rather difficult at times too. It is, for instance, quite a chore to move into a badly bombed and crowded town and find working and living space for a large group, find sources of supplies (legal or not), arrange for pay, and do a thousand and one other things, and at the same time keep work under way just as rapidly and efficiently as if set up in an office in New York. We come pretty close to doing it, too. I'm going to be the demon executive when I get back and I bet all my subordinates will be sorry I have had this army training! I have quite a large group under me, both British and American, mostly officers, who are of course much harder to manage than enlisted men. They aren't all the flower of the two armies, but they are a good bunch on the whole. I was warned one or two were trouble makers and authorized to return them if they cut down efficiency, but I would hate to have to do that, and I don't think I'm going to have to. I seem to be hitting bottom [of the V-mail form] even though I can say so much more by typewriting.

So— I love you all—

*June 19 & 22 just came. Thanks!

[Tunisia]
26 July 1943

Dear Dad:

You draw the letter to California tonight, in honor of yours of the Fourth of July, just received with jubilation. I think they may be photographing these [V-mail forms] now, and if so, I hope you have a magnifying glass. I often have to use one to read letters coming this way, all of which

(V's, I mean, of course) are now photographed. Your writing happens to be very photogenic and I read it without a glass easily. But imagine, for instance reading [my daughter] Helen's scrawl when reduced so much. Incidentally, she writes me often and very interesting, happy and loving letters. I am very much pleased with her. But then I'm pleased with my whole family, ancestral, descendant, and collateral!

Your moving sounds pretty complicated, and I expect you hate it as much as I do. I am getting a bit of it too, of a particularly intricate sort, having to assemble, house, equip, etc. a whole unit more or less from scratch and under very adverse conditions. Of course I have a lot of help and delegate a great deal, but there is a limit to the amount that can safely be left to someone else and whoever pulls a boner I am responsible, so I can't afford just to say "Do so and so," and forget the whole matter. One of the most irksome but most necessary things in the army is this rule that [if] whatever is done is wrong, it is the commanding officer's fault, even if he had no knowledge of it and in fact even if it was directly against his orders—in such a case he naturally raises hell with the real culprit, but he also catches hell himself.

Well, such are the cares of command. Doubtless I'll take it more in my stride when more used to it. As I realize, anyone including me who is used to assisting someone else is likely to be too fussy and anxious when he first takes over on his own, as I just have.

—Just there, speaking of cares, I had a long interruption. I've already worked ten hours today, which isn't long the way things have been going, but some other matters came up that required attention. You can see that I don't have to worry about the devil finding work for my idle hands! But it's a pleasure to work when you know that what you are doing is helping directly to end the war victoriously.

The town we are in is pretty lousy, besides being considerably blown up. I have made arrangements to transport my officers and men three evenings a week to a nearby town where there are a few amusements. I haven't had a chance to get over there myself yet and may not, but I already know the town from having been there some time ago. At present my principal relaxation is just talking for an hour or so in the evening, over a mess-cup of wine (which is a pretty big slug, incidentally) when I have the wine. At the moment I haven't, but I have a spy out searching for some.

Actually, as you have probably guessed, I am getting a great kick out of my new assignment in spite of the hard work, difficult conditions, and burden of responsibility. I will certainly have plenty to talk about when I get home! This is a rotten letter, all me and my burdens, but I'm tired and they're on my mind.

—Here's my spy, with wine! Goodnight, and my fondest love to all.

Gg

[Tunisia]
27 July 1943

Dearest Mother:

Remembering that I should remember birthdays a month ahead is one of the hardest things in my present circumstances, especially here where I have been so completely head over heels in work. Now I suddenly realize with a pang of regret that I have put off writing you a birthday letter until so late that it cannot possibly reach you in time. I know, however, that you will realize that there are some extenuating circumstances and that I have not forgotten you and will be thinking of you particularly when the day comes.

Anyway, even though belated, this is a birthday letter, and it carries the greatest love from your favorite son to his favorite Mother and the conviction that you will have very many more happy birthdays and that he will be there to share them with you. I am sure that the girls will help you celebrate this one and it is nice that you will have so much of your family there.

When I get back—unless still in the army—I am going to make a beeline for California as soon as I have my land legs again and just visit and relax for some time before I settle down to civilian work again. That will certainly be a happy time. It seems pretty remote now, but it is getting nearer every day and the tide of war is really on our side now. Under the circumstances, I am very glad that I am in the army, but I never could enjoy army life for its own sake and I certainly want out the day the fighting is over.

There is, as usual, little news fit to write here. I have just come back from a pleasant jeep ride through the surrounding countryside—business not pleasure, I need hardly say, but a pleasure nevertheless. I am quite the demon jeep driver and I love to drive one. I have three British drivers on hand, but I drive myself whenever practical. I certainly hope that I can get a jeep after the war because they are wonderful cars and give you such a feeling of freedom, almost like flying, in comparison with conventional cars.

And they will go anywhere, almost like a tank, which certainly will be handy in field work. My jeep is named "Anne" by the way—we sentimental soldiers always paint our girl friends' names on our jeeps.

Speaking of Anne, it really looks as if she might be cured of her brucellosis after all these years and her letters are so delightfully optimistic. It will certainly be wonderful for her really to feel well again. Her constant illness has never stopped her, but it has been a terrible burden for her.

Your crossword puzzle book is proving to be a lifesaver. I have to get my mind off work somehow by half an hour's relaxation before I can get to sleep, and that is the way I have been doing it. I get absolutely furious at the constructor before I am halfway through one, but even that does me a world of good. Probably an outlet for my temper and saves some poor lieutenant a good blasting. And it certainly takes my mind off more serious problems! Much, much love, and again happy birthday and many of them!

Gg

[Tunisia]
4 August 1943

Dearest Mother:

As a small boy with a memory none to[o] good (it has steadily gotten worse with the years), I used to remember your birthday because it came exactly a month after the 4th of July and [sister] Peg's because it came exactly 4 months before Christmas. I don't know why that is or was any easier than just remembering August 4th and 25th, but I started thinking of them that way some 35 years ago and I still do it. Therefore I know that today is your birthday and naturally have been thinking of you even more than usual all day.

I hope you got my birthday letter, although I expect it will arrive late—I have a terrific time trying to get such things off in time. I even have a little date book in which I have laboriously entered "Three weeks to Mother's birthday," and so on, but of course you have thought of the hitch in that—I never remember to look in the little book! Anyway I expect that you all know that I do love you and think of you, and I certainly have the best excuse of my life for having other things on my mind as well, with a spectacularly serious war pretty much on my doorstep.

Anyway, I now officially send you additional birthday greetings from very, very sunny Africa, where I still am.

I expect you and Dad followed the [Sicily] invasion news very excitedly and perhaps had visions of my landing on the beaches there, Tommy-gun in hand. Such, however, was not so. I don't even have a Tommy, only a .45 pistol which I haven't fired since leaving the U.S., although, under orders, I occasionally wear it belligerently for the moral effect. I don't expect to be quite as sedentary as I was, but I'm no commando (although my outfit is sometimes derisively called the Junior Commandos or Simpson's Horrors—or Hirsute Horrors, the latter because I am one of the few C.O.'s who can't order his men to shave off beards [because I have one myself] and several are taking advantage of this. We do look horrible, too!). We do have a little something to do with the invasion, naturally.

Happy to hear of the successful birth of Edith's Susan.[14] I do envy them at times, being able to keep up even a migrant home life in spite of his being in the navy, yet I know I'm more needed and more directly useful here. I am sort of lonely, I won't deny.

I envy Dad, too, with his watermelon, fresh corn, cookies, and cheese. Those are among the simple luxuries of home that we just don't have.

I expect you're all moved by now, and Bob and Vivian [Roe] are settled in, too, and Marty is starting on her second million, and things are humming generally. For me, I'm well despite being a bit ground down by heat and work. All is well and I send much love.

Gg

14. Edith Farrington, who was wheeled in a baby carriage by Simpson while his friend Bob Roe wheeled his baby sister, Ethel Mae ("Pat"), back in Denver when both boys were around eleven years old.

[Tunisia]
15 Aug. 1943

Dearest Mother:

I have your letter of July 18 & 25 & Dad's of July 25 to acknowledge with joy. Last night I got Anne's first letter direct to my new address, which makes me feel an old timer.—About a month round trip in fact, which isn't so bad. Your "shack" sounds quite nice & both you & Dad seem pleased with it now that you are actually in it. You hardly seemed enthusiastic before. I can hardly wait for the photos coming through Anne, but I'll have to, for a month or so. We have a lot of fruit trees about, too, but not a drop to eat. Oranges, lemons, & grapes in our front yard, none ripe, & figs at the former school where we eat, but G.I's are quartered there & get them all as they ripen. I'm surprised you were all picking oranges late in July—the season here is so short & has been over for months & the new ones won't start for more months. I hope I won't be here for the new ones—but you never can tell. I wish I knew as much about the progress of the war as you will when you read this. Just now we are still fighting hard in Sicily but that, at least will be over when you get this. I continue well although hot & a little thin. My best love to all of you—Gg

[Sicily]
7 Sept. 1943

Dearest Mother & Family—

Your wandering son is now in Europe, Sicily to be exact as the censor allows. Address remains the same, as it is often a forwarding address anyhow & I'm not necessarily right where that A.P.O. is.

I have Dad's letter of 1 Aug. & yours of 9 Aug. I know other mail is chasing me in Africa & hope some from you is included. You all sound busy & gay, with your birthday parties, Marty's business, new houses & all. Of course I am coming to see you as soon as I possibly can. I'm a little more optimistic about the end of the European war than I was & think now it might end next year, so perhaps it won't be quite so long before I get there. Of course the Japanese war will take longer, but I hope they have a heart & give me a trip home when the fighting ends here.

I finished my old job & haven't started the new one yet, so I'm having a little rest in a delightful camp in an olive grove. A tent is so much more comfortable than a house in a bombed town that I am really luxuriating. Believe it or not, I have electric light & telephone in my tent! (It's really a colonel's tent, but he's sharing it with me & he's away just now).

My dearest love to you all—Gg

[Sicily]
21 Sept. 1943

Dearest Father—

The last two weeks have been my poorest for writing since I came over-seas. On top of the usual business I was flitting hither & yon & then set-ting up a new headquarters & also had a slight attack of my old pal bacillic dysentery—I am well recovered again, so think nothing of it.

Now I am established in my new place "Somewhere in Sicily." It is much nicer here than where I was for the preceding couple of months. I have a comfortable apartment which I share with three congenial old friends (army friends) from whom my last job separated me: Bollag, Case, & Costanzo. The first two are back under my command along with Clark, whom I brought from the last job. The rest of my boys are scattered & not at this same place, although most of them are still in my gang. These are very stirring days, as you can imagine, & homesick as I am, it is interesting to be here.

I received last night Mother's letter of 29 & Dad's of 31 August, both with great joy. You all sound busy & happy, & that is certainly the best news I could have.

In reply to your question, I don't want anything for Christmas, honestly. I have all I need, except my family, & can't think of any luxuries that would also be practical under the circumstances.

Love to all—Gg

[Sicily]
11 Oct. 43

Dearest Mother & Dad—

I've dropped behind in writing again because I have been on the move for quite a period under conditions that made writing just too difficult. However, I am quite all right except for a passing cold & am now settled down again for a while in a place where I had been before, still somewhere in Sicily & quite remote therefore from the present fighting, so do not worry.

I've just written a long letter about this part of the world to Anne & doubtless she will pass it on in due course. My last letters from you are 5 & 11 Sept. from Dad & undated from Mother of about the same period. I'm sorry you worry, Mother, when there are gaps in my writing & there is tough fighting here. Don't be alarmed. I can't always seem to be regular in writing, but I'm no commando & do not lead charges on machine gun nests! The only Germans I have seen were either prisoners or dead or in airplanes, & some of the latter died as I watched, I'm happy to say.

So don't worry about me, or about the war. We are winning it slowly & steadily. Just don't be too optimistic when we happen to go rapidly for a while or too blue when we slow down, & it'll be all over in time—next year, possibly, for this European end, & I hope to get home after that for a while, at least.

Meanwhile I love you both & all, very dearly & I am getting along fine although I do get homesick & blue once in a while—

Gg

[Sicily]
19 Oct 43

Dearest Dad:

My calendar warns me that I must write you now to reach you on your birthday, which I hope this does. Even if this should be late, I'm sure that you will know that I am thinking of you on that day & that I love you very much. Although I have always loved you, I think I had to be a little older myself to appreciate fully what grand parents I have & how much they mean to me—If my own children ever think half as much of me, I will think I am a great success.

Happy birthday! We will try to celebrate the next one together.

Here there is the normal press of work & coming & going, although I have been in one place now for a couple of weeks & so am an old settler. This may cure me of my love of travel, although I doubt it. I am usually quite well, although like everyone else I have mild recurrences of bacillic dysentery. This will clear up easily when I get to more healthful places, & doesn't really incapacitate me.

My best love to all there, in addition to birthday wishes for you.

I hope you can read this—I have an Italian pen that writes too fine a line.

[Sicily]
Gg

[Sicily]
27 Oct 43

Dearest Peg:

A dam in the damn mail system broke somewhere & a courier suddenly appeared & handed me yours of "Sept. 3 or so," Mother's of Sept. 9 with pictures, & Mother's V-mails of Aug. 12 (that only took 2½ months— "Use V-mail for speed!") and Sept. 26 (only 1 month—someone will get fired for letting that through). The delay took nothing from my enjoyment & I have read each a dozen times & everyone in the U.S. Army here has seen the photos. (No civilians—I dislike the local civilians so intensely that I wouldn't dream of doing anything good for them. I regret the armistice [between the Allies and the Italians] because I would honestly enjoy shooting Italians.)

I'm sorry to hear (from Mother) that [your husband] Mac hasn't been well, but hope he's over it now & I'm so happy that everyone else is thriving. Photos of your new estate look wonderful, & I'll probably move right in with you after the war. You'll be crowded a bit, because I can't manage now without 4 or 5 lieutenants in the same billet to take care of me, but I'm sure you won't mind. They're nice lieutenants. Other photos also much appreciated. Mother & Dad look so well & happy, ditto for [sister] Mart who looks 10 years younger than when I last saw her. But the only photo of you is so distant I can only just recognize you. How's about a better one?

Also enjoyed Mother's confession of guzzling champagne. Save some for my return & we'll all roll in the gutter one big happy family.

I am well and still fighting—They call us the Chair-borne Infantry, but that's a libel—we sometimes fight lying down.

Much love to one & all—Your kid brother

[Sicily]
6 Nov 43

Dearest Mother—

Thanks for your dear letters of Oct. 3 & later (again no date—shame!). The former came 10 days or so ago, but I have been ill & couldn't answer sooner. My illness is nothing to worry about, just a local variety of jaundice which is painful & I'll be good as new in a few weeks. I'm in a hospital, of course, & (of course) don't like it but I'm not really in a great hurry to leave until I'm able. Meanwhile don't worry at all because I am cared for & doing very well. I would come off here where they have strange diseases.

I confessed I'm a little apalled—or however you spelled it—at the prospect of 5 Christmas bundles! I have to be mobile, you know, & have trouble meeting the low baggage limit as I usually go by air. However, I bet I find them delightful & I may well be stationed somewhere for a while when they arrive. So stupid they wouldn't let you send magazines, as that's one thing most appreciated here—especially noticeable in hospital where anything from Plumber's Gazette up gets read literally to pieces in a few days.

Enough for now. I'm getting well & I still love you all very much.

—Your devoted son

[Sicily]   V-Mail Letter
6 Nov. 1943

Dearest Marty:

Thanks for yours of 9 Oct., just received. I am happy at your news of impending marriage, because I am sure it will make you happy. I gather that the lucky fellow's name is William Eastlake? It's a new name to me, actually, although you did drop a broad hint or two. I gather that he is known as Gary—or are you studying to be a bigamist?[15] I do wish I could

15. Simpson's confusion here is justified, for Martha's previous suitor was named Gary.

be there, but you'd better not wait for me! Anyway, greet & congratulate him for me. He's in the army I gather? A few details, please. Don't be so damned offhand—or is it maidenly shyness?

Your $15,000 business leaves me aghast. My impractical little sister turns out to be the only member of the family to build up a business! Cripes, now anything can happen.

I'm enjoying an attack of jaundice just now. I'm not really yellow. The disease is a virus that attacks the liver. It hasn't got a name and sometimes turns people yellow, so they call it jaundice. It was sort of rugged in there for a week, but I'm going to get well all right. May be in the hospital a while yet.

Much love, as ever

Gg

[Sicily]

15 Nov 43

Dearest Mother & Dad—

Mother's two letters of 18 Oct. & (again) no given date arrived just now to cheer my bed of pain—Not that I have much pain any more. I am really doing splendidly, eating by mouth instead of intravenously, no fever, & generally well along the road to recovery, so that's that. I only have to get a little strength back now & I can go back & finish the war.

Do felicitate the Eastlakes for me! I am a little bewildered by the whole affair and can only hope that Marty will be half as happy as she deserves to be. I don't even know if her Gary is in the Army or Navy & have never seen a picture or read a description of him, so he is completely unreal to me, but I expect he's a nice guy if Marty wanted to marry him.

Now all your children have been married twice & at least 2/3's of them did all right the second time around, although of course no one could have as good luck as I did on my 2nd trial.

Give my regards to Aunts Lil & Martha.

So the strawberries will be ripe when I get back? Do strawberries start bearing in 1 year or 2? Do those bear all year round?—I just want to get a line on things. From where I sit (lie, really) chances for getting back still don't look good &, boy, how I want to!

Much love to you all—

Gg

[Sicily]
18 Nov 43

Dearest Dad—

Yours of Oct. 27 just arrived yesterday to cheer me enormously on my bed of pain. As a matter of fact I have very little pain now & am practically well again. I hope to be released any day now & to get back on the job. It seems as if I'd been down for a century, although it's only 3 weeks.

It is wonderful that you & Mother are so active & busy—I hope not too much so. Your schedule sounds like a tough one for anyone.

Marty's business must certainly be booming. I really can't imagine it, & her married to someone I haven't even seen a picture of. I do feel sort of out of touch & remote, in spite of Mother's and your faithful letters. Isolated over here you get an irrational feeling that everything should stand still & await your return!

That return seems remote as ever, but of course is drawing nearer steadily. I still think it will be next year.

Aside from a wild crop of rumors, which are mostly ridiculous & which I couldn't repeat even if I wanted to, little news penetrates here in the hospital & of course nothing happens here except just lying around.

I love you all very much—

Gg

[Sicily]
24 Nov 43

Dearest Marty—

I'm ashamed to say that I don't know whether or not I've written since I heard you are married—my records & memory are confused (even more than usual). Anyhow, I need hardly say (& of course like everyone who uses that cliché I will now proceed to say at length) that I wish you the greatest happiness & that I heartily congratulate William or Gary.—By now I have figured out that Gary is William Eastlake. I hope that's right. I would be interested in a lot more details, a lot more, probably, than you have time or inclination to write. One of the hardships of my circumstances here is that I feel so ill-informed & so out of touch on such occasions, so momentous & exciting for those I love, so distant & nebulous for me.—This is not a complaint, but just an explanation of my possibly peevish curiosity. Anyway, I certainly do wish you the best. No one else could have the luck I had on my second try, but I wish you the nearest possible!

I'm still in the hospital & my last blood test was not so good, so I may be another week or two, but I'm really much better & doing very well. Just a matter of convalescence now, no complications & no permanent effects, so nothing to worry about.

I suppose you still use maiden name in address?

Note my return to old address—I haven't moved, however.

Much, much love—Gg

[Sicily]
8 Dec 43

Dearest Mother—

Your letter of 31 October arrived yesterday which certainly is not good time for V-mail, but appreciated none the less. And that reminds me that even now I may be too late for Christmas greetings, which nevertheless I send with all love to all of you & whether they get there or not, you'll know on Christmas that my thoughts & love are with you. This will be our 3rd war Christmas. The first, I was at home & a civilian. The second, I was in the army but luckily enough to get home nevertheless (I was then at Fort Hamilton in N.Y. you may recall). This third I will be far away. The fourth, who knows? I'll still be in the army, for don't let anyone kid you that the war will be over then, but I might have the luck to be home any-way as the European campaign may just possibly be over & besides they do rotate officers on overseas & home duty, although not always so rapidly as we would like. (Not a trick—a fixed limit to overseas service, such as the 6 months recently suggested, would certainly interfere with the war effort & so would not be supported by us, much as we would like it personally).

I don't remember whether I wrote that I'm out of the hospital—I am. I'm not yet back at work, however, & still spend most of my time in bed, although I can now get up for a while each day & eat one meal a day at the mess, which I hope soon to raise to two. I'm really doing very well, although it is a long drag. My organization has been completely changed during my illness & I will go back to quite a different job, a good one although I won't have so many people to boss around.

Much love to my dear parents & sisters—Gg

[Sicily]
12 Dec 43

Dearest Mother & Dad—

All at once, here are Mother's letters of 1 & 7 Nov & Dad's of 7 Nov & 2 Xmas packages. Regarding the latter, I had a brief struggle with my con-science in which as usual my conscience lost—don't you think it speaks well for me that I have such strength of character that I can almost always win struggles with my conscience?—so I opened them at once instead of waiting for Christmas. One contained a fruitcake & the other contained 2 decks of cards, soap, puzzle, & candy. Thank you *very* much. It must be very hard to think of things to send under these circumstances, & I think you are very clever to think of things so welcome here. I couldn't have done it if I had tried to say what I wanted, & now I see that these are just what I wanted.

You both sound so busy & active & I only hope you don't tire yourselves too much. I don't think either of you has has much talent for taking it easy—Just look at me: I haven't done a lick of work for over six weeks!

How come I have Italians for buddies (if you want to call them that, &
in reply to your question) is that I naturally know some, being in Italy, that
some work for me (at least that's the idea, but it can be questioned
whether they work, & also whether they are for me), & also that a number
of my officers are of Italian descent, which of course makes them particularly useful here in all sorts of ways.

I still spend a lot of time in bed (I'm there now), but I get stronger every
day & will soon be back on the job.

Much love to you both—

Gg

[Sicily]
19 Dec 43

Dearest Mother (& dearest Dad, too)—

Yours of Nov. 12, 21, & 28 arrived within the last couple of days. I'm so
glad you all seem to be doing so well. We could use a little of your nice
weather—it's rainy & raw here & I rather dread the knowledge that I'll not
be really warm again till next summer—a good deal of not minding winter
at home depends on occasionally having a fire or at least warm water to
wash in, both of which are among the things we do without. Not that we
do without so much, except the basic deprivation of doing without our
families. Where I am now I have a fairly good billet, plenty of nourishing
hot food, & not even any serious air raids lately. They did make a pass at
us one night while I was in the hospital, but were driven off so efficiently
that that not a single bomb hit the town & all in all the war has been relatively easy for us lately. No complaints.

I try, with imperfect success, to visualize your home with a cat & a
canary. I guess it has to be seen to be believed!—And I certainly intend to
see it.

Write to Anne yourself about the possibility of sending [daughter] Bets
to Calif. By the time I have exchanged letters about such a situation, the
circumstances have changed, so I usually just rely on Anne's judgment, in
which I have perfect confidence & which, indeed, almost always agrees
with mine.

As I hear you are copying my long letters & passing them around I recall
that they usually contain some love-making & occasionally are a bit ris-
qué—I hope they are censored before circulating!

I am up & about & think I can declare myself well at long last.

Much love—

Gg

Dearest Marty:

In case you are wondering why the deuce I don't answer yours of no date but apparently October something, the reason is that I just now received it. It made me very happy, because you sound so happy, pleased with life & everything coming right for you at last. You have no idea [how] pleased I am—quite as much, I think, as if it was all my own good fortune. (Indeed I am equally fortunate, except for this dreary & painful separation, which cannot last forever). I did felicitate you on your marriage, & you should have received it by now. Anyhow, I repeat that I wholly approve & wish you the best possible. If I do not send congratulations, as requested, it is because the congratualations are for the lucky man. Thanks for the picture of him—I won't agree that he's pretty, exactly, but he certainly is very good looking, which is good in a man, as you say (although I never found it indispensable!). I hope he doesn't get sent overseas. Of course I have no respect for a man who tries to avoid overseas duty, but it is a weary drag & if the army finds him more useful in the U.S., & even reasonably near home, he should feel that he is doing his duty fully & has a lucky break that he do so without going overseas. By the way, is that a Service Command patch he's wearing? I don't know the U.S. patches any more. (Mine is a red, white and blue (AF) , which doesn't mean "Air Force" as everyone guesses, but "Allied Force" as I belong to a combined British-American headquarters; my unit is highly interallied & interservice & I have everything from American yeomen to British captains in my office).

Also, [there] arrived a carton of Chesterfields [cigarettes] from you, & thanks a million. They saved my life as our ration is very inadequate & I run behind even in that while in hospital.

Much love,
Gg

[Sicily]
31 Dec 43

Dearest Dad—

Your letter of 7 Dec just arrived to cheer me up. Not that I especially need cheering up, since I'm up & about, back at work almost full time, & all going well, but as usual I miss you & all the rest of my family terribly & as usual a letter is a very cheering event. About the best news I can have is that you & mother are well, busy, & happy, as you both seem to be. Some time ago I had a letter from Marty which was really lyrical with delight over her marriage, having you & Mother there, her home, her business, & life in general. In fact the whole California contingent seems to be thriving, which of course delights me.

Helen, Anne, and Joan all seem also to be happy & doing as well as possible with me away (not to be modest about it). I am somewhat perturbed about Gay & Betty, but not seriously so. They certainly seem happy & well & it is only the general situation there that disturbs me a little.

I'm so glad the wood carvings arrived & that you like them. I originally planned to send the whole batch to Anne & for her to forward on to you, but then I picked up another & sent a couple direct to you. I'm not sure I told her of change of plan, so she may still send you one. If so, keep it, also, of course, if you like it.

Much love to you, Mother, & all—

Gg

[Italy]
17 Jan 44

Dearest Mother:

Your Christmas Air-mail letter must really have come by air because it arrived yesterday & such letters usually take about 6 weeks to reach me. It was delightful to have a longer letter and to have your handwriting full size. You seem to have had a wonderful Christmas & I am sure the day ended as well as it was beginning when you wrote. Your prophesizing sounds like a very amusing touch & I look forward to being around when the prophesies are opened this year—In fact I bet that was one of the prophesies & I hope it is a better one than it was for last Christmas.

I haven't seen Ingersoll's book but understand it is very good. I have seen Ernie Pyle's & recommend it for what he calls a worm's eye view of war in this theater, what it really looks like to an individual not concerned with strategy but just with daily life. Pyle is back here now doing the same thing for Italy as he did in Africa. I think Lardner's dispatches are pretty good, too, if you see them.

I'm doing fine & actually getting some work done. Much love to all of you, even the cats—

Gg

[Italy]
22 Jan 44

Dearest Mother & Family—

I am basking in early morning sunshine which, regardless of what the tourist come-ons say, is a rarity in the south Italian winter. The lady who brings me my morning coffee remarked gaily that it must be wonderful for me to be here where we have sun every morning instead of fog as at home. I told her sternly (a) that we do *not* have sun here every morning as she would know if she could remember more than one day back, (b) that we *never* have fog at home (like most dumb Italians she has a vague idea that all Americans come from London), & (c) that it is *not* wonderful to be here and all any of us want is to polish off our job here & leave as rapidly as possible. However, I courteously thanked her for bringing the coffee & I gave her a graham cracker which she will treasure for a few days, show to all her friends, & then finally serve for dessert after dinner to her family of five.

Your much appreciated letter of New Year's Day arrived yesterday. I'm sorry several of my recent V-mails have arrived very dim—others have complained, too. This is the fault of the photographic reproduction & I trust they have fixed it now. This letter, for instance, is written in pitch black ink & with a fairly coarse pen & should be perfectly clear in reproduction.

I'm doing very well. Don't be excited—I won't be sent to U.S. for a couple of months & maybe much more. Much love—Gg

[Italy]
5 Feb 44

Dear Mother—

I'm sorry some of my V-mails are illegible when you get them. The fault is in the reproduction. This [air mail] is written in the same ink & with the same pen as I use for V-mails, & you can see that it is about as clear as possible. I'll try air mail in hopes it won't take too long—once in a while these go even faster than V-mail & then at other times they take two months, so we never know where we stand with them.

Yours of 8 Jan just arrived, which is almost a month for V-mail, but most welcome.

I'm rather sorry I ever mentioned the possibility of returning to the U.S., because everyone jumped to the conclusion that I was coming right away & will just be upset & disappointed. I *probably* will return during the 1st half of this year, but even that is not certain, & that is all I know myself. I definitely do *not* now have orders to return, so it can't be very soon. I did think I might be on my way about now, but I'm not. All depends on factors too complex to summarize, even if I were allowed to.

Marty's ceramics must really be something and I am crazy to see some of the recent ones. I was extremely attracted by the few I did see. Her success in this business is wonderful. It is in a way a pity that it is the ceramics rather than [her] paintings that have become so successful, but the ceramics are just as beautiful in their way & must give almost as much satisfaction to produce—and obviously they fill a greater need & do more good, because people buy them & live with them & do not buy paintings. Within broad limits I have always felt that one should not persist too long in an occupation for which society is unwilling to pay a decent living, even if one can afford to ignore the matter of pay. Naturally I hope that Marty does still paint, which may have more permanent value than the ceramics—or maybe not; we can't judge.

I manage in spite of hell & high water to see a little of Italian art in the flesh & a lot in reproduction. It's interesting & some is beautiful, but by & large I am not impressed. Perhaps I'd change my mind in Florence, say, but from here I consider Italian taste through the ages extraordinarily bad, with only an occasional artist or work of art overcoming the bad influence of being Italian. But then, I dislike Italians & everything they do so intensely that I am prejudiced. What *this* country needs is a good five-cent earthquake.—Which, in a way, it is getting, but alas! Italians, civil or military, are so good at looking after their own skins that most of them will survive. No, I'm not being brutalized by war. It's just that if people are going to be killed, I'd rather they were sneaking, worthless people, like Italians, rather than nice decent people like most of our American boys.

As for me, I plug along. My job is shifting by degrees & I am turning over most of my operational duties & short-range stuff to others, mostly a very nice British major, & am going onto strategic staff work, for the time being, perhaps working up to being relieved in this theater altogether. It is still very fascinating & quite important & it is odd what ramifications you can get into by starting out in censorship—very little that I have done, even when designated a censor, has been what you think of as censorship.

Frankly I am not 100% well yet & my full recovery has been set back a little, but I am at least 95% well, have nothing grave the matter with me, & continue to improve. In a month or so I'm scheduled to have a complete checking & renovation, when I can get back to where there are full facilities, possibly in Africa, so all is well. I assure you that there is no need for *any* worry—I just don't want to lie to you & say I'm bursting with health when I'm not. I work every day, on a somewhat reduced schedule, & don't get too tired.

I've got to dash off now, so will just add love to you & to all—or rather, express it, because it's there anyhow & doesn't need to be added—

Gg

[Italy]
20 Feb 44

Dearest Mother & Family—

For some little time I haven't been where my mail comes to, & none has caught up with me lately. I look forward to a feast when I do connect with my mail.

I am having quite an interesting time, doing a lot of traveling and seeing many very unusual & picturesque things. Unfortunately I have also encountered some unusual weather—like "unusual" weather in your sunny California—such as a terrific blizzard we battled through yesterday & which I thought may be really the end of us at one point, but we pulled out OK & after 12 hours in a warm bed I am back on the job today, well as ever but not one of the warmest admirers of the Italian climate. At least I get back to a comfortable billet, while the troops live & fight in that weather week after week. I'm afraid no one back home has any conception what that means.

I am pretty well, having an interesting time, & getting along fine. I do miss you all terribly & still have hopes of seeing you this year—

Love—Gg

Dearest Mother & Dad—

I just got back to one of the places where I get mail, & found Mother's letters of 23 Jan & 6 Feb & Dad's of 29 Jan waiting for me, so I'll answer them together.

First item: you are already on the distribution list for my two books when they appear.[16] They are only now really getting to press and I finished them before I joined the army, but these things are slow, especially now. They are both very technical (I'm not sure I understand them too well myself), but I know you enjoy seeing such things & bragging about them.

I'm glad to hear of Dad's reservation & hope the dentistry wasn't too painful. Of course I am also pleased & proud at Marty's continued & increasing success. We certainly are an outstanding family, aren't we? (Nice, too).

I hadn't had Anne's report on Kansas yet [Anne visited Simpson's daughter, Gay, who was being cared for by his ex-wife's mother there], & so am undecided in that matter but have just passed my ideas for her decision when she has seen & can judge the situation at first-hand.

As for me, I have another new job, but my plans & future here are quite uncertain. I am quite blue that the promised trip to U.S. has not materialized. It may yet, but time drags & I dislike all this uncertainty, but that is inevitable in a changing military situation.

I still love you both dearly—Gg

[Algiers]
5 Mar 44

Dearest Mother—

I have yours of 13 & 20 Feb—I may have acknowledged the former, before—I've been gadding about so that I'm a little tangled on correspondence. Now I am back for I have no idea how long in Africa, right where I started over a year ago. It is really quite nice to be back. This is the nicest town of all I have seen & I like the French & Arabs so much better than I do the Italians. My old room is taken, but I just called on my old landlords, The Belpaumes, & had a really rousing reception. I am very fond of them, as they seem to be of me, & in a diluted way it was a bit like coming home from the wars to rejoin true friends who have missed me. I made no real friends in Italy. I just do *not* like Italians, no matter whose side they happen to be on at the moment.

I've had a bad couple of months, now ended, because a damn fool doctor told me I was going blind. Now I've had a more thorough study by a

16. The two books are *Tempo and Mode in Evolution* and the *Classification of Mammals*.

better man who finds that it was only a passing toxic condition, probably related to the jaundice, & my eyes are really OK. What a relief! I may have to wear glasses, but why not?

Much love, Dad too,—

Gg

[North Africa]
21 Mar 44

Dearest Mother—

I have been so very busy that I have gotten well behind in correspondence. In the meantime, I have received your V's of 26 Feb & 6 Mar and your air mail of 28 Jan—snappy service, nearly 2 mos. for an air mail.

Very sorry to hear of Tieje's death. I hadn't seen him for about 10 years, I think it must be, but had an occasional note from him & I do feel I owed him a good deal in the way of encouragement & early training.

I was checked at hospital again today. Eye has stopped improving, but is now static & pretty good. Now I will be measured for glasses & will be good as new.

I just read that one of the wonders of post-war would will be plastic eyes, & Lo! Dad has one—always ahead of the times.

Don't be so cheerful about giving me to Uncle Sam—It's only a *loan*. He has to return me. If you don't insist, I do!

How did the rain hurt Mart's business? I don't get it.

Why didn't Bill [Marty's husband] call or visit Anne? I suppose he was badly rushed. Give me his APO address when you know it—overseas one, I mean.

No news on any move for me. Much love—

Gg

[North Africa]
28 Mar 44

Dearest Mother—

I am off tomorrow for a week of *rest*, & will drop you this line before I go, in case I do not wake up all week. I'm really tired & delighted to have this chance to recuperate.

I received yours of 12 Mar, with the usual pleasure, in spite of its report of Jack's [Simpson's nephew] relapse, which is bad news after so much encouragement, & of Marty's many unspecified annoyances, countered by the statement that she is doing well anyhow. I hear that Anne did have a visit from Bill, & liked him. Don't forget to give me his APO address when you get it (not the 4-number one you may have now, but a 3-number [overseas] address he'll send later), in case he should turn up in my part of the world, although I think this unlikely.

I still have no news, just lots of hard work & no indication of when, if ever, I may be transferred—I'm rather fed up because I was definitely told I would be, but nothing happens. Anyway, I'm doing well, although still pretty homesick after all this time.

Much love to you, Dad, & all—

Gg

[North Africa]
12 Apr 44

Dearest Mother & Dad—

Your air-mail letters of 19 Mar both arrived together yesterday. I was distressed to hear of all the accidents & upsets, but I guess everything is fixed up again now. I hope you two don't decide to break an arm or something to stay in style.

I am still amazed at the way Marty's business has grown & of course very delighted. I always considered her one of the family's great successes but never expected her to eclipse us all financially, also. Also I naturally expect Dad to be peppier than I am, but I do think he's bragging when he complains that he's no good for any other job after "only ten hours" in the shop! I certainly look forward to seeing some of the new products of this wonderful institution, & I certainly am interested in details of it, & of everything else that you are doing.

I guess I did tell you that my possible trip to the U.S. is all off. I'm very blue about it.

I have you down for a copy of the book on evolution, when it appears. I don't expect to understand it myself, now, but I know you'll enjoy seeing it.

I'm well & getting along OK—

love
Gg

[North Africa]
19 Apr 44

Dearest Mother—

Yours of 2 April arrived not long since & was much appreciated as usual.

Sorry to hear that Pixie [the cat] was so impure that he had to be operated on to teach him a lesson. I hope the idea doesn't catch on for humans, although goodness knows I'm pure, darn it.

I hope Peg's troubles do turn out to be thyroid, for then at least there would be something definite & hopeful to do.

I wonder if your weather has remained good? We hear horrendous tales about weather in the U.S., but California hasn't been specially mentioned, to my knowledge. Our weather is wonderful lately. We are enjoying our brief Spring—Alas! It only lasts a couple of weeks in these parts, & then comes steaming heat.

I'm doing fine, really quite well although very sourpussed, mainly because of disappointment about getting home, or *not*, rather. However, the war can't last forever—or can it?

This is my afternoon off & work is momentarily a little slack, so I'm getting it off. On finishing this I'll stroll down main street in the Spring sunshine & get my week's supply of cigarettes & candy at the PX.[17]

Much love (also to Dad, Peg, Marty, & all)—

Gg

[Algiers]

6 May 44

Dearest Mother & Dad:

Mother's V's of 7 & 24 April & Dad's of 23 April have delighted me this week. You all sound so busy & active that I almost feel as if I were idling away the war in a quiet backwash—no gardening, no potting, no income-tax figuring (Anne does that—Whoops! as Helen would say).

I have had a little more social life than usual this week. Let's see. Monday had the Belpaumes to dinner at our mess & then for a spot of eau-de-vie in a smoke-filled room. Tuesday I dined at the Royers' (he's Dean of Science in the local university & a delightful man). Wednesday Larry & I dined far into the night at Belpaumes'. Thursday I cocktailed & dined with Jeff Wright, a British colleague (a professional soldier, veddy military, in the army all his life, but a good egg). Friday (that's last night) I went to bed early with a good book in Arabic which I read for a couple of hours (at about 2 pages per hour, I am *not* brilliant at Arabic). Tonight Larry & I are calling on [illegible], our Arab mentor. What a social butterfly! So it goes some weeks, while others I just go to bed at 8:30 every night & am very bored, unless I have to work late. (I *do* work in addition to being a social butterfly).

Weather has turned hot, & we are still in wool but go into cotton on the 16th (we even change our clothes by order).

Much, much love to you both, & to my two kid sisters (that'll flatter 'em!)—

Gg

[Algiers]

14 May 44

Dearest Mother—

How Mother's day has crept up on me & I am writing on the day, instead in time for you to receive it then. Somehow I didn't hear a thing about its imminence until it was actually here. I'm not so keen on such special days, anyhow. I think of you & I love you all the time & do not

---

17. More army talk for the "post exchange," which sold a variety of merchandise.

need a Mother's Day to remind me of it, or to tell you about it. I suppose such occasions have a use in making us think about things we might take for granted—But I didn't really take it for granted that I have such a wonderful mother. I know it is very unusual & I appreciate it very much.

Tell my old man he's not bad, either.

Yours of 30 April arrived to cheer me. You ask about my birthday. I honestly & truly do *not* want a thing except you, Dad, Anne, & assorted daughters & sisters, and I'm afraid you can't send me them.

I'm plugging along, as ever. Health satisfactory.

In these parts this is Joan of Arc day & we had a very snappy parade, flags & bands & one thing another.

Much love to you all.

Gg

[Algiers]
24 May 44

Dearest Mother—

Yours of 6 & 13 May arrived more or less in cluster, if two things can cluster. Cluster or not, they were welcome, as always.

It seems funny for you to be having elections already.[18] I'd like to vote this year, particularly, but I think you Republicans have done me out of it for fear I'd vote for the Commander-in-Chief [President Roosevelt]. He's certainly a very good C.-in-C., but I wouldn't vote for him *on that account* any more than I would vote for MacArthur because he's a good general.[19]

I've finally finished study of my 1st year Arabic grammar & found a real gem right at the end. With a lot of rhetoric omitted, it is: A camel was heavily burdened with a double load of salt & dates. The camel-driver said: "Which do you prefer? Going up hill or down?" "Allah damn them both!" said the camel.

I laughed for five minutes (I really did). That story has everything, local color, Arabic idiom (in the original), native psychology, & native humor. Don't spend time hunting for a hidden point—There isn't any. That's what's so funny.

Now I'm starting 2nd year Arabic, trying to work up to camels with *four* loads. Some fun.

I'm OK. Still have hopes of getting home in a few months, but only hopes so far. Much love—to Dad too.

Gg

18. Presumably the California State primary.

19. Franklin D. Roosevelt, of course, ran for an unprecedented fourth presidential term that fall. General Douglas MacArthur was often considered a potential presidential candidate, a point of view that he did little to discourage.

[Algiers]

2 June 44

Dearest Mother & Dad—

Yours of 21 & 14 May, respectively, arrived yesterday. I'm a bit behind in correspondence, as I didn't get my ½ day off this week & that's when I usually write. Now I'm catching up by writing this at the office. I think I have unanswered letters from you at home, but don't remember for sure.

Your main news (now that the scandalous Earle Carroll episode has died down a bit) seems to be acquiring a pup, & I can imagine that's there never a dull moment with a dog, a cat, & a canary sharing your house. I did come within an ace of having a Sardinian donkey—they are very cute little things about the size of a collie dog—One of my officers actually got one to present me, but couldn't get transport to bring it back (fortunately—I have enough trouble without a mascot).

Everything is as usual here. We are busy & also having a lot of reorganization trouble. Our reorganization is really continuous, but this is an acute phase. I have a new job (as usual) & am tearing into it. I am quite well, & still love you both very much—

Gg

[Algiers]

21 June 44

Dearest Mother & Dad—

Your letters of 3 & 4 June did reach me for my birthday—mail service has noticeably improved of late. That's almost a week ago, but I've been a little under the weather—nothing at all serious—& have not written in that interval except a couple to Anne. I know you don't mind if I place her first on the letter list, & I know she passes on news of me, & I try not to neglect you too badly. You're first on the parent list!

I celebrated my birthday by an afternoon off, spent at a nearby beach, & dinner at a hotel there, run by the army as a summer rest camp—All this with Larry Bollag & Commander Gates, a naval friend. Since then I've been to one evening party (stag, military) & last night a couple of British pals & one American dropped in. (Larry & I have acquired a bottle of gin somehow, & this is a better recipe for winning & influencing people than Lifebuoy soap).[20] That's about all my social life, which is about all I can write about. Still in the same place & job & all under control.

I love you both very much—

Gg

RECEIVED TWO BOOKS FROM YOU—SWELL—THANKS!

20. Lifebuoy soap was one of the first deodorant soaps; it smelled more like antiseptic than perfume.

[Algiers]
28 June 44

Dearest Mother—

Yours of 11 & 16 June ambled in together a day or so ago.

I'm glad that you're glad I was born. I'm quite pleased about it, myself, but I should think you must have had frequent misgivings on the subject! Anyhow, balancing one thing with another I think I've done more good than harm in the world, & I know had more pleasure than grief, so I guess you can say I've come out on the credit side so far.

In spite of all the stirring events on all three European fronts, [Normandy invasion, capture of Rome, and the beginning of the Russian summer offensive] life goes along in quite a dull way. I've really finished up all the difficult parts of my job here, & for a wonder there is no new trouble to make me the trouble-shooter (my usual assignment), so I am in pretty dull routine that doesn't really need me seriously. Therefore I hope, & half, or say a quarter, expect that they may re-assign me or send me back this summer or fall, but it's impossible to tell what will happen. The end of the war is in sight—but only with my far-sighted eye! (I guess I told you my eyes have fairly well settled down with one far-sighted & one near-sighted, an annoying anomaly).

Much love to you all—

Gg

[Algiers]
4 July 44

Dearest Dad—

Thanks for your letter of 18 June. For once I have a little news, which is that I will be home sometime this summer. I don't know just when, or what will happen after that, but that much is, at last, entirely definite. Next week I will have been overseas for 18 months. That isn't long in comparison with many of the British, some of whom are in their 5th year away from home, but it's a long, long time just the same, & I'm very glad that my exile is drawing to an end. I'm pretty well polished up on my various jobs here & have bright-eyed successors working at those that are still required, as the war effort will not slacken, even in my very small department of it, by my leaving.

I'm so excited at the prospect that I don't sleep well, but I'll soon get over that, for of course this is not immediate.

There isn't any other news, of course. Still leading a quiet life. I do get an occasional good meal now (my current mess is terrible), at the local Red Cross mess for their own personnel—wrangled through a friend there. That's how I'm celebrating the 4th. Love—

Gg[21]

21. Simpson returned home soon after this letter was written, was given six weeks accumulated leave, and placed on inactive duty in October 1944.

New York City

4 Mar 45

Dearest Mother & Dad—

This is a warm & sunny Sunday, for a change, but the weather man says more cold is coming. Anyway, I'm lying in the sun in our living room (which is very pleasant) & taking it easy, as usual.

I guess Anne has told you of my good luck in that the Rockefeller Institute happens to be doing a big research project on virus liver diseases & has taken me on (for free) as a patient-guinea pig. They have finished all the fancy diagnostic tests and are very encouraging. The liver damage can be completely cured, which is unusual in liver disease, & I will be as well as ever in a few months. In the meantime I have various medicines, weekly injections, etc. My diet includes meat twice a day (plus a large breakfast with eggs) & this is unhandy now meat is so extremely hard to find even though the OPA [Office of Price Administration] promises to crash through with extra red [rationing] points. Aside from this, I naturally think a meat-rich diet is a good idea! However I do not think it a good idea (although you will) that I cannot drink at all (I mean alcoholic drinks, of course).—Although the Dr. says it will be OK to drink again as soon as my liver takes a hold once more. I have to rest a lot, including two one-hour periods at the office. Since I can't possibly rest in my office, where interruptions are inevitable, I sneak off to lie down in Roy Andrews' office, where there is a couch & I can lock myself in. He seldom uses the office these days, & so far no one has discovered my retreat & I rest in peace.

Well, so much for me.

Anne is well & vigorous, but as usual has far too much to do, with the house, her job, & also a great deal of work for the N.Y. Academy of Sciences, where she's now organizing a big symposium in addition to the regular psychology programs. She's also running out to Minneapolis next week, to speak at a conference on alcohol problems, & in general is on the run all the time. I don't see how she manages to be a perfect wife & mother along with all of this, but she does.

Bets [16] is very blooming & sweet. She doesn't care much for her school, but is doing all right & adjusting to it slowly. She has matured a good deal and has gotten over her tantrums & other babyish traits that were sometimes hard to take, & she is good company & a real pleasure to us both. After shying away from science in school she finally had to take chemistry & is so intrigued by it she's talking about becoming a chemist, although I do not expect this to last long! Her interest in the Guild at church continues & tonight she's preparing the supper for the group! She's feeling set-up now anyhow because she has acquired a new costume, navy blue complete with hat & handbag (two accessories she has hitherto resisted), in which she looks very cute & sweet, as she admits.

We dined with the Peets last night, & they asked for you. The boy is very nice & Anne & I are both crazy about him, and he about us, I may say. He's rambunctious, like any 6-year-old boy, but in a nice way.

Creighton & Bertha Ann manage to make ends meet, but only barely, as usual.[22]

Our uncertainty about living quarters continues & we have no idea what may develop. Our undesirable neighbors are still with us & no decision yet as to when or whether they'll be evicted or whether we'll have the whole house, our present quarters, or nothing.[23] Latest development, which amuses us, is that the carpenter we had in to fix our bell wants to buy the house! The prices he charges, he can afford to! If he does, he wants us to take it on a 3-year lease, which we would probably do although it would be a squeeze both financially & in the work of operating it.

We are sorry about Bill's being injured & do wish he would break down & give details, which can't be as bad as unbridled imagination. By the way, one of my most brilliant colleagues & one Anne & I both are very fond of is missing in action in Germany & we are upset about this—Bryan Patterson, my opposite number at the Chicago Museum (i.e. the former Field Museum).[24]

I must close, as some friends are about to drop in for cocktails. (I still serve them though I don't drink them—it doesn't bother me at all & I like to see others enjoy things even when I can't.)

Oh, report to Dad on research: Naturally this has slowed down a lot as my shortened working hours are largely swallowed up by administration, but I have finished my long study on penguins, recent & fossil, & the origins of penguins.[25] Illustrations & typing not done yet, but that's routine & I can now turn back to Eocene Patagonian mammals which I hope finally to complete this year. Charlie Lang (head preparator) is just finishing a mount of one of these, *Thomashuxleya*, the first mounted skeleton ever of an Eocene mammal from South America, & I have a girl (Antioch College student) who is making a plastic (small statue) restoration of the beast. So work does go on.

Much love to you all from us all—

Gg

22. Creighton H. Peet (b. 1938) was for years a special favorite of Simpson's. Simpson treated Creighton like a son, sometimes to his daughters' annoyance.

23. The Simpsons had the two top floors of a New York brownstone apartment. The lower two floors were occupied by a large, loud family.

24. Bryan Patterson (1909–1979), English-born American vertebrate paleontologist who was for a long time at the Chicago Field Museum of Natural History, then went to Harvard's Museum of Comparative Zoology, where he was a colleague of Simpson's in the 1960s. Patterson was captured during the Battle of the Bulge, December 1944; he escaped twice from prison camp but was recaptured each time. He was finally freed by advancing Allied troops.

25. Like several other Simpson projects, this one started out, simply enough, years earlier when he came upon some fossil penguin bones in Patagonia in 1934. Work on these bones led to a further consideration of penguin history and evolution that resulted in a technical monograph, *Fossil Penguins* (1946), and a popular book, thirty years later, *Penguins: Past and Present, Here and There*.

FINAL
MUSEUM YEARS
1950-1959

B Y THE end of the war Simpson was resettled at the
American Museum and chairman of the newly estab-
lished department of geology and paleontology. He was also appointed professor of vertebrate paleontology at Columbia University, whose graduate
students in zoology and geology now had the opportunity to study formally
with him at the museum. Years later he referred to the subsequent decade
as his "halcyon period." In the summers he undertook fossil-collecting
expeditions to the San Juan Basin of New Mexico, where important new
discoveries of early Cenozoic mammals were still being made. He and
Anne built a summer home—"Los Pinavetes," or "The Ponderosas"—in
the nearby mountains as a base for his fieldwork and as a retreat for writing
and relaxation.

Besides continuing his collection and description of fossil mammals,
Simpson pressed forward with more theoretical research and popular scientific writing. His publications include a variety of successful books: *The
Meaning of Evolution* (1949); *Horses* (1951); *The Major Features of Evolution*
(1953), an updated and expanded version of *Tempo and Mode*; *Life of the
Past* (1953), a popular synopsis of *Major Features*; *Evolution and Geography*
(1953), which codified Simpson's ideas about historical biogeography and
rebutted claims for continental drift based on fossils; *Life: An Introduction
to Biology* (1957), a basic college biology text coauthored with the distinguished zoologist Colin S. Pittendrigh and botanist Lewis H. Tiffany; and
Evolution and Behavior (1958), a book conceived and edited by Simpson
and Anne Roe.

The approaching centennial of the publication of Darwin's *On the Origin
of Species* in 1959 stimulated a number of symposia, conferences, and celebrations. Simpson was often an invited participant or honored guest.
Thus, he received medals, awards, and prizes from the Geological Society
of America, the Philadelphia Academy of Sciences, the University of Chicago, and similar distinctions from scholarly societies in Belgium, France,
Germany, and Great Britain.

The modern evolutionary synthesis begun in the mid-1930s, to which
Simpson had contributed *Tempo and Mode in Evolution*, was well consoli-

dated by the mid-1940s. One strong manifestation of this development was the founding of the Society for the Study of Evolution in 1946, with Simpson as its first president. He soon arranged a grant-in-aid from the American Philosophical Society so that the new society could publish a journal, *Evolution*, which survives today.

Anne began to spend more and more time pursuing her own professional career. She worked for the Veteran's Administration as a clinician, then began to branch out into the study of the psychology of professions, eventually publishing two classic works, *The Making of a Scientist* (1953) and *The Psychology of Occupations* (1956). In the late 1950s Anne held an adjunct professorship in psychology at New York University.

In the late forties and early fifties Simpson's daughters graduated from college, married, and had children. Three of the girls later received advanced degrees: Helen in botany, Gay in library science, and Elizabeth in social psychology. Joan went to work for a New York publishing house after college and began to write.

Martha and Bill Eastlake settled down on a ranch in Cuba, New Mexico. Both became successful writers. Martha published a popular art appreciation book, *Art Is for Everybody* (1951), and Bill produced several novels.

Despite this good fortune, life dealt a series of blows to the Simpson family. In 1949 Simpson's father died. His mother lived another ten years, first with Anne's brother and his wife, later in a nursing home in Albuquerque, relatively near Martha, who could visit her. Gay, who had always had precarious health, died in 1958 not long after she had married. And there were several divorces among Simpson's daughters. Then, in August 1956, Simpson himself suffered a near-fatal accident while fossil-hunting in the upper regions of the Amazon: he was struck by a tree being felled to clear a camp. He received a severe concussion, dislocated his left shoulder and ankle, and broke his lower right leg. After a week-long, painful trip back to New York City by dugout canoe, motor boat, surplus Navy plane, and commercial airliner, Simpson entered a two-year period of multiple operations, repeated hospitalizations, and slow recovery of the use of his leg. Indefatigable nevertheless, he continued his research and writing, whether lying prostrate in the hospital, reclining uncomfortably at home, or hobbling to the museum.

In 1959 Simpson apparently disagreed with the director of the museum over Simpson's ability to resume his duties as departmental chairman. Neither party ever gave a public explanation of the incident, but as a result Simpson resigned in a huff and accepted one of the Alexander Agassiz professorships at Harvard's Museum of Comparative Zoology. Thus the fall of 1959 saw the end of his thirty-two-year association with the American Museum of Natural History.

There is a five-year gap in the letters for reasons that I have not been able to determine. Thus the first letter in this section dates from 1950.

La Jara [New Mexico]
17 Jun 50

Dearest Marty—

Happy birthday to you, too! Your very nice letter arrived the day before my birthday, which is perfect timing, especially in consideration of the local mail "system." The l.m. "s." is under worse handicap than ever at the moment, as Pete Fisher is at a postmaster convention in Gallup & Mrs. Freelove, Sr., had a misery & has gone to Albuquerque to doctor it (nothing serious, I fear), so both Cuba & La Jara are worrying along with substitute postmasters.[1]

Anne, George Whitaker[2] [museum field assistant], & small (but growing) Creighton drove out ahead of me, as I had a meeting to attend, & I came by train, arriving on the 14th. All is well here, house OK except for a bullet through a window & into a wall, window repaired before we came, & what is a home without a bullet hole? Jack Lents had finished our furniture, Lucy Lucero (Zia Indian) had finished embroidering our curtains, & we really are done & fixed up inside, at least. It really looks wonderful! Probably no one else would like it just this way, but to Anne & me it's perfect, & we spend a lot of time just wandering around the house liking it.— Joan arrived safe & sound on the 15th.

We celebrated my birthday by having a spare-rib barbeque, very good & all gorged near or beyond the point of discomfort. The whole Freelove clan came, also Jack Lents, his wife, & their Spanish help, Panlive (good old Spanish name). I *think* there were 16 of us, although there was never a chance to count all at once.

Although I am perfectly well, I arrived sort of worn from a hard winter & long hours right up to the moment of leaving, also feel the altitude as I do for a day or two, so I am just taking it easy for a few days, resting myself & catching up on a few chores in leisurely fashion. George W., full of beans & 20 years younger, has the labor under control. The Society of Vertebrate Paleontology Field Conference is on the 20–24 June.[3] The whole mob (75–100 of them) will spend one night here. After the conference, Geo. & I will go out & camp in the Regina Rincón & work out a *Coryphodon*[4] prospect we found last year, then about mid-July we'll probably go up to Wyoming for a couple of weeks & stop ten days or another

1. Cuba and La Jara, New Mexico, are small towns in the northwestern part of the state on the continental divide. Martha and her husband's ranch was in Cuba; the Simpson's summer home was in La Jara.

2. George Whitaker, about twenty years younger than Simpson, was his museum assistant who prepared the fossil specimens he collected during the summer field expeditions under Simpson's supervision. It was Whitaker, trained as a U.S. Army medical corpsman, who saved Simpson's life in Brazil.

3. The Society of Vertebrate Paleontology, which Simpson helped found and whose first elected president he was, holds regular field excursions in North America for its members to examine significant fossil-collecting localities. In 1950 the Society visited the San Juan Basin.

4. Anne found one of the first fossils of the San Juan expedition, a primitive herbivorous mammal of the genus *Coryphodon*, thereafter known among the museum preparators as "Mama's first fossil."

fortnight in Colorado (Huerfano Basin, near Walsenburg). Then rest of summer here or hereabouts.

Our present plan is to come out to California the latter part of September, although of course we'll drop things to come sooner if that should seem better.

By the way, we plan to sell our hybrid station wagon ('48 Chevrolet, on pick-up chassis) in the West next Fall & return to N.Y. by train, as we won't be using a car for a year thereafter & it would just be an expensive nuisance to store an aging one. Do you happen to know how second-hand prices are out there now? The body has been refinished, motor completely overhauled, 4 new tires this year, & all in first-class shape now & should still be next Fall. Car will have about 35,000 miles then.

Well, as I started to say quite a while back, happy birthday to you! I hope you & Bill get out here for a visit again this year. There will probably be someone here all the time, as Joan & Mrs. Roe will stay when Anne & I are in Wyoming & Anne will probably return when I stop over in Colorado, but of course I hope you come while I'm here or hereabouts, too. So drop in when you can, but take two days to drive.

We are very excited about your book [*Art Is for Everybody*, 1951] & can hardly wait to see it. We haven't seen any of the ms., so are especially curious, but confident. I'm sorry it bothers you a little to have the book interest some people more than your painting, but I do not find this surprising. There are many more people (of whom I am one) more nearly literate in writing than in painting, & writing normally reaches a far larger audience. The book will also increase literacy & interest in paintings, including your own.

I'm very fond of my sister, too, & very proud of her.

Anne also sends love & birthday greetings—

G

Cruzeiro do Sul
13 June 1956

Dearest Mother—

You'll be surprised to receive a letter from just about the most remote settlement in the whole Western Hemisphere. I am here, way up in a corner of the Amazon Basin, & soon to take off up the River Juruá into still more remote wilderness. We are delayed because our outboard motors for our boats have not arrived, but everything else is here, they should come any day now, & we'll soon be off.

Remote & primitive as it is, this is quite a large town, about 2,000 people, & is legally qualified as a city in this region.—In fact it is much the largest & most civilized place in an area considerably larger than the whole state of California. Perhaps its greatest charm is that it has no traffic problem—there are no streets. Everything is either carried by hand, or on pack saddles on cows (not horses!), or in boats.

There are six of us from outside, three Brazilians & three Americans,[5] & we are hiring six people here: two men to run the large boat, two canoe-men, a cook, & a general servant. We from outside are billeted with a friendly lady, Dona Raimunda, who happens to have a larger house than she needs, & we eat in the town's restaurant, which is not exactly like the Ritz. The climate is terrible & the insects drive us crazy, but the town is really very attractive & the great Amazonian forest all around us is beauti-ful & grand beyond description.

Please give my love to Peg & Mac[6]—I shan't write them separately—& of course much love to you—It is not likely that a letter will catch me much if any before I come back down the river—But try!—G

Address: Cruzeiro do Sul, Territorio do Acre, Brasil

Cruzeiro do Sul [Brazil]
13 Jun 56

Dearest Marty & Bill—

And a happy birthday to you, too! Your letter reached here most expedi-tiously, about 10 days. Once we leave here, things will be rather hit or miss as mail goes only from town & we will be some days or weeks of travel out of town.

Until reaching here, we could hardly believe that Cruzeiro d. S. existed & now we are here we can hardly believe that any other place exists. At best it seems extremely improbable that C.d.S. & Cuba, N.M. are on the same planet. The only resemblance that rises to mind is that it hasn't rained here for ten days—beginning of the dry season. Everything is still soaking wet, nevertheless, for it would take far longer than 10 days to dry out.

C.d.S. is indeed a grand town to exist, as the Brazilians themselves con-sider it just a bit beyond the end of the world. It has no streets, only human & pack ox traffic, no running water (in pipes, plenty on the ground), only absolutely minimal supplies of any sort & mostly bartered, cash being of little use. But it has a light [generating] plant usually in oper-ation from 6–9 or 10, two clubs (one very select & one moderately so— the degree of selectness is usually settled by whether one does or does not own a coat; if you don't have shoes, you aren't eligible for a club), a bank (which cannot change the equivalent of a $5 bill), & a hotel (which makes Young's Hotel, before its remodeling, look like the Ritz). We are six & the hotel could not accommodate us (it has only one room & there were six or seven people already in it), but a most hospitable & voluble lady, Dona Raimunda, has taken us in. She & her retainers already numbered 6 or 7 (they won't hold still to be counted), so we are cozy here, too, with at

---

5. The three Americans were Simpson, his field assistant George Whitaker, and a vertebrate paleontology graduate student from Columbia University, David Bardach. The three Brazilian scientists included Llewellyn Price, a vertebrate paleontologist of American ancestry, and two collectors of living animals.
6. Simpson's sister Peg's second husband, Duncan McLaurin.

least one hammock & net in every room—It looks as if enormous tent cat-
erpillars had invaded the joint.

We have been stuck here for ten days, but such delays are expectable.
Amazingly & through haphazards too complex to relate, all our other
equipment arrived on time, but not our outboard motors & while we can
rent boats we cannot find motors to rent for so long a time. One of these
days we'll be off up the river, which is wonderful—we have already investi-
gated it for a day's travel each way.

Perhaps the most appealing of the traits of this city is an inordinate
fondness for firecrackers. In our ten days here not an hour has passed
(including the small hours) without a fusillade. We were told that firecrackers indicate the arrival of a personage, the birth of a child, or an
access [sic] of drunkenness. Of the first we had one (a new mayor), of the
second several, & the third is continuous. We were not told—for who
could have thought us such barbarians as not to know—that last night was
also St. Anthony's Eve, which calls for redoubled efforts in the firecracker
line as well as many large bonfires, exactly what was needed to make an
unbearable climate worse. Points of light even appeared every mile or so
along the margin of the black, brooding forest across the river—a romantic
sight.

We will really get away soon, even if we have to do so temporarily at
first using canoes, or perhaps we can get a deal to swap our putative motors
for real ones at hand.

Love & regards, to be distributed as appropriate. All is well on the
Juruá, & I look forward to working my way up the Puerco come October.[7]

G

[New York City]
13 Oct 56

Dear Marty & Bill—

Home two weeks [from hospital], taking considerable nourishment, &
finally feeling up to writing briefly in person. I've appreciated your frequent
notes very much; they cheer me up no end.

Just this morning came one from Huxley[8] *most* enthusiastic & grateful
for your hospitality & full of admiration for you both. I *told* you he was
bright! He'll be here next week & it will be nice to talk of N.M.

To[o] early yet to make plans, but it is possible I can get to N.M. next
summer. Probably still in a cast, but by then at least a smaller one. I can't
budge the one I have now, although I can walk on crutches to living room

7. The Juruá River is a tributary in the headwaters of the Amazon; the Puerco
River is in New Mexico, and Cuba and La Jara are on its upper reaches.

8. Julian Huxley (1887–1975), distinguished British biologist and evolutionist
whose book *Evolution, the Modern Synthesis* (1942) contributed to the consolidation of
contemporary evolutionary theory and provided the label for it. The Huxleys had
planned to visit the Simpsons at Los Pinavetes but Simpson's illness prevented the
meeting there. Thus, Marty and Bill acted as surrogate hosts and struck up a friendship
with the Huxleys that lasted for years.

if someone picks up my leg for me. Everything else practically well, & we'll have a better prognosis on the right leg when I go back to the hospital in December.

A Peruvian geologist (Benavides) who has been working in the same area where I was, just across the border in Peru, came in to see someone who broke a leg there & is still alive. He says they've lost more damn geologists that way! Actually several of their men broke bones & none were gotten out alive![9] They (the survivors, who *didn't* break a bone!) are going back in 1957 or 1958, but with a 2-way radio & a helicopter-ambulance standing by. If they go in 1958 I may join them, but not, my Dr. warns me, in 1957.

Incidentally my surgeon has the perfect name for working on a leg: Stump.

I've really about caught up on murder mysteries & science fiction. Now I'm practically off dope, I'm beginning to take notice again & read a little more edifying literature, such as catching up on "Journal of Paleontology," "Evolution," etc. I should be working on two books but don't feel quite *that* ambitious yet.[10]

Let's not talk about the Dodgers. I got in front of a TV just in time to see the débacle.[11]

Parabien, felicidades, & sanadades, as we say on the Alto Juruá.[12]

G

New York
Sunday, 10 November 1957

Dearest Mother:

Thank you for your recent letter, which Anne and I both enjoyed very much. We are interested to hear of your writing, and we look forward to seeing it when Marty has had time to type it. It should be quite fascinating! Keep up the good work!

Now I will answer your question about what I did when I had to quit college for a year. That was in 1919, after my freshman year at the University of Colorado. I went to Chicago where Uncle John got me a job at the Board of Trade. I did not like it very much, so I quit just as soon as I had found a better job with the Cable Piano Company. I was in the advertising department, where I helped out in general & also had charge of distributing two books for the Cable Company: "The 101 Best Songs" and the "101 Best Poems." You used to have copies of them—I don't have. Next year, in 1920, I traveled down South & spent some time at Sam Crosby's place

---

9. Accidents as serious as Simpson's are usually fatal in such remote tropical regions because of the quick onset of infection before proper medical treatment can be arranged.

10. *Life: An Introduction to Biology* and *Behavior and Evolution.*

11. The Brooklyn Dodgers lost the seventh and deciding game of the 1956 baseball World Series to the New York Yankees, 9–0. The Yankees had ten hits, including a grand slam home-run while the Dodgers were held to three hits and made an error.

12. "Greetings, good luck, and warm wishes."

near Port Arthur, Texas, then returned to Colorado where I worked for a few weeks at a private summer resort (at Scholl, Colorado) & then went back to the University of Colorado (in the Fall of 1920). You were in Chicago while I was with the Cable Co., but also went back to Colorado in 1920.

Now as to more modern news—there is not very much. Everything goes along fine, as usual. Anne is not yet working full time at New York University, but keeps very busy with other things. I seem to do mostly editing lately. Anne & I have done more, & nearly final, editorial work on a symposium "Behavior and Evolution" being published by Yale University Press, & I have also worked over a monograph on some fossils from Cuba [Central America], being published by the Museum. These & odd jobs keep me out of mischief. I am also writing various book reviews, and of course check or supervise a lot of things—exhibits & others—for the Museum, although I work mostly at home.

We have not yet seen our youngest grandchild, Helen's third boy, but Joan and all her family visited there (near New Haven) yesterday & gave us all the news. All is well there, too. Helen is back at work, healthy & happy, & the new baby is good, big, & growing fast. Wolf [Vishniac, Helen's husband] was in to see us a week or so ago, on his way back from giving some lectures, & Helen is coming down next week, but they won't try to bring the boys for a while yet.[13]

Next week Anne is off to Detroit for a meeting & a short consulting job (with an oil company—very well paid!) and she will have time for a good visit with Gay & with Betsy at Ann Arbor, near Detroit. Gay's husband, Frank, had a little accident a while ago but is well now, & Betsy & her brood are all well.

I know you have more recent news of Marty & Peg than we have, so won't attempt to tell you how they are.

Creighton Peet (who often asks about you) & a colleague of mine, Bryan Patterson (I don't think you know him) are due here right now for a visit, so I had better close.

Much, much love from both Anne & me—

G

[Anne adds:]
G is doing very well—he goes to the Museum once a week now.

Anne

[New York City]
22 Nov 57

Dearest Marty—

Herewith Mother's ms. Thanks for doing it, & for sending to me. I have written a highly appreciative comment to her, & encouraged her to keep up the good work. Between us, I am a little shocked because this seems

13. Wolf Vishniac (1922–1973), German-born American microbiologist, then at Yale University. Vishniac was later killed in an accident while doing research in Antarctica. He and Helen had three sons.

considerably less coherent & even less accurate than her conversation on the same subjects last summer. I suppose, however, that this is at least partly due to the difficulty of writing. On the other hand, her remodeling of the past is fascinating, as you noted, & also has a good aspect—In conversation she tends so to dwell & harp on old grudges & unhappinesses & there's little of that here. On the contrary she makes things, on the whole, more pleasant than they actually were.

Anne's in Detroit for a few days—a conference, a short consulting job, & a chance to see the kids [in Ann Arbor] & especially to learn more precisely what gives with Bets & what can best be done to help her. (Meanwhile I am well looked after by friends & our nice & devoted Swedish maid.)

Helen & Wolf both in this week—separately as the housefull [sic] of boys prevents both coming to N.Y. at once. Both happy, Helen blooming again & very pretty. They, at least, & for the moment, at least, have none but usual & not unpleasant problems, which is certainly not true for any other branch of the family.

All as usual with me. I'm off to hospital on 2 Dec., operating room on 5 Dec., but we won't know till then just what's to be done in O.R.

Much love, sweety, & my best to Bill. We are both looking forward so *eagerly* to your visit.

G

[New York City]
1 Nov 58

Dear Marty & Bill—
You'll be glad to see the enclosed, if you haven't already. H.-B. [Harcourt, Brace and Co., Publishers] is finally getting some of the lead out. Probably means the book is making some money for them, so they'll spend some.[14]

Anne's off in Washington for a couple of days—She's away about half the time lately. (Much of this is compensating for the two years of restricted activity when I wasn't fully able to care for myself, as I can now.)

She probably told you what a scare Joan gave us—Ear "infection" proved to be a tumor involving mastoid & tympanum. It took them 6 days to be sure, but it finally proved *not* to be malignant. Joan is back home now & almost as good as new. Her hearing is not affected, or at worst only very slightly.

So all's well there & the only serious family problem now is Edna Roe [Anne's mother], who *must* leave Ed & Ruth's but who is hard to arrange for anywhere else. However, Bob is doing most of the struggling with that problem.[15]

14. Like all authors since the invention of writing, Simpson felt that his college text (biology) had not been sufficiently promoted and advertised.

15. Simpson's mother was eighty-eight years old. She had been living with Anne's brother and wife but eventually had to be put in a nursing home in Albuquerque, where Martha was able to visit her from time to time.

I'm fine & still improving—No brace, no crutches—but not "no hands," as I do hobble rather heavily on canes & the pain doesn't seem to improve much. I'm up to almost anything but foot-racing or wrestling, & daily grateful.

This is still uncertain and confidential, but I want you to know of the possibility: I *may* leave New York & the Museum next fall. It's a hard decision after 31 years (exactly, today, now I think of it)—But as you know things have not been happy for me there since my return, & I do have an offer of a better job: Agassiz Professor at Harvard. More money & literally *no* duties—just to sit & think if so disposed, & occasionally to say a kind word to students (but no *teaching!*) & other faculty. Free, too, to come & go as I please. The professorship explicitly does *not* require even residence in Cambridge, although I would plan to live there as base of operations. Anne feels she can continue her work at least as well there as here. There are drawbacks of course: leaving the collections I'm working on—but Ned [Colbert] & Bert Parr would probably be so glad to see me go that they'd lend collections to Harvard for my work.[16] Less assistance & facilities—but Ned is happily cutting down on services for me here, & I think I can get research grants to cover my needs.

(Bill will be amused at my being an *Agassiz* professor, as his researches in Darwiniana will have taught him that I'm not the Agassiz type.[17] There are 3 Ag. professorships—The two already there are old & good friends of mine: Bryan Patterson & Ernst Mayr.)[18]

Well, time will tell—& so, when I know, so will I.

The best to you both—

G

[New York City]
Thanksgiving
[Nov. 27, 1958]

Dearest Mother:

You're gadding about again, so I'll send this to Martha to forward to you wherever you are.

This is Thanksgiving & is a big day here. Bets & Garth [her son] are here staying with us for several days. Helen, Wolf, & their three boys are staying with Wolf's mother, Luta, & they'll soon arrive for Thanksgiving dinner, & so will Joan, Al [her husband], Trina, & Peter [their two children]. Quite a mob! All are well & blooming. The children delightful,

16. Edwin Colbert was chairman of the department of geology and paleontology; Albert Parr was director of the American Museum.

17. Alexander Agassiz (1835–1910), Swiss-born American naturalist and engineer who made a fortune in copper and used some of it to endow professorships at Harvard's Museum of Comparative Zoology, which had been founded by his father, Louis Agassiz (1807–1873).

18. Ernst Mayr (b. 1904), German-born American zoologist and evolutionist, formerly a colleague of Simpson's at the American Museum, whose book, *Systematics and the Origin of the Species* (1942) is another of the pillars of the modern evolutionary synthesis.

although rather strenuous, as they should be at their ages. This is more family than we've had together for years. It is actually 2½ years since I saw Bets or Garth or Helen's older two boys, & I had never seen her youngest, born while I was laid up. Everyone is out watching the big Thanksgiving parade, a big event for small fry here in New York, but will be back in an hour or so for the festivities. We all think of you very much, miss you, & send our collective love—a large package!

Anne & I have decided to leave New York next year, in September 1959. I've been here over 31 years & am sort of tired of it, and I have been offered a good job at Harvard University—Agassiz Professor of Vertebrate Paleontology, pays well & has no duties except to write & do such research as I please. (No teaching, which I do not like—I mean, I do not like to teach & do not have to though I will be a professor.)

We will of course continue to spend summers in New Mexico, & this makes no difference to our plan to be there next summer. We will have to come back a bit early in order to make the move, which will be quite a job, especially as I have to move my library of several thousand volumes & re-install it at Harvard—They are remodeling & equiping [sic] quarters for me in the Museum of Comparative Zoology, which is part of Harvard College & was founded about a century ago by Louis Agassiz, the man for whom my professorship is named.[19]

Not much other news here. I'm dashing down to Washington tomorrow for a meeting of the Secretary of Interior's committee on Geological Survey, & Anne leaves at about the same time for a week in Miami, Florida, at a conference on psychology.

Much love, again—

Gg

New York
21 December 1958

Dearest Mother—

Merry Christmas! Another year has rolled around & although we have to be far away we think of you & send our love.

This is a far merrier Christmas for us than the last two, when I was in the hospital & doubtful whether I would ever walk again. Now I am walking, not even using crutches, & I have a new job to look forward to, & all is going well with us.

In fact, although thankful for those blessings, we do not feel very Christmasy this year. We had our family reunion, & a very good one, on Thanksgiving, which seems a more appropriate occasion anyhow. Anne & I will be alone on Christmas Day, although we will have Christmas Eve with Joan & her family. I think the commercialization of Christmas has gone so far that it is spoiled for us & we seek our own occasions for celebration without commercials!

19. Simpson is wrong; the Agassiz professorships are named after Louis's son Alexander, who endowed them.

Anne has been sick for a couple of days—nothing at all serious, just a severe cold—& that has put us behind in writing & various odds & ends. Fortunately I am now able to take care of meals & so on when she is under the weather.

We are both working along. Anne's research project is making great strides—She has a good assistant & efficient (also very pretty!) secretary. I am working on the collections I made on the ill-fated Brazilian expedition, as I want to finish them before I leave the Museum. We are also finishing up the revision of our book "Quantitative Zoology"[20]—Very technical. I'm afraid you won't care much for it! I also have five new lectures to write for April, & have to prepare for a new course I'm giving at Columbia University, which I had agreed to give before I decided to leave there. That will be my last teaching there, & perhaps anywhere, since I am not required to teach at Harvard & probably won't.

Marty send[s] us some Christmas greens from New Mexico—cedar, piñon, & sagebrush—& they make us homesick. We do look forward to getting out there next summer.

Merry Christmas, again, & much love—

G

[New York City]
11 Jan 59

Dearest Mother—

There really is not much news this week. Anne has been under the weather all week, but is better today—Some flu she's been having trouble shaking off. I'm fine. I've started a new course of hydrothermy & massage, which is enjoyable & does me good.

Helen is coming down from New Haven to see us this afternoon. It's been quite a while since we saw her & so of course we are looking forward to it. Joan & her kids usually spend Sundays with us, but aren't today.

Now that I get around more easily, we go out a little more or have company here. All the Peets were up a while ago. Young Creighton is now 20 years old & over six feet tall. He is studying at Tufts College. That is near Harvard, so we hope to see more of him when we move to Tufts [sic] next fall—He's a sophomore & will be there a couple of years after we move.

Margaret & Ned Colbert, whom you know, are in Brazil, on the same sort of exchange professorship Anne & I were on 5 years ago. (Not exploring the jungle, as I was on my latest trip there.) Rachel Nichols, whom you also knew at the Museum, is going to retire next year. (She is older than me & is one of only 3 in the Department who were there when I went to the Museum. The others are Carl Sorensen & Edith Marks, both of whom also retire soon.) Of others you may remember there, Jerry Walsh

20. Richard C. Lewontin, then a geneticist at North Carolina State, now at Harvard University, collaborated with Simpson and Anne on the revision, which was published in 1960.

& Charlie Lang are doing well—both retired some years ago. Barnum Brown has been retired for a long time but still comes to the Museum frequently & is writing his memoirs. (He's older than you, but also quite sprightly.)

We are so pleased that Marty & Bill could have some time in Hollywood & that this may mean considerable to them financially—as you know, a little extra will help them greatly. It must be lonely for you not to see them for a while, but it is a needed chance for them & they'll be back soon. Bill's new book keeps selling, not enormously but steadily, & we hear many good comments on it.[21]

I noticed that you had an unusually heavy snow in Albuquerque a week or two ago, & I hope you enjoyed it—or don't you like snow anymore? Here it has been unusually cold, but mostly clear. There has been practically no snow. I like New York with clear, snappy winter weather. What I hate is the long, slow drizzles we sometimes have, but so far we have not had much of that, either.

Much love from us both. Keep well-occupied as you can, & we'll visit together again before so terribly long.

Gg

N.Y.C.

8 May 59

Dear Marty & Bill:

I venture to address myself to you even though I am not a Big Cow Man & have no herd to innoculate against blackleg, blackjack, faro, & fantods.[22]

I have just ordered a gadget for measuring relative humidity—something we've always wanted for our weatherwatching. So sure it's dry, but just *how* dry? I've had this sent to me there at Cuba [New Mexico] in care of Bill because we don't care how dry it is in New York. So open it up & have fun until I get out there (probably on 9 July) & take it away. Actually, this is a birthday present for Anne, who loves weatherwatching even more than I do.

My appointment—also Anne's—at Harvard is now finally official—no surprise, as word of it had already reached literally as far as Red China.[23] I leave with a bit of nostalgia but no present regrets. Things get worse & worse at the American Museum, & I'm lucky to be getting to hell out. The

21. *Bronc People* (1958), set in contemporary New Mexico, is about a rancher's son who is determined to become a bronc buster, his black friend, a born intellectual, and assorted Navajo Indians.

22. Blackleg is a disease of sheep and cattle; one of the symptoms is swollen legs. Blackjack and faro are popular gambling card games of the Old West. Fantods as used by the novelist Galsworthy is nervousness, restlessness, or uneasiness.

23. Anne's initial appointment at Harvard was as lecturer and research associate in Harvard's School of Graduate Education. In 1963 she was made full professor—the first woman ever in the school and the ninth in all of Harvard's history—to hold that title.

Director has just been fired (in the big business way of giving him a meaningless new title, "Senior Scientist" no less, & taking away all authority). New director not yet named. This would have done me nothing but further harm if I had been staying on. It's a general crackdown from the Trustees who run the lives of the Museum staff as a hobby.

Anne moves to Cambridge on 15 June or thereabouts. I stay here until I go to Arizona about 6 July, & Anne will join me at La Jara, of course.

We saw the Claxton's briefly—here a month en route from winter in Tucson to summer in Vermont. They are aimless & seem unhappy. Doing nothing whatever is fun only when it's an interval between doing something. They hadn't intended to look us up, but we heard they were in town & looked them up.

You of course heard that Creighton (Sr.) had a coronary. He's up & about & O.K., but serious problems with legwork for his kind of writing (less & less in demand) & with having to move. Thank God B.A. [Bertha Ann, his wife] has a fair job—but no pension coming. We may be able to help Creighton (Jr.) stay in school, which is extremely desirable at this point, as our own (or other) children may need less help next winter.

I get around as pert as a grasshopper (one with a slightly busted leg). Anne overdoes, you'll be amazed to hear, but is enjoying it. She's away again at meetings hither & yon. I'm winding up as much as I can at Amer. Museum & also have a lecture, a conference, & an institute (Baltimore, Cold Spring Harbor, & Tucson) before I relax among you cattle folk.

Much love, & keep your fences dry & your powder up.

G

Do you remember my colleague Horace Wood?[24] He's had a tendency to pinch strange girls & generally act queer over recent years & a week ago suddenly became violent & was shoveled into a loony bin. This is not an occupational disease.

[New York City]
12 June 1959

Dear Marty and Bill:

This is just a quick note to let you know about my arrival in Albuquerque. I am scheduled to arrive there on 9 July, by Frontier Airlines flight 82 from Tucson, due in Albuquerque at 12:03 p.m. Please do not bother to meet me, because it is really a terrible nuisance and I can perfectly well come out on the afternoon bus. I'll call you from Cuba when I arrive there. I won't attempt to go up to Los Pinavetes [the Simpsons' home] that same day, so if you still have guests please reserve me a room in one of those motels in Cuba.

The plans for picking up the jeep and the camp equipment up at my

24. Horace Elmer Wood II (1901–1975) was a vertebrate paleontologist at the University of Newark (later part of Rutgers University) whose specialty was fossil rhinoceroses and their kin.

place are still somewhat nebulous. A couple of weeks ago, George Whitaker suddenly had a severe hemorrhage from an ulcer and he has been laid up ever since. It is quite doubtful whether he will be able to go west this summer. Even if he does not, some of the other lab men, including Walter Sorenson [sic], will get out there some time in June. I'll try to see that you are notified, but don't be too surprised [sic] if they drop in on you. It is, of course, alright to let Walter use the keys to my place and to remove any of the museum's equipment.

Sincerely,

G

Thanks for check—I'll write you about that from home in a day or two. Love to you both—G

[New York City]

14 June 59

Dearest Marty—

Many thanks for the check. I must explain that I have not divided it just as was originally suggested. (I, at least, had made no commitment that I would do so, & it is my money, after all.) What I have done is to give each of the girls the amount Anne had said they might get—$300—& the same to young Creighton Peet. I have kept the remaining $500. (Our trip to Albuquerque cost more than that.)

I have a somewhat guilty feeling as if I was cheating the girls, which God knows is not true & I resent feeling so (especially as each has had considerably more than this from me recently). I must confess I got a bit angry, too, when Bets wrote all bursting with plans to spend *her* inheritance from Grandmother, proposing to quit her job & go back to school on the strength of it—which even if I gave her *all* the money would mean that I would shortly be providing full support for that family instead of part as at present.

However, I have simply explained to the girls that this is a gift, not an inheritance, & how I am dividing it. If any of them kicks, I'll throttle her.

The Peets are not doing at all well & I am quite worried about them & glad I can ease things by giving a little to young Peet. (It is of course impossible to give anything directly to his parents.) Creighton is tired & feeble & can't do much. B.A. [Bertha Ann, his wife] has a job, thank goodness, but it's not enough for the whole family. Young Peet is working as a barker at Coney Island, the best pay he could locate for a few weeks when he has the promise of a more interesting & also well paid summer job with a sugar company. I thought his parents were kidding when they complained about the social status of his Coney Island job, but found to my amazement that they are perfectly serious in deploring this! They are still not reconciled that young Creighton won't (& can't) be a writer—the only gentlemanly work! Bertha Ann is also upset because young Creighton takes a predatory attitude toward young females—I told her that if a healthy 20-year old male did *not* have such an attitude she would have a right to worry! In short, the Peets are just generally upset, & with good reasons, which they camouflage by worrying about the wrong things.

We'll be moving all this week, God help us!

Much love—

G

[New York City?]

19 Jun 59

Dearest Marty—

Enclosed a bit more on Grinell that should be kept with certificates, I guess. Peg sent this separately to me about a month ago & I misplaced it in the furore of moving.

We are all moved. Anne is in Cambridge—11 Farrar St., Cambridge 30*, if you haven't noted—& I'm going up for a couple of days next week. All my office stuff has been moved, too, but I won't settle in there till Sept.—My office books, & files filled 108 boxes! (And Ned saw to it I didn't take anything to which I couldn't prove title—Even had my secretary, without telling me, list everything I took & left!)

All well. No more news. Oh, yes, I'm to get another Sc.D. ([University of] Chicago) in November—not announced yet. I need more doctor's degrees like I need the hole in my head.

Love—

G

*Phone now installed—Easy No. to remember: University 4–5566.

[New York City]

25 Jun 59

Dearest Marty—

Thanks for your most reassuring letter. Anne thought I was being stingy, but she's recovered (after sending a bit more on her own!), & the girls were all most appreciative, including Bets. So all is well.

Like a stinker I neglected to say Happy Birthday in time. I do now, & many of them.

Things are in a hooraw here, as you can imagine. I've been in Cambridge most of the week, seeing our new home, ordering office equipment, attending a conference, & so on, but now am back in N.Y. till I leave forever on 6 July.

Our place in Cambridge is delightful. I really like it better than our apt. here, even apart from the couldn't-be-quieter contrast in neighborhoods, shopping facilities, etc. Everyone at Harvard is most cordial, too, & I actually can hardly wait to get cracking there.

You may have noticed that the boys didn't make it on 15 June to pick up jeep & camp stuff. George is now certified OK by the Museum Dr., & he & Walter Sorensen are flying to ABQ [Albuquerque] on 13 July, probably reach Cuba [New Mexico] on 14 July. I'll be there then myself, of course. If Bill gets the Buick out before I come (which is not really necessary) he may have to shift some camp stuff & leave Buick out of garage until the boys come, but that doesn't matter.

I'll see you soon. Much love,

G

HARVARD
1959-1967

THE SIMPSONS moved to Harvard in the summer of 1959 and took an apartment in Cambridge near the university. Simpson held joint appointments at the Museum of Comparative Zoology as an Agassiz Professor and at Harvard itself as professor of vertebrate paleontology. Anne was affiliated with the school of graduate education, first as lecturer and research associate, then as full professor. In the mid-1960s Anne started the Center for Research in Careers at Harvard and became its director.

Simpson's reputation was international and he continued to receive honors and invitations to visit and lecture worldwide. Anne too had many opportunities to lecture and consult. Consequently, the Simpsons, who were enthusiastic travelers anyway, intensified their professional pace. Yet they often made side trips to visit members of their far-flung family. Gradually, however, health problems began to slow them down. In 1964 they both had what Simpson later called "his and her heart failures," and later that year they took a long South Seas cruise to convalesce.

Martha and Bill were still living in New Mexico. In 1965 Martha published her second book, *Rattlesnake under Glass*, a cookbook of western United States recipes combined with ranchland-inspired anecdotes of the New West. Bill continued to write novels and occasional screenplays for Hollywood; for a time he also taught writing at the University of New Mexico in Albuquerque.

Simpson's mother died in 1960, shortly before her ninetieth birthday. Because both of Simpson's parents were dead and because Anne conducted most of the correspondence with the daughters, Simpson's letter writing to the family was mostly confined to Martha.

Simpson's professional energy never slackened. He published a major treatment of the ways in which humans organize and classify organisms, *Principles of Animal Taxonomy* (1961), a lineal descendant of the earlier 1945 classic, *Principles of Classification and a Classification of Mammals*. He also published two collections of essays. The fourteen essays in *This View of Life* (1964)—the title is derived from the last sentence of Darwin's *Origin of Species:* "There is grandeur in this view of life . . . "—present Simpson's views on the impact of evolution on nineteenth- and twentieth-century culture, the nature of evolutionary inquiry itself, the lack of preordained goals or purpose in evolution, and humanity's place in the universe. Simpson long considered this book his favorite, until near the end of his life when a new book shared that distinction.

In 1965 Simpson published a second collection of essays, *The Geography of Evolution,* which he quickly regretted, for these essays argued for continental stability at a time when the new theory of plate tectonics was gaining ground. This is the only book that Simpson clearly wished he hadn't published.

[Cambridge, Mass.]
[10 Oct. 1959]
Saturday

Dear Marty & Bill—

Thanks for your recent letters. We are always eager for news of N. M. & do appreciate your supplying it so well.

I do not know whether you have heard that Bela Mittelman[1] died last Sunday & Perry Claxton last Thursday—Bela in N.Y. & Perry in South Londonderry, Vermont (their summer home). They died in precisely the same way—Both sitting in their living rooms & just suddenly ceasing to live from one instant to the next. Well, I guess that's a better way than most.

Young Creighton has settled in at Tufts & drops by to see us occasionally—had dinner here Wednesday. He has a cold water flat which we haven't seen & of which his parents disapprove violently, although they haven't seen it either. (You know B.A. & Creighton have an unexpected snobbish streak, which is the main reason why young C. is in college at all—He is completely nonintellectual in spite of being intelligent & has great difficulty with classes.) C. will be 21 on 30 October & we hope to get B. A. & (old—one can't say "big" any more because young C. is bigger) Creighton up for the weekend but are uncertain whether they can get away.

Anne is in Chicago—left yesterday noon & will be back this afternoon. She's away all over the place half the time—So much for her intentions to take it easier this year! She is well, however, except for an occasional cough & she is enjoying the new setup very much. It's exactly what she always wanted.

I, too, like the new setup & have no regrets, au contraire, even though I am finding the readjustment more of a problem. I really am taking it easy, however. My fears of too much social life in a more tightly knit community are not being realized, although we do have enough. For example we

1. A psychoanalyst friend of Anne's who had collaborated with her on alcoholism research.

recently spent the evening with Paul Levine, wife, & in-laws in Lexington. He is a young geneticist (on Harvard faculty) & his wife (very nice) is daughter of Carl & Ruth Epling, old friends of ours at U.C.L.A. (He's a botantist.) Tomorrow we'll spend the day at the Mayrs' weekend place 60 miles north of here in New Hampshire. Ernst Mayr (ornithologist) is another fugitive from the American Museum who preceeded [sic] me here by several years. The fall color in the N.H. hills should be near its peak, & clear weather is promised, so that should be delightful.

Well, off to watch football on TV. (Did A tell you that our old TV that I was about to chuck out works *perfectly* in its new location? It just didn't like New York.) By the way, Bill, I sat up to see the fight last night & regretted it. Wasn't that a lousy one?

Much love—

G

Cambridge
17 Oct 59

Dearest Marty & Bill—

I am finally getting with it here at Harvard, & it is quite simple if you only pay attention.[2]

I think I did tell you that I am two professors. One is paid but has no tenure. The other has tenure but is not paid. But if the professor without tenure is fired, the one with tenure has to be paid. I am on the faculty, but I do not teach. Everyone who teaches is on the faculty, with certain exceptions. If you are not on the faculty, you have to pay dues to belong to the Faculty Club, but if you are on the faculty you do not have to pay dues.

My title is, among other things, professor or rather professor-professor. My rank is also professor, but you do not have to be a professor in order to have the rank of professor. Anne is a lecturer but has the rank of professor, but naturally there are lecturers who do *not* rank as professors. Chairmen & directors have titles but no rank as such, of course.

Now obviously *precedence* has nothing to do with either title or rank. With certain exceptions (established in 1652, 1717, & 1812) precedence is based on year of *first* academic degree. So of course my precedence— Anne's too—is as of 1923. This is all-important at football games, symphony concerts, & state dinners. 1923 puts us above the salt, but well below Don Quixote ("donde esta la cabecera")[3] because there are still plenty here with precedence from the '90's, & perhaps some from the '80's for all I know.

The departments of Biology & Geology have elected me a member

2. What Simpson is about to describe is the sort of complex organization typical of an academic institution more than three centuries old, parts of which grew like Topsy.

3. "Where is the salt." Simpson of course is referring to the old saw that one's position in the social hierarchy is reflected by one's seat assignment at dinner. Above the salt puts you closer to your host, below the salt puts you farther away.

(rather members—I'm two professors each in two departments—four so far), but of course I'm *really* in the Museum of Comparative Zoology. That has no connection with biology or geology, but is *at* Harvard College. It is not in, with, by, or notwithstanding Harvard College but just at it. It is

1959–1967affiliated with Harvard University, which of course is not *directly* related to Harvard College. The University agreed in 1889 to pay M.C.Z. $1000 per year for use of a lecture hall, which it does not use. Otherwise the University contributes nothing to M.C.Z. M.C.Z. (*of course* not Harvard University or College) owns its building & land. But it also owns the land on which is Peabody Museum, which *is* owned by Harvard—University, that is, not College, natch. Harvard does not pay rent for that land, or for use of the well in our patio.

My salary as the salaried one of us is not paid by any of the Harvards, but by the Bay State Trust Co., trustee for the estate of Alexander Agassiz, deceased. Agassiz professors do not get fired by Harvard, because then Harvard would have to pay them in their tenure positions. But Harvard does pay *curators*, who of course do not work for Harvard at all. (The Bay State has something to do with the revolution.)

That's all simple & straightforward, but it becomes more confusing when it comes to The Faculty. Most people who teach, & many who don't, belong to the faculty, but only a few faculty members belong to The Faculty, & some nonfaculty members are Faculty members. (Some but not all Faculty members & faculty members are eligible for the Faculty Club.) Members of The Faculty have the privilege of telling the President to go to hell, & he has the privilege of firing them if they do, unless, of course, they are also faculty members. In the latter case they can only be fired for raping a dean's wife in public. (Only wives of husbands below the rank of dean & precedence below 1930 can be raped in public.) One of my fellow Agassiz professors did something worse than rape: he went to a meeting of The Faculty (to which he did not belong; all Agassiz profs. are on the faculty but not all are on The Faculty). Next day he received a polite, icy letter from [Nathan] Pusey (that's the president—of Harvard University, that is, of course nothing to do with the *College!*) telling him to cut off his buttons. But he only has zippers.

Some time between 1690 & 1710 (I never could remember dates) it was decreed that only Masters of Arts of Harvard College (*not* University, & I do wish you'd pay attention) could belong to The Faculty. Consequently when they appoint someone to The Faculty who is *not* a M.A. of H.C. they quickly drop a diploma in his mail box. (This is not necessary for faculty members.)

My mail box, by the way, is a large mahogany literally box built in with the Museum (which will have been built just 100 years on 15 November of this year). Since I am not likely to be here more than 15 or 20 years, my name is just stuck on a strip of adhesive, over the name of a predecessor who predeceased in 1892. Every morning the mail clerk sorts the mail just outside my office, then takes it downstairs to my box—I follow him, get the mail out of my box, & we walk back up together.

Anyway, they made me a member of The Faculty with all privileges (*vide supra*, p. 2, lines 20–25) thereunto pertaining, hence also Magister Artis, & so completed what is known in unmentionable circles as the stuffed shirts' (or squares') hat trick:[4] degrees from Harvard, Yale, & Princeton (in order of antiquity & prestige to cognoscenti) & fellowship in Amer. Philosophical Soc., Amer. Acad. of Arts & Sciences, & National Acad. of Sciences (same order).

Smoking, chewing, & spitting are especially forbidden in M.C.Z., but in the Harvard Herbarium (no connection with the Arnold Arboretum or with the Dept. of Botany, in the same building), which belongs to Harvard—one of them, anyway—or maybe doesn't, but is *at* Harvard—or is it really? It's on M.C.Z. *land* but on Divinity Ave., which belongs to the Divinity School, but is administered by the R.O.T.C. In, to repeat, the H. Herb. is a large spitoon (spitton?) engraved by Tiffany: "Presented to the Harvard Herbarium for the use of the Director of the Museum of Comparative Zoology." Which is the only privilege the Director (A. S. Romer) has.

Well, I've omitted all the really good parts, but in briefest outline that is the situation here in a nutshell. It really isn't so much *what* they say as *how* they say it. I've got to run—an assistant dean's wife just started up the steps of the Widener Library (that's *in* the yard, *at* but not *in* Harvard College.) Cheers!

G

P.S.—We've just had a visit from a colleague who is native to East Africa & who says he's a Kikuyu, but who is awfully pink & who's [sic] name is Leakey.[5] He enthusiastically proposes to show us Kikuyu country, Olduvai, Rusinga, the Ngorongoro lions & petrifactions if we'll go to Kenya—Tanganyika—Congo next summer. So, inshallah,[6] we will. (We aimed to anyhow, as you know, but having the red carpet laid on clinches things!)[7]

We have to be different, though. We will *not* go to see A. Schweitzer. (Remark inspired by a Sch. record, organ—ugh!—on radio just now.)[8]

Leakey does speak Kikuyu, but not Arabic & has just given a new find a

4. "Hat trick" is hockey slang for the scoring of three goals in a single game.

5. Louis S. B. Leakey (1903–1972), Kenyan-born anthropologist of British parentage. Fossil excavations by him, his wife Mary, and his son Richard in East Africa have made major contributions to our understanding of early human evolution. Early in life Louis Leakey was adopted into the Kenyan Kikuyu tribe; speaking their language fluently, Leakey served as translator at the trials of alleged Mau Mau leaders during the uprisings against the British colonial government in the 1950s.

6. Anglicization of the Arabic expression "if Allah wills (it)," a frequent pious exclamation among Muslims. Appropriate here, for much of East Africa was explored by Arab sea-traders and a large part of the native population today practices the Muslim faith.

7. The Simpsons did visit Leakey in East Africa in 1961 and while on a field trip with Louis were present at the major discovery of a fossil ape, *Kenyapithecus wickeri*, about ten million years old and, according to some interpretations, close to the line of human ancestors.

8. Albert Schweitzer (1875–1965), Alsatian-born theologian, philosopher, African missionary doctor, and organist famous for his interpretations of J. S. Bach. He received the Nobel Peace Prize in 1965 for "his efforts on behalf of the brotherhood of man."

wrong Arabic name. He thinks ظمأ means "Africa"—imagine such ignorance! (As you know, it means "to be thirsty," "metal castenets," or "ethiopian.") Anyone thinks I'm kidding will be mighty sorry, too.

G

P.P.S. (later, as they say, the same day)—Above was a reaction on returning from 1st faculty meeting in a daze (not, alas, an alcoholic one). I've just had yrs. of 15 Oct., & will forward enclosure to Dorry [Claxton]. (Anne's away again.) Anne, by the way, saw Ruth, who's in bad shape, & Joan, who's in excellent shape, more realistic (it says here) & on good terms with Al & infants.[9] Much love—

G

[Cambridge]
13 March 1960

Dear Family:

I get to write the family letter this week (there wasn't any last week). I'm using Anne's electric typewriter and it scares the—well I'll spare Edna Roe and not say what it scares out of me.

Anne was up and somewhat about for a bit after her probable virus pneumonia, but about a week ago she either relapsed or came down with something else and has been in bed again ever since. The pattern is highly original and the doctor is really not quite sure what is wrong. Tomorrow, if Anne is not much better, he and we decide whether to put her in the hospital for a few days to see whether exhaustive tests will come up with something more definite. She does feel a little less exhausted today, but still with a great deal of chest pain.

I am working at home, which is agreeable and efficient in my fine study, writing a book entitled "Principles of Animal Taxonomy"—for ages 4–6. Watch the best seller lists. Starting later this month, I will go and present an expurgated version of the book to postgraduate audiences (which will dwindle rapidly as they get a load of it) at Columbia University, for which they will give me some advance royalties and then print it if they remember to.[10] (Iiifff you hesitate with this typewriter or breathe hard it takes the bit in its teeth, and I don't know how to backspace either.)

All winter long big storms avoided Boston—I figured the snow didn't want to fall on the Irish, and I wouldn't want to either. Then last Thurs-Fri, or rather a week ago, whammmmyyyy, as this typewriter wants me to say. The heaviest single snow ever recorded in the Boston area. It crippled traffic and the mayor, who by an oddity happens to be crippled too and by no oddity is Boston Irish, came on TV and delighted me by announcing

9. Ruth, here, probably refers to the wife of Anne's brother, Ed Roe. Simpson's daughter Joan had recently divorced Alfred Meyer, with whom she had two children.

10. Simpson gave the prestigious Jesup Lectures at Columbia University in 1960 in which he discussed the main points more fully developed in the *Principles of Animal Taxonomy*.

that "packing of cazz is bond."[11] It speaks well for my increasing familiarity with this area that I finally realized that an alternative spelling of that would be "the parking of cars is banned." I wish parking had been barred, which comes out "bad," but I have to report accurately.

(I find I can backspace by sheer brute force. Did you ever wrestle an electric typewriter?)

I had a wonderful idea that when the weather is bad (not Boston accent, I meant Boxstonian for barred of course) I'd just go everywhere in taxis. Of course it never occurred to me that when the weather is bax— thataway, whatever I mean—you can't *get* taxis. Anyway it's much nicer just to stay home, and I get more done too. (Brilliant machine doesn't even know a margin when it gets to one.) Even so I have gotten out for a couple of sahshays and even got Anne to the hospital for X-rays. It's really lovely here with the deep snow, if only it weren't for the bad going for my gimpy leg.

I should have gone out and shot someone before beginning this letter so I'd have some exciting news, but this is about the only place left where there isn't anyone I really want to shoot. Give me time.

Much love to you all—

G, George, Geege, Daddy, etc. etc.

[Les Eyzies, France]
[Late August, 1960]

Dear Marty & Bill—

We are somewhat at wits' end about Los Pinavetes, sorry it has been such a worry to you, & worried as to what can be done about it now & in the future. The immediate solution of renting to your friends is satisfactory to us, but is not a long-range solution & may not be practical now. If it is practical, we do authorize you to follow it up. We do not quite see how a family with small children can spend the winter there what with heating & plumbing problems & the at least occasional impossibility of getting out readily. It seems to me impossible to winterize the present water system— tank, terrace drain, & some other parts could almost certainly freeze & even getting water into the tank would seem impossible in coldest weather. We have always figured that a safe winter system will require a well, pump—hence electricity—new piping to the house, & also some revision of the house plumbing, certainly of the cold-water drain thru the terrace (which did actually freeze & burst on us once). That is all a very complex operation, the supervision of which we would hesitate to saddle upon you, & also a rather expensive one that we had not budgeted for this year, although we could manage to borrow or scrape together the money for it if necessary. And then there is the fear that if & when we do put in electric- ity, pump, & so on, we may just be providing more things to damage or

11. John S. Collins.

steal. It also worries us about the long-range plan of further building &
moving our (quite valuable) libraries there, as we will never be able to stay
all the time on the place.

We are ensured against theft, etc. with the E. A. Steidly Agency in

310

HARVARD
1959–1967

Albuquerque & they should be notified if anything really worth much has
been stolen or serious damage done.

This is all extremely upsetting—as much for you as for us & we do
deeply regret the nuisance for you. I am also sorry to be so indecisive, but
can't seem to see a good solution from here. If your friends really want to
& can move in—frankly I would not if I were they—that could be a satisfactory temporary & compromise solution.* We entirely trust your judgment there, but of course could not consider you responsible if things went
wrong as could conceivably happen despite the best of judgment.

Anyway, much love from our Cro-Magnon lair.[12] As she has noted,
Anne will soon be back in the U.S. & a little more accessible—

G

*I suppose token rent should be charged, but the amount would be imma-
terial to us.

[Cambridge]
28 Oct 61

Dearest Marty—

Life in your two residences sounds like a social whirl. I hope you aren't
letting them wear you out.

Our lives in the big city are much quieter—mostly just going to the
office, & in my case not always that. Although the weather is still reasonable I do much of my work at home. Anne, on the contrary, goes to the
office more & more—her new office is so near that she finds it simpler to
drop in there Saturdays & Sundays than to bring work home.

We have both been refusing most lecture & other travel arrangements,
but a few are unavoidable. I have to go over to Worcester for one in a couple of days (my secretary will drive me) & have two more lectures in
Nov.—Ann Arbor & New York. The African slides are ever popular &
make lecturing relatively painless.

Next weekend we're going to Rochester to see Helen, Wolf, & the
boys—we haven't visited since they moved.[13] In spite of the fact that I am
extremely fond of Helen, I confess that this is as much duty as anticipated
pleasure. Joan & her two will spend Thanksgiving with us.

The news of poor little Pete Whitaker is bad—He has pulled through,
but it might have been better if he hadn't.[14] He is blind & almost uncon-

12. Letter was written from Les Eyzies-de-Tayac, a village in southwestern France
where fossils of early *Homo sapiens*—Cro-Magnon Man—have been found as well as
cave paintings and Paleolithic sculpture.

13. Helen's husband, Wolf Vishniac, moved from Yale to the University of
Rochester.

14. Son of George Whitaker, Simpson's field and laboratory assistant at the American Museum, who had had a serious fall into an empty swimming pool.

trollable, goes into spasms of rage & it takes a couple of people to hold him down. Anne has put them in touch with some neuropsychiatric aid that may at least make it possible to live with him. However it sounds as if he had permanent brain damage.

Our fall royalty checks have just come in, & they were several hundred dollars more than last year. We want to share the windfall, & so enclose a small part of it—I'm sure you'll be willing to make us happy by accepting it. As you know, there's always more if needed. This is just a windfall & has no reference to need.

(We are also sharing a bit of it with the Whitakers, who *do* need it.)

We are both feeling fine. No illness yet, & we're going to have flu shots as a precaution we haven't tried before.

Love from us both to you both—

G

[Cambridge]

9 Dec 61

Dearest Marty:

On our way back from lunch yesterday we had the [Harvard] Coop send you two records. The Coop isn't the most efficient store in the world & the clerk was extremely skeptical when I insisted that New Mexico is in the United States. (I'm partly resigned to the fact that almost no one in New England ever heard of New Mexico. They say either, "Oh! You mean Arizona!" or, "Don't you have to have a passport?")—So please let us know if & when they arrive.

They are the Forellen Quintet (augmented Budapest quartet) & the Italian Symphony [of Felix Mendelssohn] (Toscanini & N.Y. Symphony), plus pretty good flip sides. I hope you don't have them, or at least these particular versions. At the moment, at least, these are my favorite music, chamber & symphonic, respectively. I always play the Ital. Symph. when I am depressed, & it always cheers me up. It is the greatest expression of pure joy of life that I have ever encountered. So I hope you like it too, & that if you ever should feel blue you'll play it & cheer up with me.

Anne & I have taken turns with the latest in viruses, but we're both OK again now. Anne is off tomorrow for a week in Baton Rouge, a sort of working vacation—extremely well paid!—as personnel consultant for Humble-Esso.[15] This is becoming a sort of annual event for her, which she enjoys because she likes to have contact with industrial psychology, finds B. R. pleasant, & also gets an evening or two in New Orleans. She'll also give a lecture at L.S.U.—*desegregated* (she called to make sure before accepting).

As for me, I just plug along. Lots of busy work—short papers, book

---

15. Anne gave advice on the recruitment and selection of new graduates in engineering. Despite some initial resistance and skepticism—she was the first woman consultant the managers had ever encountered—her detailed advice was enthusiastically followed.

reviews, an occasional lecture (I turn most of them down but give 4 or 5 each winter just to keep my hand in), & beginning a revision of the African fossil prosimians (bush babies, pottos, & kin). We both continue to love it here. Everything about it extremely congenial, stimulating and

relaxing in the right proportions. Anne has her finger in a dozen pies, of course. I do just enough professorial or organizational work not to feel isolated: a couple of committees, informal student contacts (I'm still unofficially working with a couple of the students I abandoned at Columbia) & the like.

Weather cold & clear. Visibility unlimited. Blue jays raising hell. Sunset at 4:52. Stock market up, mixed. Much love to you both from us both—

G

[Las Vegas, N.M.]
Tuesday [25 Sept. 1962]

Dear Marty & Bill—

I'm going to write now instead of telephoning later in the week, principally because there is no place to telephone without a large audience. I just don't want to discuss my health, let alone my opinions, in public. My health happens to be fine, even better than when you left. If it should turn otherwise, I *will* call.

My opinions are a different matter. This place & everyone in it are simply *lousy*. Mrs. Farmer is a partial exception, as she continues gracious & solicitous, but that hardly suffices, especially as she hasn't time for much during the day, at least. Did you realize that she does *all* the work in the hotel, without even a chambermaid?

The faculty & student attitude is one of utter indifference. They obviously don't intend to have anything to do with me if they can help it, & as a rule they can. Dean Meyer,[16] who arranged the lectures, didn't bother to get in touch at all except by a message relayed through Dean Farmer, who got it wrong. Result that while I was waiting for M. to call on me, as F. said he was going to, my first lecture audience sat apathetically awaiting at a time & place of which I had no knowledge. When I did get there, I found that there was no projection equipment, for which I had specifically asked & had been assured it was there. The second lecture, today, was set for 8 A.M. as I was told, but when I got there, there was no audience at all. Finally I found that it had been transferred to a different place, without telling me. (There was no projection equipment there, either.) There were about 50 at my first lecture & about 15 at my second. I hope soon to be talking only to myself. Everyone absolutely expressionless. No questions. No discussion. Everyone gets the hell out quietly without a word as soon as I stop.

16. E. Gerald Meyer (b. 1911), a chemist and dean of the graduate school at New Mexico Highlands University; he was later vice-president of the University of Wyoming.

My only human contact, outside of a kind word from Mrs. F. in passing, I forced by asking a junior faculty punk where I could get a cup of coffee, & to my intense surprise he took me out & bought me one.

This is going to be the most thrilling fortnight of my life, & I hope it teaches me a great big lesson.

There are three cafés within walking distance—the De Luxe, where we had breakfast, & one a little worse & one a *little* better, so I won't starve. It turns out that the Sigma Xi meeting I am to address does not include dinner, as I thought, so I'll eat only at the three cafés.[17]

It baffles me why the hell they invited me when they don't even pretend to be interested in me or anything I have to say, but it baffles me even more why I accepted.

Aside from that I am really fine. I moved to the back of the hotel to a less comfortable but much quieter room where I am destined to spend a *lot* of time. Now I'll go out & stir up some excitement, like mailing a letter, if there is mail service in this town. (Of course I've already walked every block in town, so there's no more sightseeing to do.)

Much love, & don't worry. I am perfectly well but just bored &, to tell the truth, damn mad.

G

[New Mexico Highlands Univ.]
[Las Vegas, N.M.]
Monday [, October 1, 1962]

Dear Marty & Bill—

Thanks for your call. It was nice indeed to hear your voices. As you surmised, I had a very interested audience & could not exactly speak freely, but I trust that you were reassured that I am well & that things are going a bit better here.

What improvement there is in the latter respect is due to Dean Meyer, for motives that certainly include a belated desire to make a better impression on me but also, I think, some equally belated friendliness. Anyhow, he asked me whether cocktail parties & football games interest me, & I allowed as how they did. So Sat. aft. he took me out to the Golf Club, a municipal affair but operated by the university & used almost exclusively by faculty. "Cocktails" were bourbon & 7-up, but with some argument & eyebrow-raising I got the 7-up omitted. The occasion was the presentation of golf trophies for the past year's play. Then we drove back to Meyer's house where he casually said we'd drop across the street for a bite to eat. He put it just that way, but in fact across the street was the President's (Donnelly's) house where a large party had long been in progress.[18] I'm not

17. Sigma Xi, an honorary scientific society, has many campus chapters around the country. There is usually an annual dinner at which new members are inducted and a distinguished scientist gives an address. In this case Simpson apparently sang yet went without his supper.

18. Thomas C. Donnelly (b. 1905), political scientist and president of the university for the previous ten years.

at all sure we were actually invited, & Donnelly had obviously not been told who I am or that I was on the campus, but he rallied nicely & was very polite to me. The "bite to eat" was an elaborate buffet of snacks to go with the bourbon & 7-up which is obviously the Las Vegas Anglo drink. (There were no Spanish at either party & as far as I know there are none on the faculty, although the "student" body is mostly Spanish.)

That reminds me to tell you that this is an almost completely segregated community. It consists of two separate incorporated towns, with entirely independent municipal officers & services, one 99% Spanish (West Las Vegas) & one 99% Anglo (East Las Vegas). They are of about equal size. Spanish can shop & eat, or work at menial jobs, in East Las Vegas, but not *live* there, except for students in the dormitories. (The "University" is wholly in East Las Vegas.) The Spanish town is the original one, & when Anglos started moving in they built a separate town. The two have never gotten together, in spite of some attempts to do so & in spite of the fact that the two are built right up to the boundary.

—To resume—that is why I said the president hadn't exactly had me to dinner. It wasn't exactly dinner, & he didn't exactly, or at least intentionally, have me. Afterwards the Meyers & their 2-year-old-son & I went to the Highland-Colorado Mines game, which was interesting but inept. In its two previous games Highlands hadn't scored, but they won this one 25–20. It was nostalgically like the high school games I used to see in the early 1910's—only one formation, single-wing right, & about four plays. No deception or surprises, & Highlands won because they were simply heavier & faster. Meyer hooted when I said that it was nice to see an amateur game for a change. They pay their players like anyone else, but they just can't afford to outbid Albuquerque & Las Cruces so get third-choice in New Mexicans & can afford only a few junior college transfers from other states.[19]

Yesterday, Sunday, Meyer drove me around for a couple of hours & then bought me some (excellent, I may say) fried chicken at a drive-in. He's an academic wheeler-dealer & gave me quite a snow job about the "university." He has succeeded in getting a shocking amount of federal money for nonexistent research. (For reasons now obvious to me, I have never been shown around the science building, but I have snooped all over it by myself & there is *not one* research laboratory.) I won't go into details, but his largest pipe dream is to get a federal "urban renewal" program that would buy up about half the (Anglo) town—2½ square miles of it!—& give it to the "university" for campus expansion.

Anyway, you see I didn't have the expected dull weekend. I have no nonlecture engagements this week, but only have 4 days to go & will live through them all right. Incidentally, there has been no heckling about evolution because my audience—now down to 7 or 8, including only one junior biologist—simply doesn't react at all. They sit absolutely deadpan, then scatter as soon as I stop talking. (The head of biology is a woman, who has attended no lectures. I have not been introduced & have spoken

19. The University of New Mexico in Albuquerque and New Mexico State University in Las Cruces.

to her only once, when she was typing in the secretary's office & told me she wasn't the regular secretary, who was away, & so didn't know where my lecture was—I had lost it, as generally happens.)—

Love, G

[Las Vegas, N.M.]
Wednesday [3 Oct. 1962]

Dearest Marty & Bill—

A final word to assure you that all goes well here. The trains are running & I should be able to make my getaway all right on Friday.

Things are being a bit less dull this week. There are some nice & friendly people, but they are slow to react & there is no particular way to meet them. In my snooping around I found a faculty coffee room where there are usually some professors at leisure. Of course no one ever told me about it or invited me there, but when I drop in on my own they don't throw me out & do speak to me politely. Then yesterday the head of the biology dept., a woman to whom I was not introduced, came to my lecture for the first time & afterward asked me to dinner. The dinner was pleasant & the Frau Professor interesting, but this, too, turned out to be a bit strange: she took me to a (very good) restaurant on a mesa out of town, hurried the service as much as possible, & while I was swallowing the last bite put me in a taxi & sent me home—!

Then yesterday a mathematics professor & his wife took me to lunch—same routine, a quick bite in a restaurant & away, but still a friendly gesture & a very welcome change from always eating alone. This evening Creely's friend Bunker, head of the English dept., had me out to his house after dinner (i.e. 7 P.M.—we dine early here) & he, his wife, & I had a pleasant chat for a couple of hours.

Although the faculty is all Anglo & includes one Jew & a handful of Protestants, all the people mentioned above, like the great majority of the faculty & specifically *all* the deans & dept. heads I have encountered, are Catholics. I find this very significant in a nominally non-Catholic institution in a state where the Anglos are overwhelmingly non-Catholic. It also seems significant that I know they are Catholics—naturally I didn't ask, but each one managed to let me know. So there has finally been some discussion of evolution & religion. Like most *educated* Catholics, these people are not anti-evolutionary but they are intensely concerned with the bearing of Catholic theology on evolution, (rather than the reverse). I am very accustomed to this, having had almost constant contacts with Catholic evolutionists for some 35 years, so have no trouble dealing with it in a friendly but firm way.

The foul-ups & general ineptitude do continue, in spades. They pulled a beauty just today: arranged as an extra that I should give a public lecture *tomorrow* night & then put a notice in the local paper (The Optic) saying it was *tonight*. The notice in the paper wasn't brought to my attention until after the announced hour for the lecture, so naturally I wasn't there. I then called Dean Meyer, who said that this was just a little mistake & of course the lecture is really for tomorrow. I've no idea how many turned up for it tonight, & suspect no one will when I actually give it, if I can find it— They still keep changing locales without notice.

I also saw considerable parts of the Dodger-Giant [baseball] playoffs on TV, so altogether life has been much less dull this week than last.

The Farmers continue to be solicitous as hotel-keepers & charmingly cordial in passing contacts, although we never have had 5 minutes of sustained conversation.

Much love—

G

[Cambridge]
Sat., 8 Dec 62

Dearest Marty—

Yours of Monday just received.

I'd love to visit Mazatlan (& San Blas) with you, & I'm sure Anne will agree. We are keen on Yucatan, too, but can do that another time or just possibly & briefly on the way to Mazatlan. I'm not quite sure about exact timing. 26 Jan to 3 Feb. might cut it a little fine, as Spring Term starts 4 Feb. & involves Anne. I've written her to check, & you'll get timing as soon as possible.

Anne's enormous grant to found a Center for Career Studies has come through, so she'll be busier than ever, Director of the Center, teaching a course or two, & finishing work on her present grant. I'm happy to say they did cut down the center commitment from 7 years to 5. That's what Anne wanted, anyway, & *they* told her to up her application to 7 years then *they* cut it back to her original figure—!

As you know, Anne's off on her one really long absence—back the 18th. I hate living alone, even though Anne has all my dinners cooked & frozen & I just heat them up. In fact it turns out that I am much occupied & actually won't be having dinner at home or alone for the next 7 days but with the Patterson's, Jimmie & Emmy Reid,[20] the "Visitors to the Biology Department" (a Harvard oddity), the Natural History Club, a group of medicos at Harvard Health Center, &—believe it or not—an ex-girl friend & my secretary. (Aren't I the rounder!) My social life threatens to become as hectic as yours. (I also just had dinner with an ex-secretary, but her husband was along. Interesting, in fact, as he's a postal inspector & was sent here to work on Boston's famous million dollar mail robbery. He says he has solved it, but can't find the money!)

20. James Reid was the editor at Harcourt, Brace who encouraged Simpson to undertake the writing of the college biology text, *Life: An Introduction to Biology*. The Reids and Simpsons were also good friends and members of the Martini and Mandolin club. Reid played the first mandolin, Simpson the second mandolin, Anne the piano and guitar, and H. J. Harris, a clinician who wrote *the* book on brucellosis, the banjo-mandolin; nonplaying spouses served the gin and joined in the singing.

I'm astonished to hear that Bill's doing an opera. Is he doing the music, too, or if not who is, or don't operas have music now? (I'm an old fogey.) Anyway, I'm glad it's coming well. What of the novel?

Much love—
G

[Cambridge]
16 Dec 62

Dear Marty & Bill:

A note from Anne confirms my fear that she must be in Cambridge on 3 Feb. Apart from that restriction, the Mazatlán plan is fine & we look forward to seeing you there. We haven't checked plane schedules yet, but will do so directly. My thought that we might just possibly go via Yucatán is out—too much time & expense.

No further news here. I've had another bug & am spending the weekend more or less in bed, but it's not serious. The weather here is not exactly good, either bitter cold or wet & raw, but the big midwest blizzards haven't hit us & we have only a powder of snow. I've had so much busy work lately that I haven't done anything interesting but hope I can get down to it this week. I've also had an unusual number of lunch & dinner engagements, which also cut into things although they help to cut the loneliness with Anne away. Some interesting & some not. Unfortunately as the bug hit me on Friday I missed the probably most interesting ones—for example with Dobzhansky[21] who was here to give a colloquium & whom I didn't even see although I talked on the phone. (We actually have 2 phones here, just to keep our average normal, plus a total of 4 at our offices!) This coming week is so far devoid of outside engagements, but Anne will be back Tues. night.

The only excitement is about a book, *The Origin of Races*, by a friend of ours, Carleton ("Carl") Coon. It's being bitterly attacked by other friends of ours (including Dobzhansky), in a way I consider unfair almost to the point of being underhanded.[22] I'm trying to set them right in a review, which is going to make me bitterly attacked as well. A main point is that we'll never solve racial problems if merely recognizing that races exist & trying to study them is considered subversive & subjects one to abuse from *both* sides.

So far, however, it is safe to stick to the current sentiment that live

21. Theodosius Dobzhansky (1900–1975), Russian-born American geneticist whose *Genetics and the Origin of Species* (1937) triggered the modern evolutionary synthesis.

22. Carleton Coon (1904–1981), American anthropologist at Harvard whose book *Origin of Races* (1962) raised controversy because he asserted that the various human races may have evolved at different rates, thereby resulting in some races reaching the *Homo sapiens* level before others. In a review of the book in 1963 Simpson said he considered the book "an honest and substantial contribution to the scientific study of races and as nearly free from aprioristic bias as any on its subject."

Indians are O.K. Just don't mention the fact that Negroes, as a rule, have dark complexions. (I haven't yet quite figured out why it is O.K. to recognize that Nordics have complexions like shad bellies but non-O.K. to recognize that Negroes are attractively brown.)

Hoping you are the same, with love—

G

21 Sep 64
Between Honolulu & Suva
(that places us within 3,000
miles, more or less)

Dearest Marty & Bill:

I don't know just how it happens, but you get a letter from me this time, such as it is. Anne wrote from Vancouver, so you do know that we are actually embarked, & now you see we are well on our way. This ship is very large & very posh—in fact rather too much so, as it has the imperial air of a big luxury hotel. However, service is superb & we are pampered as far as possible. The food is generally excellent, *not* middle class English but eclectic. To be sure I did have a brief but very severe upset that *I* attribute to food poisoning, but the doctor does not agree & no one else who ate the same food was sick so it probably was just me cutting up again. Because of that I was back in bed for a few days & was unable to go ashore in Honolulu. However, Dorothy & Earl & Dickey & his wife came on board & we had a visit[23]—They took Anne to a beach club for lunch & [raced?] about a bit, then had dinner with Anne at the posh grill on the ship. Now I'm fine again & planning to go firewalking in the Fiji Islands day after tomorrow. Anne is as well as she has been lately, or better, & I do think the trip is doing us both good. We are happy to learn (by Marty's letter received in Honolulu) that your plans are becoming more definite & do hope everything works out.

Much love,
G

[Southwest Pacific]
[10 Oct. 1964]

Dear Marty & Bill:

We are rolling along between the Great Barrier Reef & the Australian coast—a journey of several days, which makes you realize how enormously long the reef is. This letter will be mailed tonight by the Barrier Reef pilot when he leaves us at Thursday Island in the Torres Strait, where we turn west between Australia & New Guinea. All of which sounds very roman-

23. Dorothy was Simpson's second cousin on the Hawaiian side of the family; Earl was her husband and Dickey their son.

tic, & feels & looks so, too. This is an unusual bit of ocean voyage because we are almost always in sight of land. There are innumerable islands inside (west of) the G. B. Reef, low coral islands & high mountainous ones, & we also can frequently see the coast of the mainland. We are back in the tropics, hot, damp, but breezy & with water of incredible shades of blue.

Oronsay [the cruise boat] is smaller & older than Oriana, really like a ship & not like a Statler-Hilton as Oriana was. We much prefer Oronsay & are having a great time—only just a little sorry for ourselves at this moment because we had our 2nd (last) cholera shots yesterday & they have upset us a bit. Be *sure* to have cholera shots before you take off, even if told (as we were) that they are not necessary. They can become necessary at any time, as they now are for landing in Hong Kong, & it is a nuisance to have them en route.

We hope you, too, are soon to be in these romantic waters, or hereabouts (give or take a few thousand miles).

Much love from us both—

G

[Cambridge]

21 Apr 65

Dearest Marty & Bill:

Marty's of 17 April just arrived (air mail, yet!) & I am replying *immediately*, not because Anne can't but because she is out on the town i.e. working overtime & I have just returned from more or less same.

We are both OK—sort of second-hand, to be sure, but still legible. Once Anne had her [heart] fibrillation corrected, plus a couple of days to recover, she started out like mad again. I worry, but she really has *some* sense. Anyway, overdoing apparently has little bearing on her problems. She is beginning to think that retiring in a year or so might be all right. Depends on many things, of course.

We are so excited & delighted at all the good things ($$$, of course, but even more the recognition, kudos, etc.)—You both have worked hard for this & it is high time. I suppose the immediate Hollywood deal means we'll see little or nothing of you next summer, & we don't like that, naturally, but it should be fascinating for you. Go, go, Eastlakes!

(Re Cadillac, Bill, I understood the need when you weren't in the chips—but now you can afford a VW!)

It's bad news about Pop[24]—But apparently you coped in good time. We just had a *very* nice letter from Rob't Pryer [sp?], about several things but including his opinion that Pop was making out very well & would be a new man—for 80 years old, that is, of course.

So we are doing just fine, & *you* go & enjoy without worry at all. *Much* love to you both—

G

24. Pop was Martha's ailing father-in-law.

Paris

13 Nov 65

Dearest Marty:

I have greetings to convey to you, & am writing them in a spare moment here in Paris although I may mail the letter in USA, where I will be tomorrow.

I had not planned to look up your friends on this brief visit, but was delighted to see some of them through a remarkable coincidence. It seems that Madeleine had long wanted to go to the opening ceremonies at the Sorbonne, having heard (correctly) that they are very picturesque. Finally this year she did go, & was thunderstruck to see your brother on the dais in all his glory! She waylaid me on the way out & asked me to dinner at her restaurant. I was fortunately able to accept & had a pleasant visit &, of course, an excellent meal. Besides Madeleine & her husband, Étienne & his wife were there. All were most pleased to have news of you, & they send love & (in Madeleine's case) kisses. They all seem to be thriving. Madeleine also invited a lady ethnologist to provide professional interest for me—a kindly thought, but not a successful one since the lady is almost comically like Maggie Mead, with the same complete faith in her own omniscience & utter scorn for everyone else in the world.[25] However, the others were so charming that she was unable to spoil the evening.

The ceremony was quite splendid.[26] No procession as such, but we entered the Grand Amphithéâtre de la Sorbonne between the ranks of the Garde Républicaine, who drew their swords with a great clatter & saluted each of us as we passed. The band of the Garde opened proceedings with the Marseillaise, & you have never heard anything till you hear a French military band belt out the anthem in a closed hall! Eleven degrees were then bestowed, each after an éloge lasting some 15–20 minutes & taking up the recipient's life year by year. Earl Warren was the other American, but in absentia, as the Supreme Court was sitting & he could not be absent there. (Oddly enough, he was represented by the French, not the American, ambassador.)

Bill will be amused that the only other scientific degree, therefore coupled with mine, was to an astronomer from the Glorious Soviet Union—the Armenian "Republic," to be exact, Ambartzoumian by name.[27] (Since he is president of the Armenian Academy, member of the USSR Academy & president of the International Astronomical Union, I gather that he is quite a swell. His democratic principles prevented his wearing a robe or evening clothes on the appropriate occasions, & he speaks no French at all although a little English.)

25. Margaret Mead (1901–1971), American anthropologist at Columbia University and a colleague of Simpson's at the American Museum. Years before, she took Lydia's side in their marital troubles, much to Simpson's great annoyance.

26. Simpson was receiving an honorary doctorate from the University of Paris.

27. Viktor A. Ambartsumyan (b. 1908), Soviet Armenian astronomer, member of the U.S.S.R. Academy of Sciences, and expert on the diffusion of light through planetary atmospheres.

We also had a very grand official dinner for all the "docteurs honoris causa," a reception for Ambartzoumian & me at the Faculté des Sciences (interestingly now located where the Halle des Vins used to be) & another, also there, for me alone. Champagne flowed like water, & whiskey (peculiarly, I thought) flowed like champagne. I have been déjèuner'd & diner'd every day & generally given the full treatment, which is of course delightful but which happily is just enough at this point, or will be after my last dinner tonight, which happens to be presented by a clerical colleague, the Abbé Lavocat.*[28]

Hoping you are the same, with love—

G

*The good abbé has a long flowing beard,
& we make a very striking pair!

[Cambridge?]
19 Oct 66

Dearest Marty—

Dilatory as ever, I hadn't acknowledged your memo on books rec'd, & not, from Okla. Press. This checks out, as I have now heard from Harvard Coop that the books you list were not available—some out of print & some not yet published (although all ordered as in stock from the latest Okla. Press catalogue). I'm pursuing some & letting some drop. I hope you are enjoying the books that did come.

The birds that flap no wings, i.e. in swoops are all woodpeckers or sapsuckers, but not (to my knowledge) just one species of either.

No special news. They're still trying to find out what my latest pain is all about. Some of the procedures are both painful & embarassing—for example, being put upside down on a violently shaking table, inserting a tube in the arse, & shaking up a barium sludge into the colon & parts north.[29] Several young to medium ladies now know everything about my plumbing except what's wrong with it.

May we sometime set foot on $5,000 per acre rancho land if we clean our shoes first?

Much love to both—

G

28. Réné Jean Marie Lavocat, French vertebrate paleontologist at the University of Montpellier and Simpson's contemporary.
29. Simpson was having trouble with diverticulitis at this time.

TUCSON
1967-1970

B Y 1967 the winters in Boston and health problems combined to encourage the Simpsons to leave Cambridge and Harvard and retreat to the year-round home they had purchased several years earlier in Tucson, Arizona. Anne retired more or less completely, but Simpson, now sixty-five, continued working as diligently as ever in the library-laboratory annex he had had built adjacent to their home. He was appointed a professor in the geosciences department at the University of Arizona, which position he retained until his formal retirement in June 1982, at age eighty. The Simpsons regretfully sold their summer home in New Mexico because it was virtually impossible for them to live comfortably year-round in so remote a place.

Martha and her husband Bill also sold their holdings in New Mexico and moved to Tucson, not far from the Simpsons. Bill taught writing for several years at the university, but in the early 1970s the Eastlakes divorced. When Martha moved to Tucson, the half-century correspondence between brother and sister ended, except for an occasional letter or card when the Simpsons were traveling.

Simpson's three daughters continued their active professional lives. Helen Simpson Vishniac became a professor of microbiology at Oklahoma State University. Joan Simpson Burns continued her writing in Williamstown, Massachusetts, where her husband taught political science. In 1975 she published *The Awkward Embrace*, a study of nine contemporary men influential in American cultural life. Elizabeth Simpson Wurr has written several books, including a novel, several studies in social psychology, and a medical diary that described her recovery from an almost fatal attack of tubercular meningitis. She lives today in Woodside, California, with her husband, an engineer.

Much of Simpson's writing from his years in Tucson falls into the popular vein. He published another collection of essays, *Biology and Man* (1969); *Penguins: Past and Present, Here and There* (1976); his autobiography, *Concession to the Improbable* (1978); two books on the fossils of South America and their discoverers, *Splendid Isolation* (1980) and *Discoverers of* the Lost World (1984); a work on the scientist he most admired, *The Book of Darwin* (1982); and a general book on paleontology for student and non-specialist alike, *Fossils and the History of Life* (1983).

31 Oct 67
Tucson

Dearest Marty:

Anne is off in Wisconsin for a few days & I am here. We're both off again in a week or two.

She did get the bird puzzle with great glee & meant to send thanks earlier but just lost out in the rush.[1] I assure you that she appreciates it very much.

Was it Ferlinghetti who advised the students at Knox [College] to ride with LSD? (I keep mixing that up with LDS, which of course means Mormon in these parts.) It definitely was he who did advise the students here to use marijuana, LSD, & other drugs. I suppose he practices what he preaches & therefore am not surprised at his rudeness to you—it obviously was such even though you may say you don't mind.[2]

Mrs. Johnstone obviously is no damn good, & it was just dumb luck that she had a ready-made buyer for us.[3] I *do* hope you can settle your affairs there soon. I feel apologetic that ours were so quickly & satisfactorily settled, & catch myself feeling as if I had somehow played a dirty trick on you by getting out when you can't (as yet), illogical as I know that is.

I am overjoyed that you are evidently planning to come here for Thanksgiving. Do come as soon & stay as long as possible. We'll be back from Yale on 18 November. I have to leave again for a meeting 26 Nov. to 2 Dec. in Denver but then back & we both leave again on 10 Dec. Anne will be here while I'm in Denver, so I hope you can stay on then, but do come as long before as you can, too.

Tucson has suddenly cooled off & is delightful. We've switched air conditioning units from cooling to heating, but in fact neither is needed at present.

Much love to you both—

G

I wrote Seligman, our lousy Albuquerque lawyer, & told him (rather snidely, I confess) that we were no longer lords of La Jara & he could return the unexpended part of our retainer ($125) in re Mrs. Meyers & the gate. He replied, predictably, that oddly enough his out of pocket expenses came to more than that, but he would call it quits & not charge us the rest!

1. Anne often occupied herself in the evenings by working on complicated puzzles or doing handwork of one sort or another.
2. In 1967 Martha's husband was writer-in-residence at Knox College, Galesburg, Illinois.
3. Simpson is referring here to the realtor who sold the house in La Jara. The Eastlakes were also trying to sell their place, but it took much longer than the Simpsons' sale.

28 Feb 68
Tucson

Dearest Marty—

Thanks for yours of 27 Feb. Anne's away again. Edna Roe had a pain & although she was out to dinner a couple of days later Anne felt duty bound to run up to Denver. In fact Edna is a great deal better off than Anne.

Of course I rejoice that Bill might get $50,000 & a nice trip, but I also am somewhat repelled (as ever) by Hollywood. I am a Burton buff, have read most of his innumerable books & several of his many lives (that is, biographies by others of him). He spent some time in Cairo, but as far as I know never went farther up the Nile. A trip up the Nile, or even to present-day Cairo, is simply irrelevant. Incidentally, I do not think Burton is at all enigmatic. Eccentric, yes, but that's not the same. He did unusual things, but his motives are always perfectly obvious.

I'm very sorry that the ranch sale seems stymied again. It's extremely annoying.

I did notice that you'd had your face plasterized, & very becoming too.[4] I just thought if you wanted it mentioned you'd mention it.

Much love to both—
G

Mid-February 1970
Argentina

[Postcard addressed to Marty & Bill Eastlake]

These creatures [elephant seals] have to be seen to be disbelieved. From a distance we thought they were big rocks, but close by they rear up & roar, which rocks rarely do. They are on Peninsula Valdés where we visited a few days ago. (Not really at Puerto Madryn as the card says.) Yesterday we drove out to Sarmiento to visit Justino Hernández, Patagonian companion of my youth, now a grandfather like me, & we had a *great* reunion—As Anne says, full of wine, women (wives), & song (raucous)—

Love—G

23 Feb 70
Antarctica

Dear Marty & Bill—

A couple of nights ago, at 2:45 A.M. midst snow & ice we crossed the Antarctic Circle, turned around & hightailed it north again. Now we are cruising up Gerlache Strait (I never heard of it before, either!), which has

4. Apparently Martha had facial plastic surgery about this time, at age seventy.

to be one of the most spectacular sights on earth.[5] Tremendous jagged mountains on both sides—islands to left & the Antarctic continent to the right—jet black where not snow-covered (only a few spots), with a glacier in every valley & even on every slope, each ending in a delicate blue ice cliff from which bergs calf from time to time. Bergs of every imaginable & many unimaginable shapes. The lower ones with seals basking on them, or an occasional misanthropic penguin. Where there happens to be a rocky slope near shore, there is generally a penguin rookery, at this season full of large, demanding chicks just shedding juvenile down, pursuing every adult who comes ashore with food (in stomach, to regurgitate).

The whole trip has been just *great*. Ship very comfortable, fine cabin—lousy service, but you can't have everything, it says here.

In spite of all we have read & all the pictures we have seen, the real Antarctic is an almost blasting revelation. One of the big, big bangs of a lifetime.

In short, we're high. Hoping you are the same—

Love—G

5. Gerlache Strait lies between the Antarctic Peninsula and large offshore islands to the west, Anvers and Brabant.

AFTERWORD

T HE SIMPSONS took another South Seas cruise in the spring of 1984. Simpson caught pneumonia after sailing a few days and became seriously ill. He improved enough to continue the trip with the hope that rest and relaxation would restore his fragile health. On the contrary, his condition deteriorated, and when he reached home he was immediately hospitalized. Throughout the summer he suffered various complications, and was in and out of the hospital. By October his strength had waned and on the evening of Saturday, 6 October, he died, aged 82. Shortly before he died he had the opportunity to see his daughters, who, of course, had come from California, Oklahoma, and Massachusetts to be at his bedside.

At the time of Simpson's return from the ill-fated cruise, I sent him the full set of these letters transcribed from the originals. In mid-July he wrote me the following:

> This is one of the saddest letters I have ever written. You know that my sister Martha has been quite ill for some time [she suffered from emphysema], although by doctors' & nurses' reports not yet crucially so. She died yesterday afternoon. The best I can add is that this was quick, quiet, & painless. . . .
>
> I have read your ms. of letters to Martha, etc. but I cannot yet take the time, energy, or emotion to make any remarks on them.
> I am slowly getting better but it looks like a long haul.
> I can't write more.

Anne Roe Simpson still lives in Tucson, Arizona.

SELECTED BIBLIOGRAPHY

1928. *A Catalogue of the Mesozoic Mammalia in the Geological Department of the British Museum*. London: British Museum (Natural History). 215 pp. This monograph resulted from Simpson's year of postdoctoral research on the primitive mammals in the collections of the British Museum.

1929. *American Mesozoic Mammalia*. Memoir, Peabody Museum, Yale University, volume 3, part 1, pp. 1–236. Although published after his British Museum study, this monograph reported Simpson's earlier dissertation research at Yale. These two studies were so comprehensive and authoritative that the field of primitive mammals lay fallow for a full generation.

1934. *Attending Marvels: A Patagonian Journal*. New York and London: Macmillan. 295 pp. Travel adventures and natural history of Simpson's year of fossil-collecting in Patagonia, 1930–1931. Reprinted in 1982 by the University of Chicago Press.

1935. The first mammals. *Quarterly Review of Biology* 10:154–180. A synopsis of Simpson's major findings regarding Mesozoic mammals; drawn from his dissertation and postdoctoral research.

1937. The beginning of the age of mammals. *Biological Review* 12:1–47. An overview of the early worldwide expansion of mammals during the Cenozoic Era, following the demise of the dinosaurs at the end of the Mesozoic Era.

1937. Supra-specific variation in nature and in classification: From the viewpoint of paleontology. *American Naturalist* 71:236–267. The first of what Simpson later called "door-opener" papers to his entry into more theoretical considerations of evolutionary processes.

1937. Patterns of phyletic evolution. *Geological Society of America Bulletin* 48:303–314. The second "door-opener" article.

1937. *The Fort Union of the Crazy Mountain Field, Montana, and Its Mammalian Faunas*. *United States National Museum, Bulletin* 169:1–287. A major monograph on early Cenozoic mammals which also introduces statistical techniques to discriminate between fossil species. A kind of pilot study for *Quantitative Zoology*.

1939. *Quantitative Zoology: Numerical Concepts and Methods in the Study of Recent and Fossil Animals*. New York and London: McGraw-Hill. 414 pp. Coauthored with his wife Anne Roe, whose experience as a clinical psychologist provided the statistical expertise for characterizing large popu-

329

lations from small samples. Subsequently revised with third coauthor, Richard Lewontin, a population geneticist, and published by Harcourt, Brace, 1960.

1940. Los Indios Kamarakotos (Tribu Caribe de la Guayana Venezolana). *Revista de Fomento, Caracas* 3:201–660. An ethnographic study of a local tribe of Caribe Indians that kept Simpson occupied while he waited for an end to the rains that were preventing him from collecting fossils.

1942. The beginnings of vertebrate paleontology in North America. *Proceedings of the American Philosophical Society* 86:130–188. Another of Simpson's interests was the history of paleontology; this paper won the society's Lewis Prize for that year.

1944. *Tempo and Mode in Evolution*. New York: Columbia University Press. 237 pp. The single most important of Simpson's works, *Tempo and Mode* contributed to the modern evolutionary synthesis by demonstrating that the microevolution of the population geneticists was sufficient to explain the macroevolution of the paleontologists.

1945. The principles of classification and a classification of mammals. *American Museum of Natural History, Bulletin* 85:350. A discourse on taxonomic arrangement and its application to living and fossil mammals; still in print and extensively cited. Although somewhat out of date, no one has yet taken up the formidable challenge of providing a revision.

1947. Holarctic [northern hemisphere] mammalian faunas and continental relationships during the Cenozoic. *Geological Society of America Bulletin* 58:613–688. One of several articles published during this period when Simpson effectively argued against fossil evidence putatively supporting continental drift.

1948. The beginning of the age of mammals in South America, part 1. *American Museum of Natural History, Bulletin* 91:1–232. The first part of a comprehensive review of Cenozoic mammalian history in South America during most of which time the continent was isolated from the rest of the world.

1949. *Genetics, Paleontology, and Evolution*. Princeton: Princeton University Press. 474 pp. A collection of papers presented by a varied group of biologists at a conference in 1947 which consolidated the previous decade's evolutionary synthesis. Simpson collaborated in the assembly and editing of the volume with Glenn Jepsen, a Princeton paleontologist, and Ernst Mayr, a Harvard biologist. Simpson's paper, "Rates of Evolution in Animals," basically reviewed the major points he made in *Tempo and Mode in Evolution*.

1949. *The Meaning of Evolution*. New Haven: Yale University Press. 364 pp. The most successful of all Simpson's books, it was translated into ten languages and sold over half a million copies in the English version. Revised and updated in 1967; still in print.

1951. *Horses: The Story of the Horse Family in the Modern World and through Sixty Million Years of History.* New York: Oxford University Press. 247 pp. A popular account of the natural history and evolution of horses. An outgrowth of Simpson's updating of an American Museum pamphlet on horses.

1952. Probabilities of dispersal in geologic time. *In* The problem of land connections across the South Atlantic with special reference to the Mesozoic: A symposium. *American Museum of Natural History, Bulletin* 99:163–176. Important nail in the coffin of the theory of continental drift among North American geologists. Simpson demonstrated that land animals moving across stable continents could as easily explain the past distributions of fossils than could drifting continents with sedentary animals.

1953. *Life of the Past: An Introduction to Paleontology.* New Haven: Yale University Press. 198 pp. Another popular book explaining the science of paleontology. Its chief interest today lies in the illustrations, drawn by Simpson himself, which have a certain amateurish charm.

1953. *The Major Features of Evolution.* New York: Columbia University Press. 434 pp. An expanded and more elaborate version of *Tempo and Mode in Evolution.* Some reviewers thought it lacked the brilliant succinctness of *Tempo and Mode.* Nonetheless, a generation of paleontologists and evolutionary biologists cut their professional teeth on *Major Features.*

1953. Evolution and geography: An essay on historical biogeography with special reference to mammals. Condon Lectures, Oregon State System of Higher Education, Eugene, Oregon. Pp. 1–63. Simpson's summation of the principles of determining past geographies of land animals—still valid—and what we can infer about the former configuration of past continents—no longer valid.

1957. *Life: An Introduction to Biology.* New York: Harcourt, Brace. 845 pp. Coauthored with zoologist Colin Pittendrigh and botanist Lewis Tiffany, this college biology text emphasized the principle of evolution amidst life's manifold diversity. Simpson revised the book in 1965 with coauthor William Beck, who is currently preparing a third edition.

1958. *Behavior and Evolution.* New Haven: Yale University Press. 557 pp. Another collaboration with his wife Anne Roe Simpson; a collection of papers by a number of researchers dealing with various aspects of the evolution of behavior which exemplified the integrative power of the evolutionary synthesis.

1960. The world into which Darwin led us. *Science* 131:966–974. One of several essays written by Simpson around the centennial of Darwin's *On the Origin of Species* in which Simpson celebrates the major contribution that Darwin made to human knowledge.

1960. The history of life. In *The Evolution of Life,* vol. 1 of *Evolution after Darwin,* edited by Sol Tax, pp. 117–180. The University of Chicago [Darwin] Centennial. University of Chicago Press. A mid-career overview of the contribution that fossils have made to our understanding of life's evolution.

1961. *The Principles of Animal Taxonomy*. New York: Columbia University Press. 247 pp. An expansion and deeper treatment of the intellectual and practical bases for ordering animate nature in a hierarchial scheme foreshadowed in the 1945 monograph on mammals.

1963. Historical science. In *The Fabric of Geology*, edited by Claude Albritton, Jr., pp. 24–48, Reading, Mass., Palo Alto, and London: Addison-Wesley. Perhaps the best of several essays by Simpson on the distinctive character of an historical science like geology or paleontology as contrasted with the experimental sciences such as chemistry or physics.

1964. *This View of Life: The World of an Evolutionist*. New York: Harcourt, Brace, and World. 308 pp. A collection of earlier published essays brought together into one volume; the essays treat Darwin, the nature of historical biology, the problem of apparent purpose in living nature, and thoughts about cosmic evolution and the human evolutionary future. Simpson often remarked that this was his favorite book.

1965. *The Geography of Evolution: Collected Essays*. Philadelphia and New York: Chilton Books. 249 pp. The only book that Simpson regretted because it brought together those articles in which he strenuously argued against continental drift on the basis of fossils just at the time when the new global theory of plate tectonics was revolutionizing geology.

1967. The beginning of the age of mammals in South America, part 2. *American Museum of Natural History, Bulletin* 137:1–259. The completion of this extensive monograph started two decades earlier.

1969. *Biology and Man*. New York: Harcourt, Brace and World. 175 pp. Yet another collection of Simpson's essays treating a variety of subjects, including the biological aspects of race, language, and ethics.

1976. *Penguins: Past and Present, Here and There*. New Haven: Yale University Press. 150 pp. Popular summary of the natural history and evolution of these highly specialized birds.

1978. *Concession to the Improbable: An Unconventional Autobiography*. New Haven: Yale University Press. 291 pp. As Simpson described it, a book about the things he'd done and seen and thought and what he considered interesting and important.

1980. *Splendid Isolation: The Curious History of South American Mammals*. New Haven: Yale University Press. 266 pp. A general account of the evolutionary history of a variety of mostly extinct beasts during the Age of Mammals.

1980. *Why and How: Some Problems and Methods in Historical Biology*. New York: Pergamon Press. 263 pp. A collection of extracts of previously published papers upon which Simpson comments in order to explain the context in which they were written and the nature of the problems he was attempting to solve. Curiously, the title recalls a children's natural history book written by Rev. Charles Kingsley, *Madam How and Lady Why*, which Simpson elsewhere remembered as having a major influence on his early interest in nature.

1982. *The Book of Darwin.* New York: Washington Square Books. 219 pp. A paean to the master evolutionist of the nineteenth century by the master of the twentieth. Extracts from various of Darwin's works with commentary by Simpson.

1983. *Fossils and the History of Life.* New York: Scientific American Books. 239 pp. Simpson's swan song about the subject he spent six decades of his life studying. Although fairly technical, accessible to the nonspecialist.

1984. *Discoverers of the Lost World: An Account of Some of Those Who Brought Back to Life South American Mammals Long Buried in the Abyss of Time.* New Haven: Yale University Press. Simpson completes the circle and ends as he began with an account of his intellectual adventures in South America.

1985. Extinction. *Proceedings of the American Philosophical Society* 129:407–416. Posthumous publication of the manuscript Simpson was working on when he died. Although incomplete, the work seems to argue that we still do not understand the causes of extinction, even though the phenomenon has been studied for two centuries by paleontologists. Simpson also suggests that he did not give any particular credence to extraterrestrial causes of extinction.

INDEX

Genetics, 4
The Geography of Evolution, 304
Geological Society (American), 42
Geological Society (British), 63
Geology, 53
Germany, 72, 91, 92–93, 95–97, 98–99, 101
Gidley, James, 42, 88, 173
Gilmore, Charles W., 42
God, 156
Goodrich, Edwin S., 69
Granger, Walter, 168, 174–175, 203, 205, 208, 214, 223, 229, 231
Granger, Mrs. Walter, 229, 231
Gregory, Herbert E., 108
Gregory, William King, 42, 49, 88, 108
Guanacos, 144, 147, 148
Gunter, Herman, 135

Hakluyt, Richard, 135
Haldane, J. B. S., 194
Harris, H. J., 316 n. 20
Harvard University, 303–321; Agassiz professorship at, 11, 12, 286, 294, 295, 306; Anne at, 11, 297 n. 23, 303; Center for Research in Careers at, 303, 316; Museum of Comparative Zoology at, 11, 12, 286, 294, 295, 303, 305–307
Hearst, W. R., 39
Hepatitis, 9, 237, 265, 266, 268, 273, 274–275, 281
Hereford, Charly, 247
Hernández, Justino, 145, 190–191, 325
Hieroglyphs, 64, 99, 103, 111
Hobbies, 117
Holland, James, 202, 209, 216
Holland, Sir Thomas, 58, 72, 79
Holmes, Walter, 115, 121, 123
Hopwood, Arthur Tindell, 56, 57, 58, 59, 62, 65, 73, 85, 87, 90–91, 93, 95, 109
Horses, 285
Horses, 28, 180
Huene, Friedrich von, 66, 70, 72, 87, 91, 93, 95, 139
Hunter, Fenley, Mr. and Mrs., 201
Huxley, Julian, 290

International Education Board, 55
Irigoyen (pilot), 152–153
Irwin, Herb, 248
Italy, 271–274

Janensch, Werner, 96
Jepsen, Glenn L., 204
Jewett, Frank B., 241
Jones, Whiskey-proof, 153
Jurassic period, 63

Kamarakoto Indians, 219, 230, 231
Keats, John, 157
Kiaer, Johan A., 96
Kindle, Cecil H., 69

Kinney, Helen J. *See* Simpson, Helen Kinney
Knox, Thomas, 211
Koch, Albert, 62 n. 5, 65
Kock, Antonio de, 153–154

Lang, Charley, 282, 297
Language, 102–103, 104–105; Arabic, 247, 248, 250, 277, 278; Egyptian, 48–49; French, 71, 150, 152; German, 101; Spanish, 145, 147, 157, 166–167
Lardner, Ring, Jr., 271
Lattimore, Owen, 207
Lavocat, Réné Jean Marie, 321
Lawrence, D. H., 48
Leakey, Louis S. B., 56, 307–308
Leakey, Mary, 307 n. 5
Leakey, Richard, 307 n. 5
Le Gros Clark, Wilfrid, 62, 66, 70
Lemoine, P., 31
Lents, Jack, 287
Levine, Mr. and Mrs. Paul, 305
Lewontin, Richard C., 296 n. 20
Life: An Introduction to Biology, 11, 285
Lindbergh, Charles A., 97
Linnean Society, 63
Lippman, Mort, 15
Literature, 109–110
Longwell, Chester, 50 n. 33
Loomis, Frederic B., 42
Lord, Mrs. Rolfe, 174, 175
Louisiana, 135–136
Lucero, Lucy, 287
Lull, Richard Swann, 7, 28, 43, 50 n. 33, 55, 62, 66, 67, 70

MacArthur, Douglas, 278
McBride, Katherine E., 140, 216
McCarthy, Mary, 132 n. 15
McLaurin, Duncan, 209, 264
McLaurin, Jack, 34, 209, 212, 214, 231, 250, 275
McLaurin, Margaret. *See* Simpson, Margaret (McLaurin)
Major Features of Evolution, 9, 285
Mâle, Emile, 33
Mammals: Cenozoic, 41, 140; classification of, 140 (see also *Principles of Classification and a Classification of Mammals*); Jurassic, 63; Mesozoic, 6, 7–8, 28–29, 43 n. 20, 56–57, 61, 66, 69, 70, 73, 88, 97, 238; Mongolian, 49; monotreme, 118 n. 4; multituberculate, 91, 118; Patagonian, 282; Pleistocene, 115; Purbeckian, 63; Tertiary, 115, 116
Marks, Edith, 296
Marsupials, 180
Martineta, 147
Maspero, Gaton, 193
Matthew, Margaret (Colbert), 66 n. 14, 229, 231, 296